Molecular Devices

Molecular Devices

An Introduction to Technomimetics and Its Biological Applications

Andrei A. Gakh
The University of Virginia, Bethesda, MD, USA

WILEY

This edition first published 2018

Registered Office
John Wiley & Sons, Inc., 111 River Street, Hoboken, NJ 07030, USA

Editorial Office
111 River Street, Hoboken, NJ 07030, USA

For details of our global editorial offices, customer services, and more information about Wiley products visit us at www.wiley.com.

Wiley also publishes its books in a variety of electronic formats and by print-on-demand. Some content that appears in standard print versions of this book may not be available in other formats.

Library of Congress Cataloging-in-Publication Data:

Names: Gakh, Andrei A., author.
Title: Molecular devices : an introduction to technomimetics and its biological applications / Andrei A. Gakh.
Description: Hoboken, NJ : John Wiley & Sons, 2018. | Includes bibliographical references and index. |
Identifiers: LCCN 2017043486 (print) | LCCN 2017055745 (ebook) | ISBN 9781119448150 (pdf) | ISBN 9781119448136 (epub) | ISBN 9780471411390 (cloth)
Subjects: LCSH: Nanotechnology.
Classification: LCC TK7874.8 (ebook) | LCC TK7874.8 .G35 2018 (print) | DDC 621.381–dc23
LC record available at https://lccn.loc.gov/2017043486

Cover design by Wiley
Cover image: © RED_SPY/Gettyimages
Illustrations: Courtesy of Andrei A. Gakh

Set in 10/12pt WarnockPro by SPi Global, Chennai, India

Printed in the United States of America

V10002096_070718

To my father -
Who introduced me to the concept of the information reality.

Contents

Preface

Human imagination has only few limitations in its own world of information domain. This is not true for technical engineering bound by the laws of physical reality. These physical limitations are environment-specific, so engineering solutions used for the construction of ships would need to be modified for the construction of airplanes, because ships are designed to operate in the water, and airplanes – in the air (Figure 1).

In the late twentieth century, human engineering entered the molecular subdomain, first with the creation of the biotechnology toolbox, followed by construction of rudimental nonbiological molecular devices. While the molecular environment is quite different from anything engineering sciences encountered thus far, the inner logic of the technological progress dictates that this development should start with the design of functional molecular systems that can mimic conventional man-made devices (Figure 2). By the analogy with biomimetics, these functional molecular systems can be called technomimetics.

It is naïve to assume that practical molecular devices of the future would simply employ assemblies of currently known technomimetics. But the engineering of these basic technomimetic molecules is a necessary step toward the development of more complex functional molecular devices. It is not a coincidence that the first automobiles adopt some technologies of a horse-driven carriage (Figure 3), and the early airplane designs resemble a kite (Figure 4).

It is very likely that current technomimetic development would follow the same script, albeit indirectly. As a historical analogy, one of the prominent goals of alchemy was the making of gold from other metals, but alchemists ultimately cleared the path toward modern chemistry – a far more valuable enterprise than all the gold bullions of the world combined. With this analogy in mind, the current field of technomimetics can be viewed with well-deserved optimism. After all, for exponential dynamic information processes such as modern human technology, attaining the right direction is usually far more important than the transient results collected along the way.

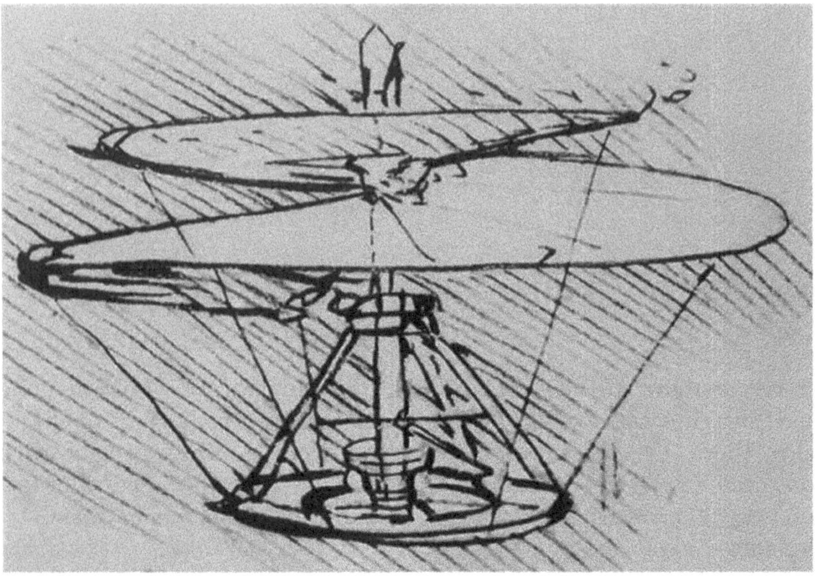

Figure 1 Leonardo Da Vinci's design of a helicopter equipped with the Archimedes' screw, even though it is intended to operate in air, not in water.

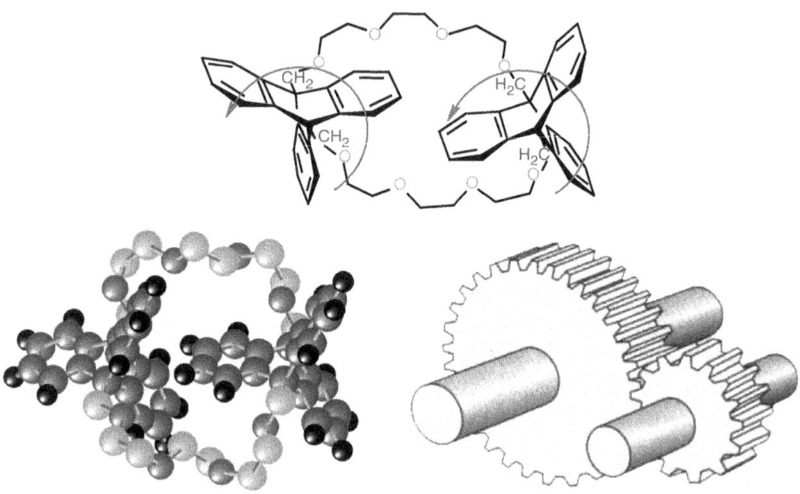

Figure 2 An example of a technomimetic molecule and its conventional mechanical prototype.

Figure 3 Daimler "motorwagen" that adopted some technologies of a horse-driven carriage. *Source*: Courtesy of the World Encyclopedia of Cars.

Figure 4 BlackBurn monoplane that resembles a kite.

As with any emerging sciences, technomimetics is a quickly growing field. Not surprisingly, the writing of this book is an evolving story, spanning more than 15 years. My initial interest in molecular devices was prompted by the needs of applied research in the area of advanced materials in the late 1980s. The discovery of fullerenes incited a hope that these spherical molecules can be used as universal molecular engineering construction elements. Unfortunately, the intrinsic steric limitations of C_{60} molecular surface precluded the synthesis of an almost perfectly ball-shaped "Teflon fullerene" $C_{60}F_{60}$, and lead to a "fluorine dance" phenomenon (facile fluorine atom migration on the fullerene surface) in crowded and deformed, partially fluorinated fullerenes $C_{60}F_{48}$ and $C_{60}F_{36}$. Better results in the field of molecular devices were subsequently achieved with triptycene crown ethers as molecular gearing systems, which were designed and constructed in 1994–1995. About the same time, it became clear that other conventional man-made devices could serve as an inspiration for fully functional molecular systems.

This book is a natural expansion of the ideas presented at the ACS meeting in 1995, as well as the follow-up paper published in 1997, in which "technomimetics" was formally introduced as a scientific term. Had the book been written in the early 2000s, it would be a very different document given the immense achievements of the last decade. This, once again, is a reminder that the essential evolutionary steps cannot be skipped, even in the book writing process. Perhaps the only reason why this book is now in print is that the content's time has come. In some way I feel that my role here is only as a transmitter of ideas mature enough to enter the information domain on their own terms.

The book would have to be impractically large should an attempt be made to cover all current areas of research in technomimetics, including molecular electronic devices. A conscious decision was made to cover (at least for now) only simple mechanical molecular devices, such as container compounds, tubes, tweezers, gears, and others. These basic technomimetics have intuitive, easy-to-understood mode of actions and are likely to become integral parts of the ultimate molecular devices of the future.

Finally, I would like to express my deep appreciation and gratitude to those who contributed to this effort. To my wife and my daughter, who enthusiastically supported my work despite the obvious disruptive nature of the writing process. To my colleagues, who were tremendously helpful in the preparation of the book. To all my friends who read the draft and provided valuable comments. To Nobel Laureate George Olah for his initial encouragement. And last, but not least, to Ed Immergut and to all of the Wiley publishing team without whom this book would probably never see the light of the day, at least in the current form.

June 2016

Andrei A. Gakh
Bethesda, MD

1 Introduction

CHAPTER MENU

1.1 Technomimetics

The term *technomimetics* was introduced about 20 years ago [1]. However, the origin of the idea to create *individual molecules and molecular systems that can mimic conventional man-made devices in function, design, or mode of operations* can be traced back to the late 1970s (e.g., molecular tweezers) [2]. Despite the early start, the major advances in this area were made only during the past two decades. Nowadays, this is one of the most prominent research areas, with hundreds of research papers published every year. The past decade also saw an increasing rate of practical applications, primarily in biomedical and material sciences fields, with a better recognition of technomimetic molecules as a distinct subclass of molecular devices [3]. However, the major practical impact of this technology is expected over several decades into the future. Construction and fine-tuning of complex functional technomimetic molecules remain the major unresolved problems.

Technomimetics represent a limited subclass of molecular devices and molecular machines. Molecular devices themselves constitute a more general family of individual molecules and molecular systems capable of providing valuable device-like functions. Many of them do have distinct conventional prototypes and therefore can be identified as technomimetic molecules. However, other molecular devices operate in a manner that is distinctively different or rare within the domain of conventional man-made devices and therefore do not fit within the technomimetics subclass [4].

Molecular Devices: An Introduction to Technomimetics and Its Biological Applications, First Edition. Andrei A. Gakh.

The early set of technomimetic molecules includes simple molecular devices, such as container compounds (see also Chapter 2) [5, 6], gearing systems (see also Chapter 3) [7, 8], belts and tubes (see also Chapter 4) [9], and tweezers (see also Chapter 5) [10, 11]. Subsequent developments in the late 1990s and early 2000s added more complex technomimetic molecules, such as molecular brakes [12] and chemically driven motors [13]. The late 2000s and early 2010s saw an expansion of this set with advanced technomimetic molecules, such as molecular wheelbarrows [14], cars [15], and scissors [16]. Some examples of technomimetic molecules are presented in Section 1.2 for illustration purposes.

1.2 Examples of Technomimetic Molecules

Containers are perhaps the simplest conventional devices that can be easily constructed on a molecular scale. More importantly, molecular containers can closely mimic the functions of conventional containers by providing an isolated environment that can be filled with guest molecules or ions. Container compounds are attractive molecular carrier vehicles and already in high demand for biomedical applications. This area of research was active since the late 1980s and culminated with the synthesis of various carcerands and hemicarcerands prepared by simply joining two bowl-shaped cavitand units (Figure 1.1, see also Chapter 2) [5, 6].

The inner space of these container compounds is described as "a novel reaction environment" [18] or "molecular reaction flask" [19] and can accommodate various molecular guests, including small organic molecules, reactive intermediates [20], and even fullerenes. Two examples of container compounds with benzaldehyde and fullerene C_{60} entrapped inside are presented in Figure 1.2.

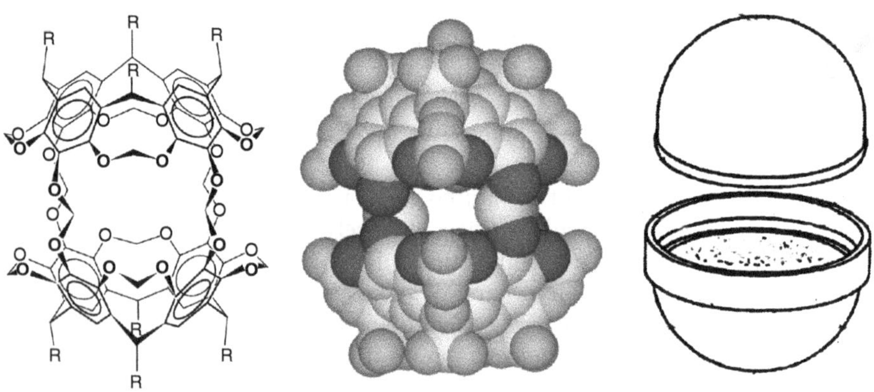

Figure 1.1 The pyrogallol[4]arene-based molecular container compound and its conventional prototype ($R = CH_2CH_2Ph$, some hydrogen atoms and CH_2Ph fragments were removed in the space-fill model for clarity) [17].

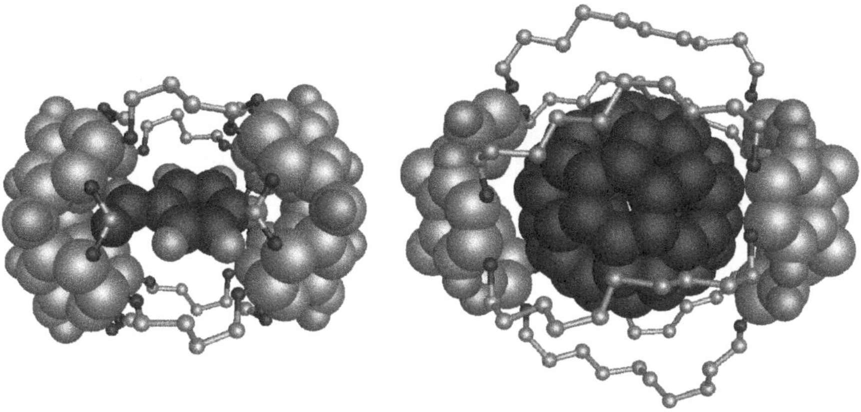

Figure 1.2 X-ray structures of benzaldehyde (left) [21] and fullerene C_{60} (right) [22] guest molecules encapsulated in molecular containers (some hydrogen atoms and CH_2CH_2Ph fragments were removed for clarity).

Further progress in the container compound research field was achieved with the advent of endohedral fullerenes [23–26], characterized by the tightly meshed all-carbon network sidewalls. A major synthetic hurdle associated with the difficulty of filling up the empty internal cavities of fullerenes was eventually resolved with the implementation of modern ion implantation technology as well as "surgical" open-and-close synthetic methodology. An additional set of valuable endohedral metallofullerenes (Figure 1.3) was prepared by the optimization of the traditional arc discharge fullerene production process. The commercial importance of endohedral fullerenes is evident from the recent reports indicating that N@C_{60} is one of the most expensive (per gram) synthetic organic compounds ever made [27]. The high price is undoubtedly associated with a proposal to use this container compound in miniature atomic clock applications due to a very long electron phase coherence time of up to 250 μs at 170°K [28].

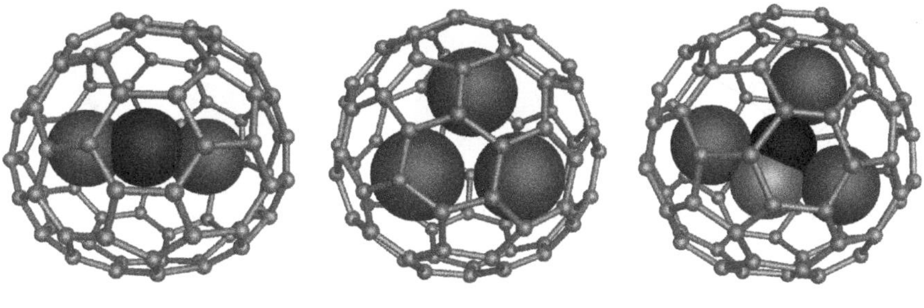

Figure 1.3 X-ray structures of endohedral metallofullerenes $Sc_2O@C_{80}$-$C_{2v}(5)$ (left) [29], $Sm_3@C_{80}$-I_h (middle) [30], and $Sc_3NC@C_{80}$-I_h (right) [31].

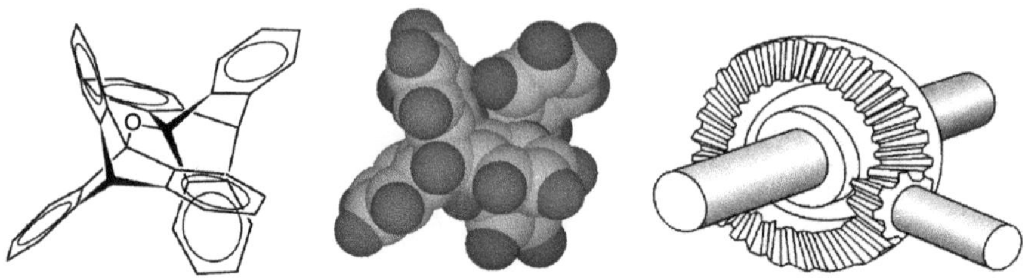

Figure 1.4 A simple bevel molecular gear, bis(triptycyl) ether, and its mechanical prototype [33].

Molecular gearing systems [1, 8] (Figure 1.4, see also Chapter 3) belong to an important class of technomimetic molecules, which clearly demonstrate the limits of conventional engineering within the molecular domain. Unlike conventional devices, molecular gearing systems are impossible to construct without gear slippage. However, the optimal molecular design allows for millions of correlated rotations between the gear slippage events. It is also possible to introduce various braking elements into molecular gearing systems, so the rotation can be controlled externally [32].

Belts and tubes (see also Chapter 4) are perhaps the most abundant conventional construction elements and are widely used in modern technology, predominantly for the transportation of gases and liquids, and also as parts of various mechanical devices. Belt molecules were actually the earliest examples of technomimetics and were accidentally discovered more than a century ago [34], but the rational synthesis of molecular tubes with precise structures remains a challenge [9]. A recent example of a simple belt-like molecule, cyclo[4]fluorene, with distinctive green fluorescence, is presented in Figure 1.5 [35].

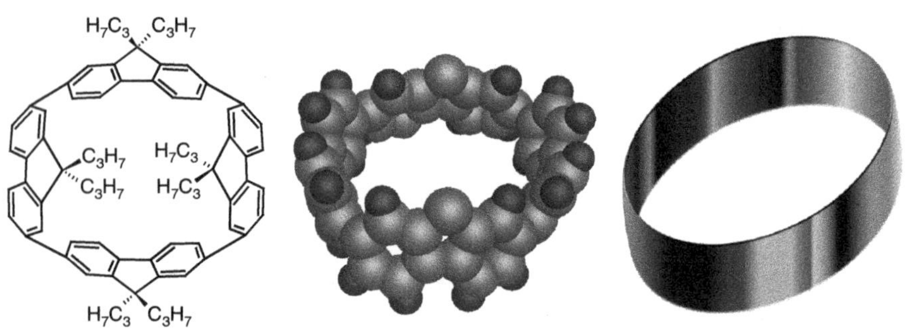

Figure 1.5 Cyclo[4]fluorene as a simple belt-like molecule (left), its X-ray structure (middle, n-C_3H_7 fragments were removed for clarity), and mechanical prototype (right).

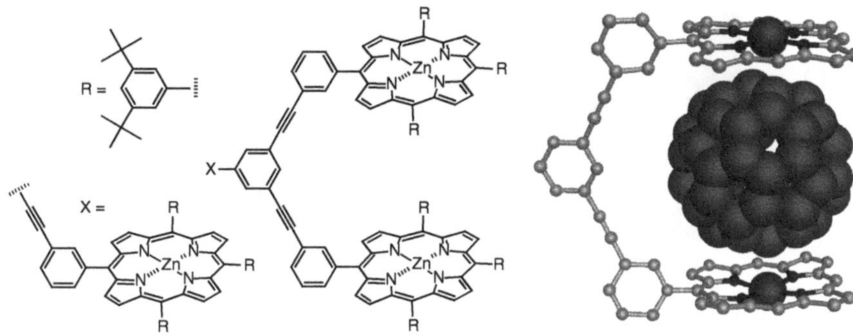

Figure 1.6 Flexible molecular tweezers with porphyrin pincers (left) and X-ray structure of its complex with fullerene C_{60} (right, R and X fragments as well as hydrogen atoms were removed for clarity) [36].

Molecular tweezers [2] (see also Chapter 5) [10, 11] represent one of the most developed classes of technomimetics with potential applications ranging from advanced chemical sensors to novel biomedical agents [10, 11]. While simple molecular tweezers typically do not have precise pick-and-choose capabilities of conventional mechanical tweezers guided by human or artificial intelligence, the desired “intelligent” selectivity can be attained in more complex systems by the incorporation of multiple functional groups. A better application potential can be expected for dynamic molecular tweezers, which can be controlled allosterically by the presence of certain chemical species in the environment or responding to light, electrochemical, and mechanochemical stimuli. An example of molecular tweezers with porphyrin pincers is shown in Figure 1.6 [36].

Molecular scissors and pliers are elegant examples of externally controlled molecular devices, which undergo scissor-like transformation upon the application of certain stimuli, typically narrow-band electromagnetic radiation [16]. Unfortunately, the preponderance of current examples of molecular scissors can mimic their conventional mechanical prototypes only in the mode of action, not in the cutting function. It is, however, expected that these molecular devices can eventually be used to perform similar actions, such as chemical bond cutting. An early example of molecular scissors is presented in Figure 1.7 [37].

Advanced versions of molecular pliers can mimic their conventional mechanical counterparts not only in the mode of action but also in the function. An example of molecular pliers that can be used to change the conformation of a heterocyclic guest molecule is presented in Figure 1.8 [38].

Molecular wheelbarrows belong to a family of sophisticated technomimetic molecules. Their design encompasses all major elements of the conventional

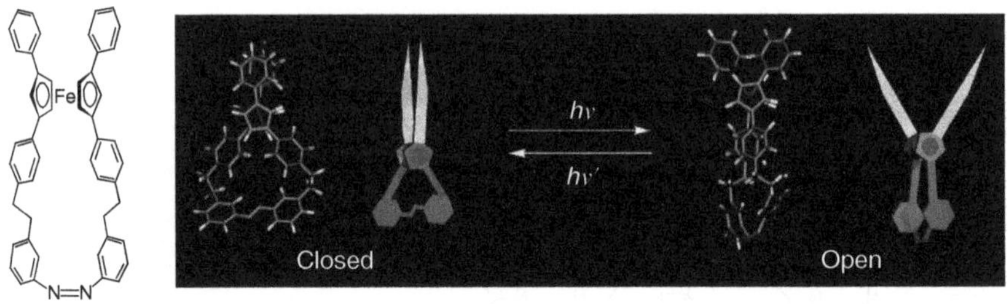

Figure 1.7 An early example of light-driven chiral molecular scissors. *Source*: Adapted from Muraoka *et al.* 2003 [37]. Reproduced with permission of American Chemical Society.

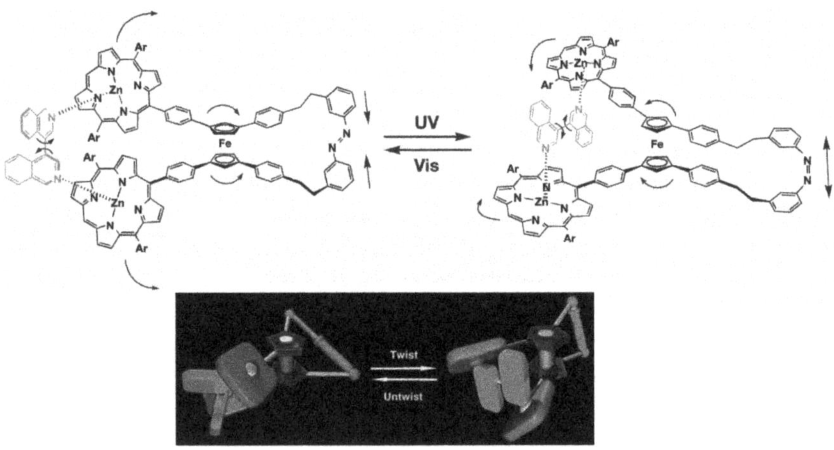

Figure 1.8 An example of advanced light-driven molecular pliers capable of changing the conformation of a heterocyclic guest (Ar = 4-decyloxyphenyl). *Source*: Muraoka *et al.* 2006 [38]. Reproduced with permission of Nature Publishing Group.

mechanical prototypes, such as wheels, axels, supporting frame, and even handlers. The structure of a molecular wheelbarrow is presented in Figure 1.9 [14].

Molecular cars and trucks [15, 39, 40] can be considered as a logical extension of the abovementioned molecular wheelbarrow design. Instead of two wheeling elements, molecular cars usually have four. This design completely separates the chassis from the plain "road" surface and allows for an unrestricted movement along the applied external force [41]. There are currently several variants of molecular car designs, which entail the use of fullerene C_{60} (Figure 1.10) [42], adamantane [43], *p*-carborane [44–46], and even organometallic complexes [47] as car wheel mimics. Light-driven [48] and electrically driven [49] motorized molecular cars are also known.

The aforementioned set of molecular cars does not permit the attachment of a useful payload, so these "nanocars" can be considered as replicas of conventional cars only in the mode of action, not in the function. Payload functionality

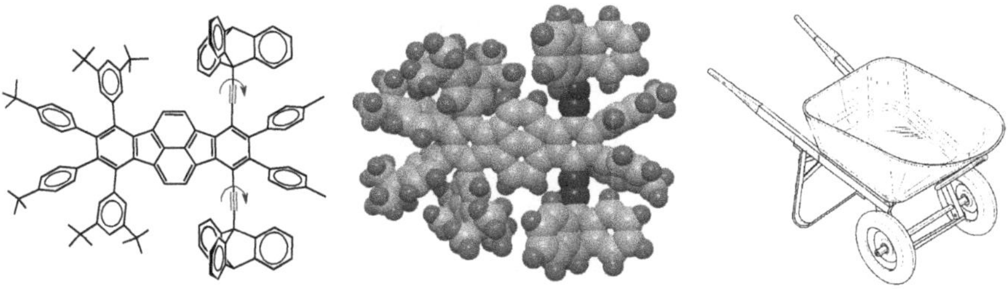

Figure 1.9 The molecular wheelbarrow and its conventional mechanical prototype. The two wheels are made of triptycene moieties attached to the molecular frame with two acetylene axels.

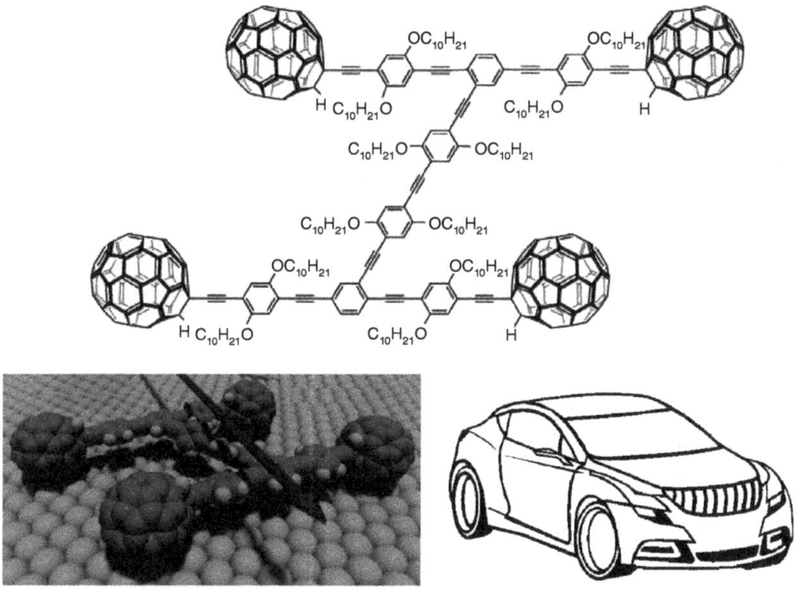

Figure 1.10 An example of molecular cars with four wheels. *Source*: Adapted from Shirai *et al.* 2005 [42]. Reproduced with permission of American Chemical Society.

can be achieved by the addition of a functionalized "cargo bay". Due to the ability of picking up and dropping off the payload, these technomimetic molecules can be categorized as molecular trucks (Figure 1.11) [50].

Molecular gyroscopes [51–57], turnstiles [58–62], and compasses [63, 64] belong to the extended family of rotary devices [65]. In a very general sense, a common element of all these devices is a rotor embedded in a stationary molecular frame. Two representative examples of molecular gyroscopes are shown in Figure 1.12 [54, 56].

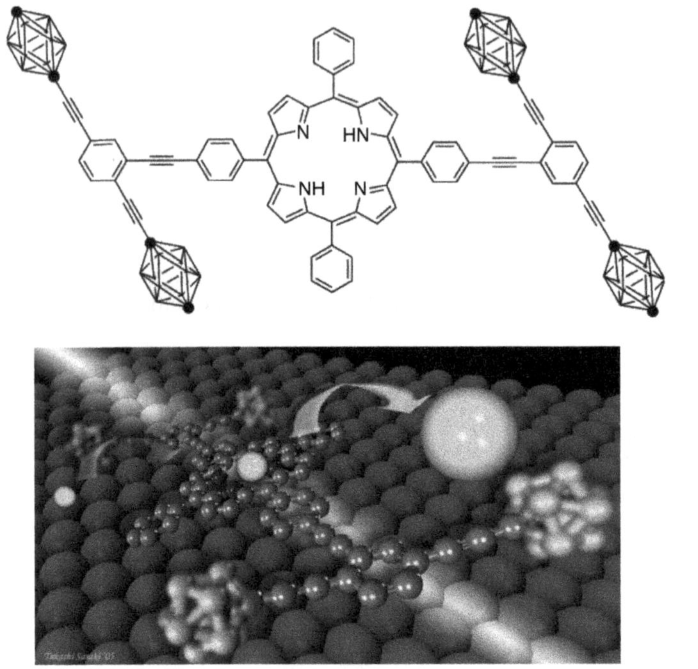

Figure 1.11 The molecular truck. *Source*: Reprinted from Sasaki *et al.* [50]. Copyright 2007, with permission from Elsevier.

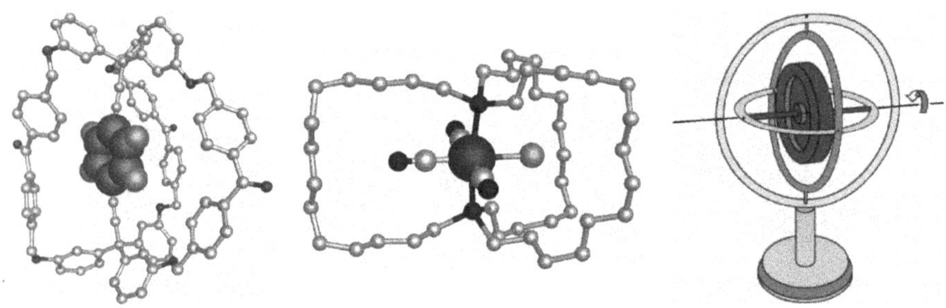

Figure 1.12 X-ray structures of molecular gyroscopes [54, 56] (left and middle, some hydrogen atoms were removed for clarity) and their conventional mechanical prototype (right). *Source*: Schalley 2002 [66]. Reproduced with permission of John Wiley & Sons.

While in conventional technology, gyroscopes and ball bearings are quite different devices with distinct application fields, it is possible to consider their molecular counterparts together since in the molecular domain their modes of operations are somewhat similar. In essence, molecular gyroscopes and ball bearings are molecular systems that allow some form of partially restricted rotation of a smaller, internal part of the molecular assembly relative to a bigger, external framework. In this respect, the only difference between the molecular gyroscopes and the ball bearings is the degree of rotation restrictions, which is usually lower in the molecular ball bearings compared to the molecular gyroscopes.

The abovementioned considerations can be applied for molecular belts as components of molecular ball joint bearings and gyroscopes. One of the examples of Matryoshka-type "gyroscane" complexes, constructed from two cucurbituril belt molecules, is presented in Figure 1.13 [67]. In this gyroscane complex, the internal cucurbit[5]uril part and associated chloride anion can rotate freely within the external cucurbit[10]uril molecular frame.

An exquisite example of a molecular ball joint bearing was recently constructed from cyclo[4]chrysenylene and fullerene C_{60} derivatives. Fullerene C_{60} and cyclo[4]chrysenylene produce a strong complex ($\log K_a = 12.3$ in benzene), in which the internal C_{60} part can freely rotate in all directions within the cyclo[4]chrysenylene framework. Subsequently, extensive NMR studies demonstrated the existence of partially restricted motion for systems with the fullerene C_{60} derivative bearing *N*-methylpyrrolidine "shaft" moiety (Figure 1.14) [68].

More details regarding this technomimetic molecule with dual-mode rolling dynamics were obtained with the use of modern computational methods (Figure 1.15). It was found that the substituted fullerene C_{60} ball underwent two types of distinctive rolling dynamics (spin and precession) with relatively low activation energies in a range of 4–5 kcal/mol [69].

Other notable examples of technomimetic molecules include molecular pumps [70], spoked wheels [71–74], pinwheels [75], molecular hinges [76],

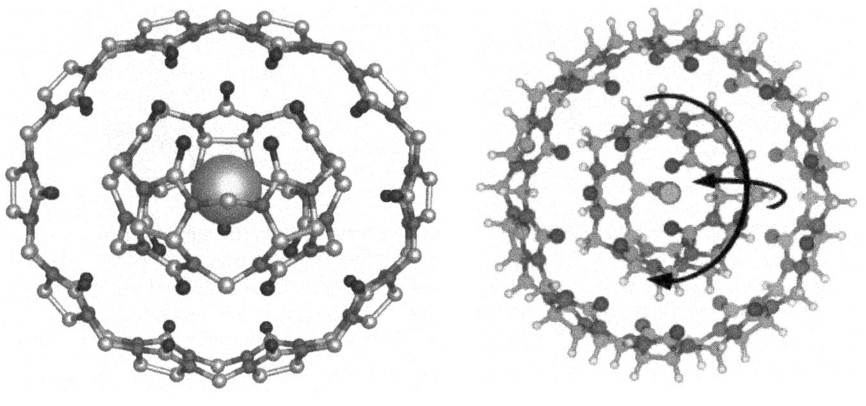

Figure 1.13 X-ray structure of the Matryoshka-type complex of cucurbit[5]uril and associated chloride anion within cucurbit[10]uril (left, hydrogen atoms were removed for clarity) and its dynamics (right). *Source*: Day *et al.* 2002 [67]. Reproduced with permission of John Wiley & Sons.

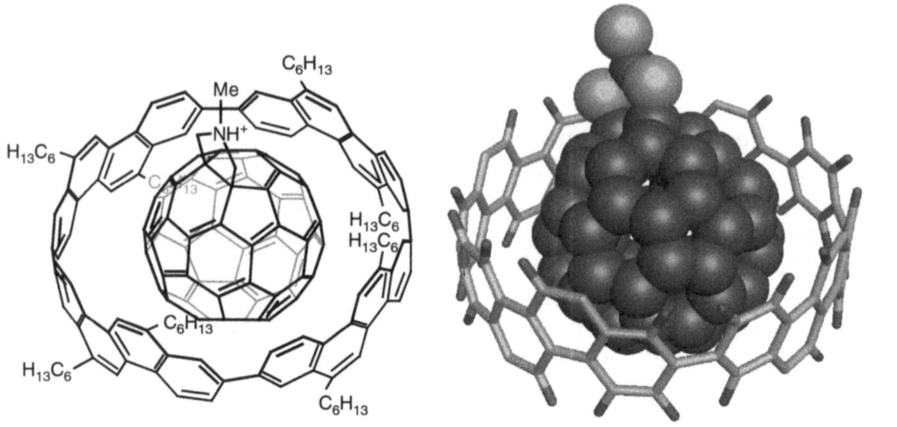

Figure 1.14 The complex between substituted (*P*)-(12,8)-cyclo[4]chrysenylene and *N*-methylpyrrolidine derivative of fullerene C_{60} (left) and its energy-minimized 3D structure (right). *n*-Hexyl groups as well as hydrogen atoms of the *N*-methylpyrrolidine handle moiety were removed for clarity (right).

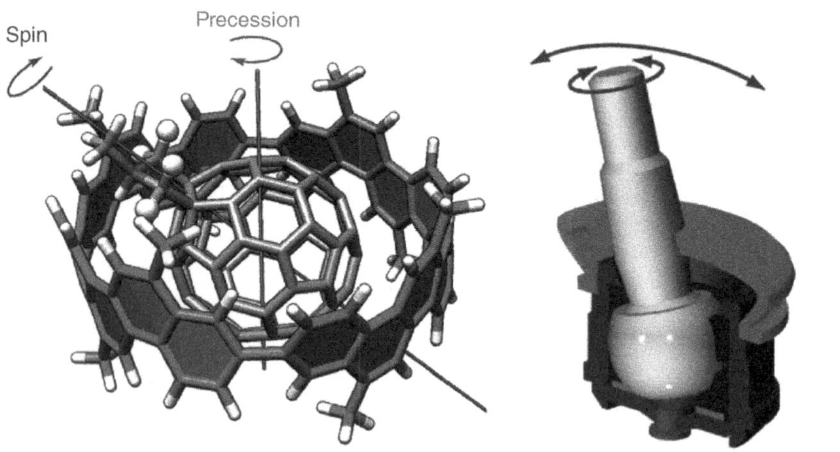

Figure 1.15 Dual-mode rolling dynamics (spin and precession) of the substituted (*P*)-(12,8)-cyclo[4]chrysenylene complex with *N*-methylpyrrolidine derivative of fullerene C_{60} as molecular ball joint bearings (left) and its mechanical counterpart (right). *Source*: Isobe *et al.* 2015 [69]. http://pubs.rsc.org/en/content/articlelanding/sc/2015/c5sc00335k#!divAbstract. Licensed under CC BY 3.0. *Additional credits*: Motor & Equipment Manufacturers Association (MEMA).

caterpillar tracks [77], elevators [78], and many others [79]. They are not as well developed as the previously mentioned examples but are important for understanding the true breadth and scope of the current technomimetic research field.

1.3 Manufacturing of Technomimetic Molecules

Unlike conventional precise engineering, which is the standard production method of almost all man-made functional devices, precise atom-by-atom molecular engineering still belongs to the realm of science fiction. More importantly, there is a growing consensus that manufacturing of molecular devices can be successfully accomplished using modifications of traditional synthetic methods successfully employed in organic chemistry and biochemistry for decades, if not centuries.

The first breakthrough in this direction was actually achieved in the late 1960s with the development of the early methods of "repair replication", [80] which were subsequently reduced to practice in the form of "polymerase chain reaction", or PCR [81]. PCR in its essence is a combination of the two well-known synthetic approaches: template-directed synthesis and repetitive synthesis. While the existing PCR protocols can only be applied for the production of DNA copies, both template-directed and repetitive syntheses are currently being used for manufacturing a variety of technomimetic molecules. These and other synthetic methods, which are specific for technomimetics, are described later in detail.

Repetitive (or iterative) synthesis is widely used in chemistry, even though it is not always named that way [82]. In essence, repetitive synthesis is a methodology where the same synthetic procedure is being repeated multiple times. This allows to "grow" the original molecular scaffold by adding chemically similar fragments in a repetitive manner. The simplest repetitive reaction is polymerization, such as the one used in the production of polyethylene. More precise version of polymerization is controllable oligomerization, which is frequently used for the synthesis of belt molecules and will be further discussed in Chapter 4.

A good example of repetitive synthesis is the preparation of [3.3]metacyclophane oligomers. These oligomers can be considered as "folding stairs" molecular ribbons. The general structure of these molecular ribbons is presented in Figure 1.16 [83, 84].

The repetitive synthesis of the [3.3]metacyclophane oligomers required a special synthetic strategy, which was developed on the basis of the key building block equipped with two sets of linker groups, CO_2Et and CH_2NHTs. The synthesis of this key building block is presented in Scheme 1.1 [83].

Controlled "growing" of the oligomeric units is initiated by the reduction of the ester groups as the first step. The second step is bromination leading to the corresponding bromomethyl derivative. Subsequent heterocyclization with two equivalents of the key building block discussed earlier yields the elongated cyclophane ribbon with the regeneration of the four ester groups required for the next iteration cycle (Scheme 1.2) [83, 84].

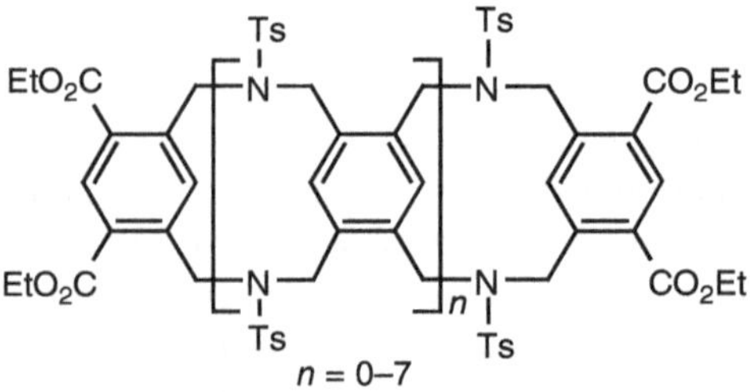

Figure 1.16 The general structure of the [3.3]metacyclophane oligomers ($Ts = p\text{-}CH_3C_6H_4SO_2$).

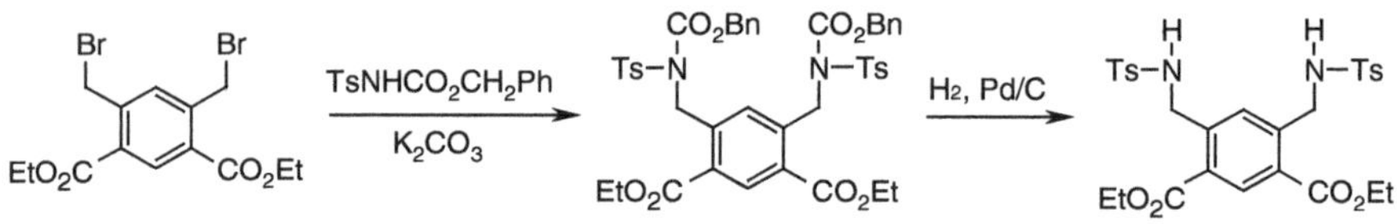

Scheme 1.1 Preparation of the key building block for the repetitive synthesis of the [3.3]metacyclophane oligomers ($Bn = PhCH_2$).

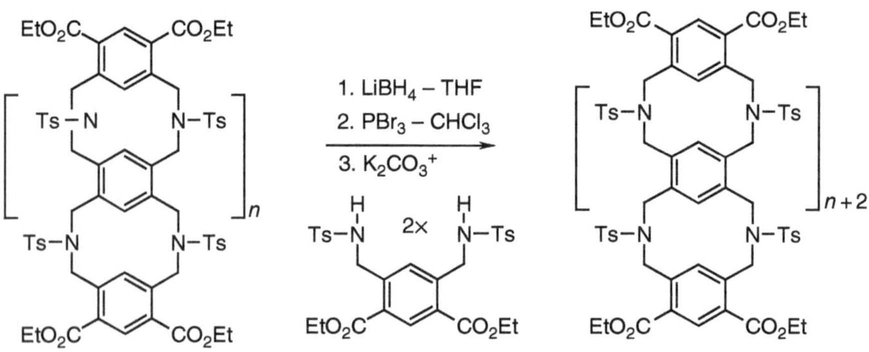

Scheme 1.2 Preparation of the stepwise-folded molecular ribbons by repetitive synthesis.

Single-crystal X-ray analysis and ^1H NMR spectroscopy confirm the all-*syn* arrangements of the repeating [3.3]metacyclophane units. Benzene rings align in a staircase-like manner almost parallel to each other. The [3.3]metacyclophane oligomer with nine benzene rings is the longest molecular ribbon prepared so far by this method (Figure 1.17). It was assumed that the synthesis of even longer cyclophane ribbons might be possible due to their reasonably good solubility, which appears to be independent of the ribbon length [84].

Further improvement of repetitive synthesis can be achieved with the addition of template-directed methodology [85, 86]. Template manufacturing is a well-known conventional engineering approach. In essence, to fabricate

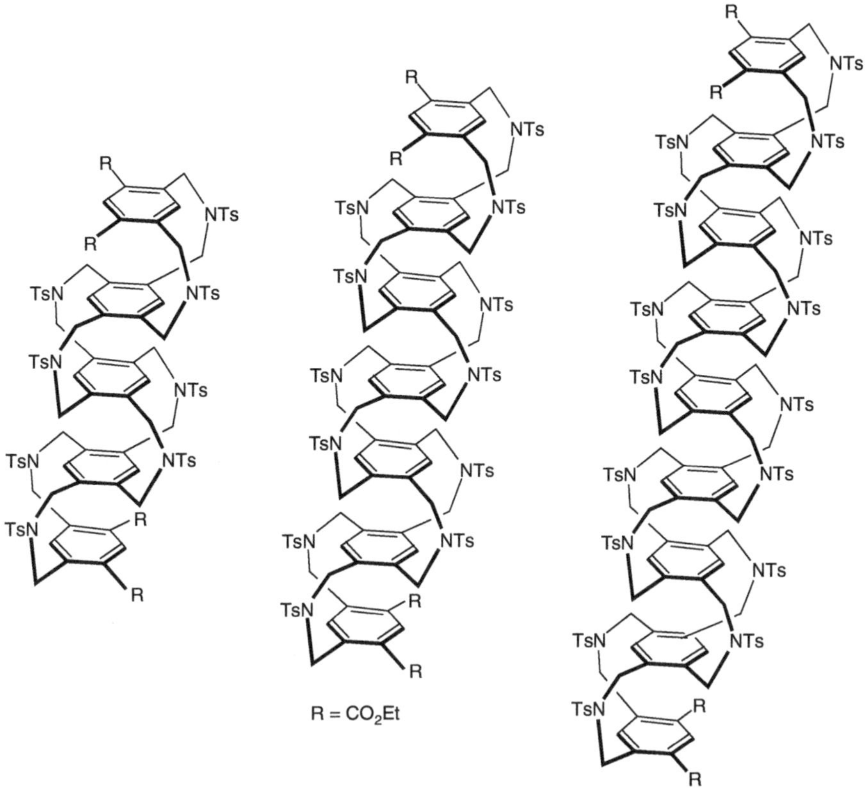

Figure 1.17 Examples of the stepwise-folded ribbon molecules prepared by controlled oligomerization.

a complex artificial object, it is sometimes easier to use a temporary support structure around (or within) the object and, after the construction is complete, remove the supporting elements during the final stages of production. Examples of the template manufacturing technology include the construction of buildings, bridges, concrete and ceramic items, and others. In chemistry, the ideas of template manufacturing are widely used in enzyme catalysis where the enzyme serves as a template to bring two (or several) molecules together in the proper position to facilitate their joining. The final product is being subsequently removed from the active site. Recently, the same idea was applied to the manufacturing of molecularly imprinted polymers, where template–monomer complexes were used to produce precise cavities within the polymer matrix (Scheme 1.3) [87–89].

A classic example of template-directed synthesis is preparation of container compounds from properly functionalized resorcin[4]arenes [90]. During the

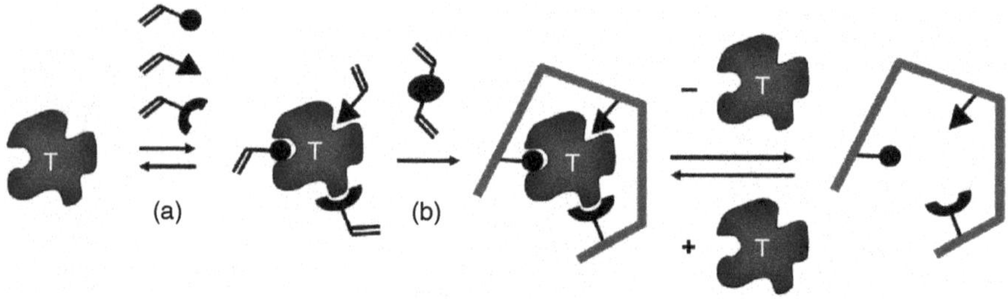

Scheme 1.3 Template-directed synthesis of molecularly imprinted polymers (T – template molecule); (a) template–monomer complexation, (b) matrix polymerization. *Credits*: The Zuse Institute, Berlin.

synthesis, the two resorcin[4]arene hemispheres are brought together with the help of a guest molecule, which remains inside during the multistep closure of the container. The yield of this closure process is determined by the nature of this template, which is typically a molecule of solvent with the proper size match (Scheme 1.4) [17, 91]. Unfortunately, in many cases, the template molecules are difficult to remove from the closed molecular containers. This problem can be resolved by constructing larger semipermeable container compounds (e.g., with four $(CH_2)_4$ links instead of four CH_2 links), which allow for the removal of template molecule at elevated temperatures [92].

Template-directed synthesis of technomimetic molecules is frequently performed in conjunction with repetitive methodology (template-directed repetitive synthesis). The abovementioned example actually represents a rudimental case of the template-directed repetitive synthesis since the closure of the two resorcin[4]arene hemispheres is conducted by performing repetitive linking of the hemispheres in several points. In advanced cases, a template molecule or an atom cluster serves as an assembly point for template-directed oligomerization of monomers in a precise and controllable manner.

One of the examples of the advanced template-directed repetitive synthesis of a giant porphyrin-based molecular tube fragment is presented in Scheme 1.5. In this particular case, the template approach allows to perform 12-ring closure via C—C bond formation in a highly controllable manner and in a reasonable yield. The template molecules can then be safely removed, thus yielding the desired template-free molecular tube fragment (Figure 1.18) [93].

Another classic example of template-directed repetitive synthesis is the production of all-carbon molecular tubes with the use of metal clusters. In this case, the metal clusters serve not only as templates but also as catalytic centers (Figure 1.19) [94, 95].

Unfortunately, precise manufacturing of single-walled carbon molecular tubes is difficult to achieve via metal cluster catalysis at elevated temperatures. This catalytic process typically leads to the formation of multiwalled molecular

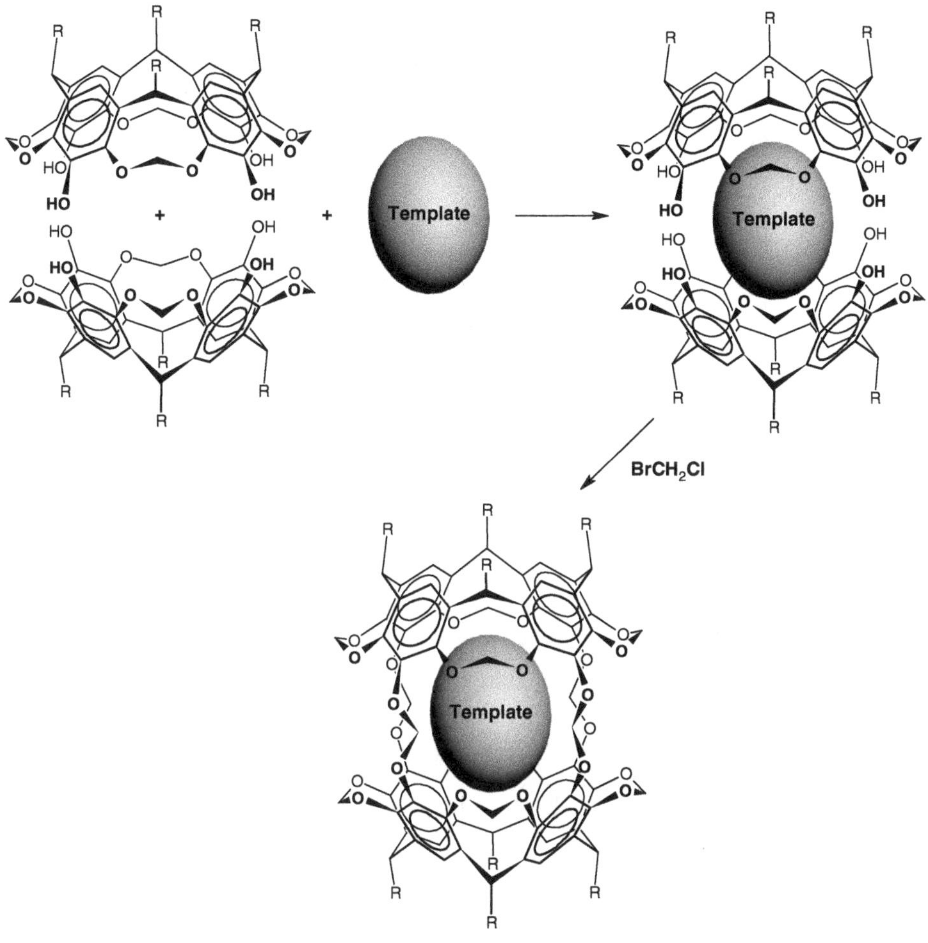

Scheme 1.4 Template-directed synthesis of resorcin[4]arene-based container compounds ($R = CH_2CH_2Ph$; template = pyrazine (87%) [91], 1,4-dioxane (68%) [91], $(CH_3)_2SO$ (61%) [17], $(CH_3)_2NCOCH_3$ (54%) [17], $(CH_3)_2NCHO$ (49%) [17], but not $CH_2(CH_2CH_2)_2NCHO$ (0%) [17]).

tube assemblies of various sizes. The size selectivity of carbon tube fabrication can be improved by utilizing carefully crafted preorganized polyaromatic templates. This organic template-directed method can provide access to single-walled molecular tubes with prearranged structural parameters. A recent example of this approach is presented in Figure 1.20 [96, 97].

Advanced template-directed repetitive synthesis allows for the addition of chemically homomorphic units in a controllable manner. The process is also known as coded heteropolymerization [98, 99] since it is possible to write (encode) information directly into the structure of the resulting

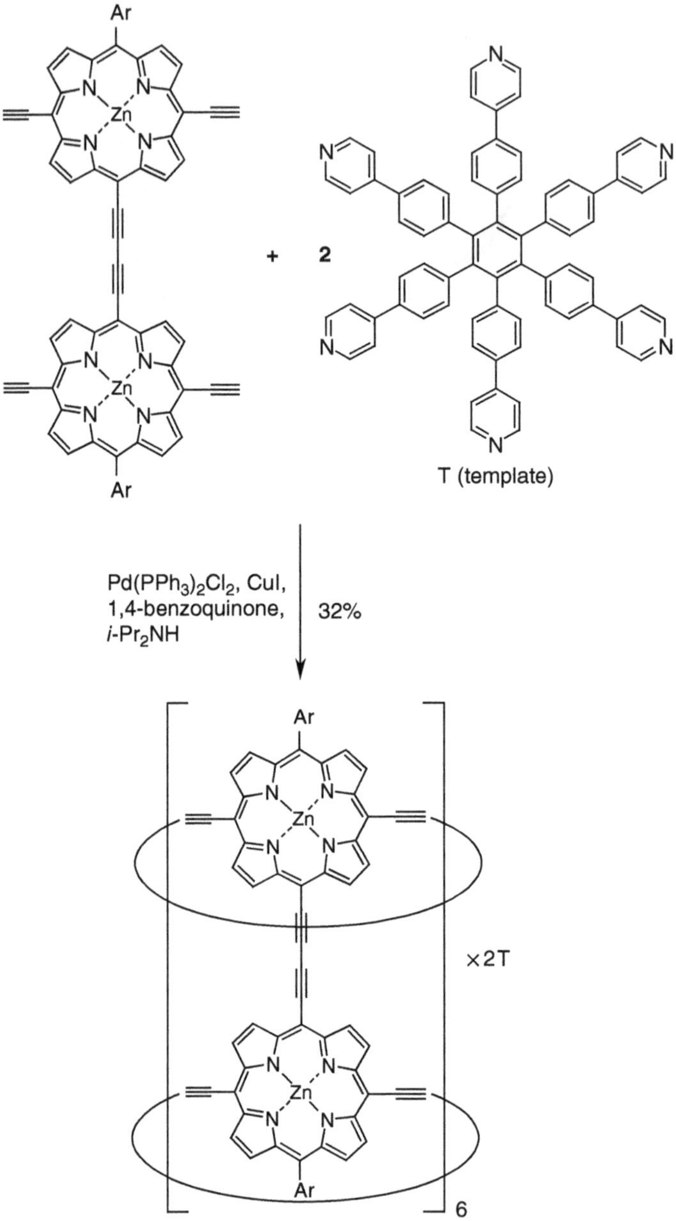

Scheme 1.5 Synthesis of the giant porphyrin-based molecular tube fragment via template-directed repetitive synthetic methodology; Ar = 3,5-$C_6H_3[Si(n\text{-}C_6H_{13})_3]$.

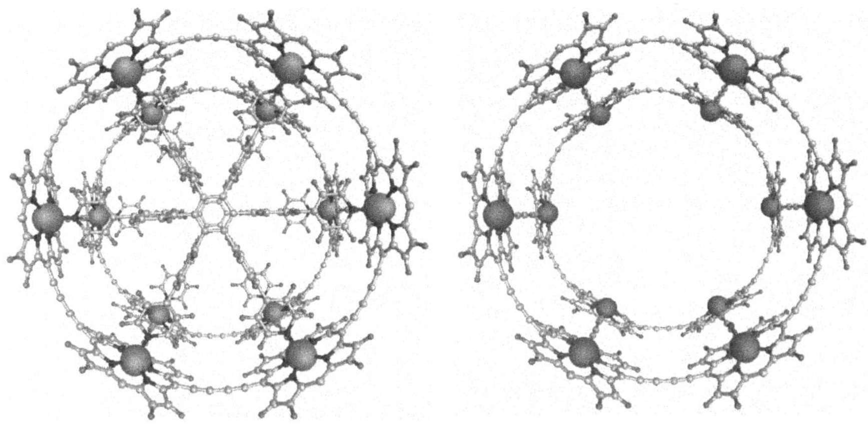

Figure 1.18 Energy-minimized structures of the giant molecular tube fragment with the attached template (left) and after template removal (right). Aryl moieties were omitted for clarity.

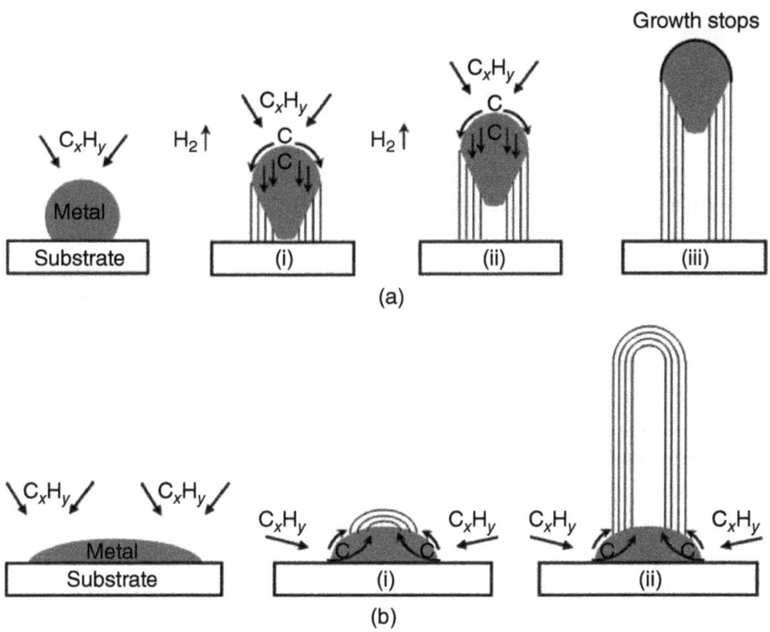

Figure 1.19 Template-directed repetitive synthesis of all-carbon molecular tubes with the use of metal clusters: (a) tip-growth, (b) base-growth. *Source*: Yan *et al.* 2015 [94]. Adapted with permission of Royal Society of Chemistry.

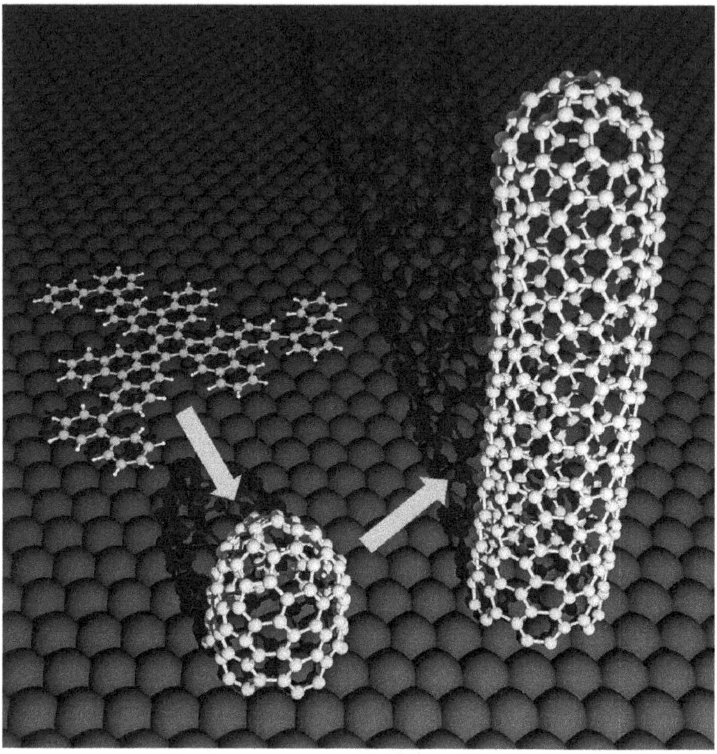

Figure 1.20 Preorganized polyaromatic template-directed repetitive synthesis of single-walled carbon molecular tubes. *Credits*: Juan Ramon Sanchez-Valencia/Empa.

heteropolymer, thus transferring chemical information from the template to the coded heteropolymer. Coded heteropolymerization requires execution of a tedious repetitive protocol and is typically done in semiautomatic synthesizers. The primary use of these semiautomatic chemical synthesizers is for the production of biologically relevant molecules, such as oligonucleotides and proteins, but they are capable of assembling other coded heteropolymers. There are also some molecular systems that would allow small protein manufacturing in a semiautomated mode (similar to conventional automatic peptide synthesizers), but the real progress in this area is still years away. It is also not very clear how complex technomimetic molecules can be manufactured this way, but at least the feasibility of this approach has already been demonstrated (Scheme 1.6) [100].

In rare circumstances, self-templating is possible, so the growing molecular fragment self-directs the proper alignment of added construction blocks. This unique self-templating mechanism is probably in play during a versatile and efficient synthesis of cucurbit[*n*]urils from simple building blocks, such

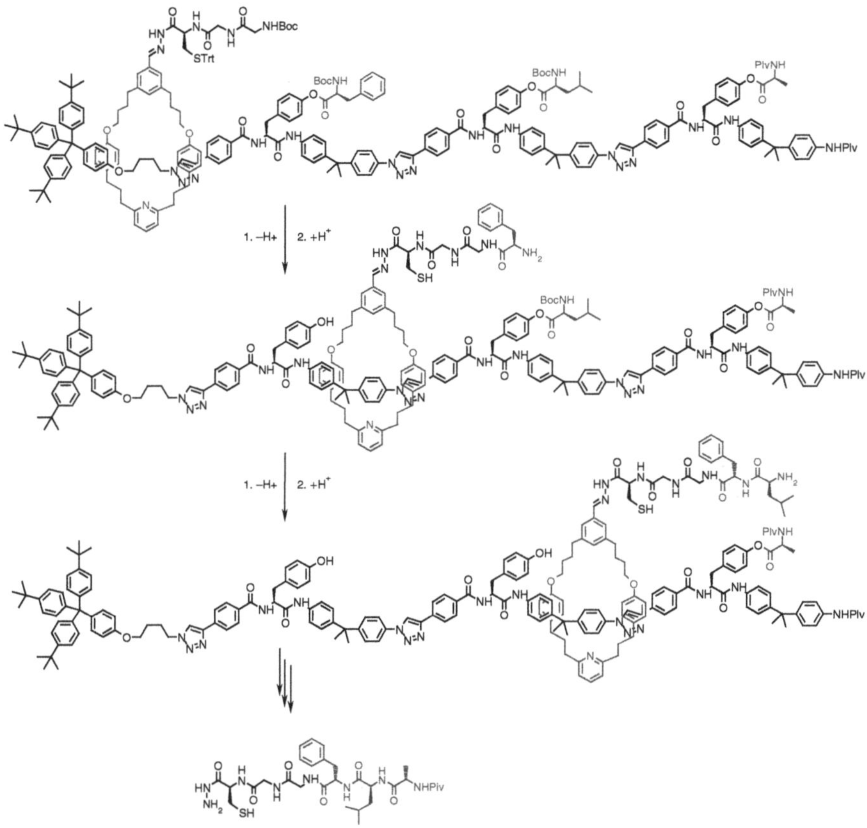

Scheme 1.6 The artificial molecular protein synthesizer utilizing template-directed repetitive chemistry methodology; Trt = triphenylmethyl, Boc = *tert*-butyloxycarbonyl, Piv = pivaloyl.

as formaldehyde and glycoluril in acidic conditions. To achieve the best results, prolonged heating is required since it would allow for de-assembly of imperfectly connected units and ultimate fabrication of the most thermodynamically stable macrocyclic belts. This process is presented in Scheme 1.7 [101]. Self-templating usually leads to a mixture of several structurally similar products, which in this particular case consists mainly of cucurbit[*n*]urils with various sizes ($n = 5–10$). It is conceivable that this unusually efficient macrocyclization reaction is also facilitated by "normal" template effects of hydronium (H_3O^+) cations [101].

One way to speed up and simplify the manufacturing of technomimetic molecules is to evoke a self-replicating synthesis, so theoretically an unlimited number of copies of the desired molecule can be prepared without a

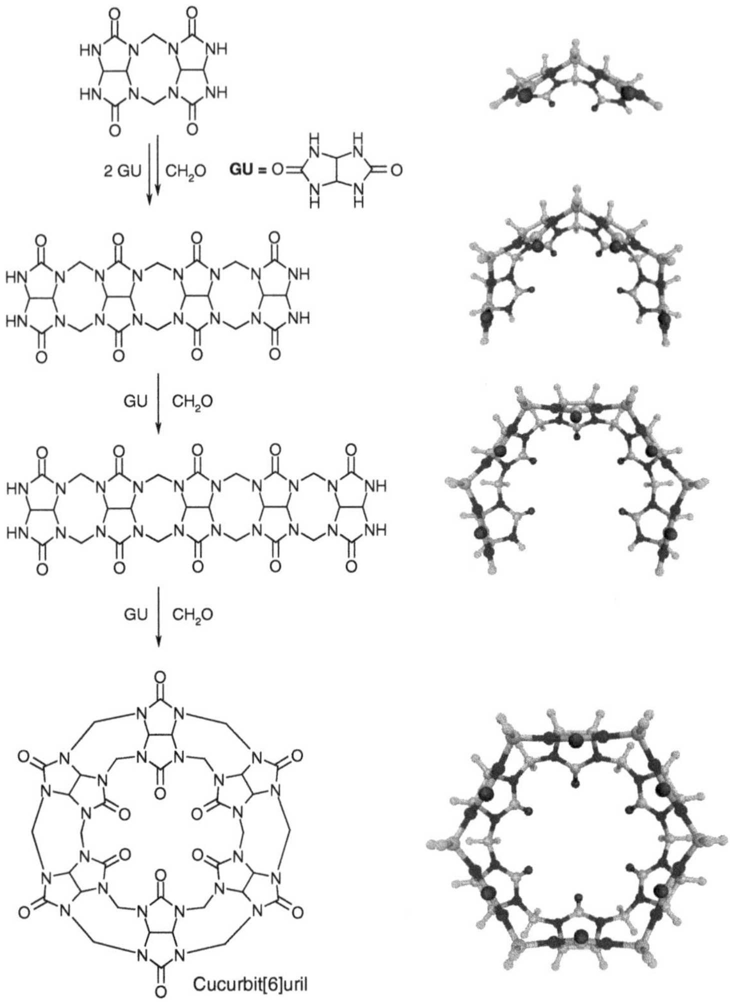

Scheme 1.7 The stepwise self-templating condensation of glycoluril (GU) and formaldehyde in acidic conditions. Only one cucurbit[*n*]uril ($n=6$) is depicted as a final product for simplicity.

conventional labor-intensive multistep synthesis. Although such artificial self-replicating chemical systems do exist [102], the utility of this approach so far was limited to experimental self-replicating molecules, which are of little value to technomimetics (Scheme 1.8) [103].

Self-replication typically involves self-assembly, which is defined as "the spontaneous assembly of molecules into structured, stable, noncovalently joined aggregates". [104] This approach so far achieved only a limited utility

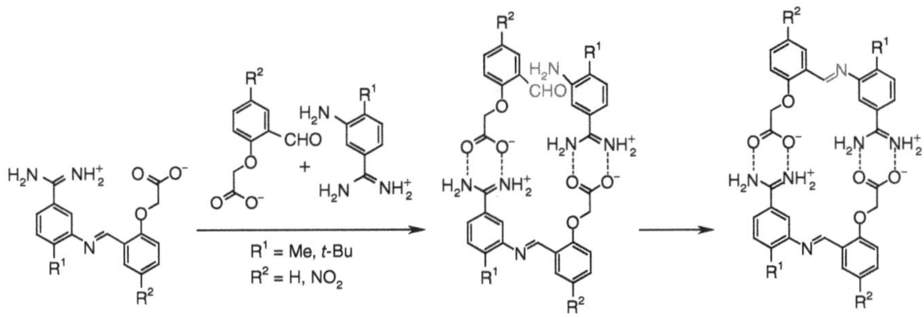

Scheme 1.8 An example of a simple self-replicating chemical system. Autocatalytic imine formation is facilitated by the template effect of the parent imine molecule.

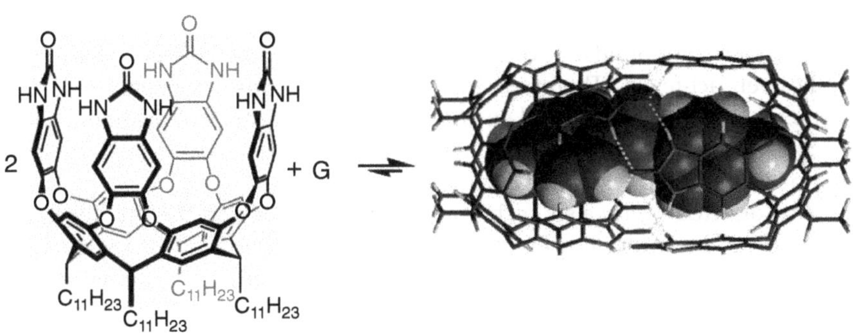

Scheme 1.9 Self-assembly of the molecular capsule from two deep cavitand units with tetraphenyleneurea "locks". *Source*: Choi *et al.* 2004 [108]. Reproduced with permission of American Chemical Society.

in the construction of technomimetic molecules since many technomimetics require a covalent bonding of functional fragments and therefore cannot be produced via self-assembly. As a representative example of self-assembly approach relevant to technomimetic field, some molecular capsules can be prepared by self-assembly of two complementary half-spheres (see also Chapter 2) [105–107]. One of these self-assembled dimeric capsules is presented in Scheme 1.9 [108].

1.4 Scope of the Book

It is virtually impossible to put all current research in the area of technomimetics in one book without breaking the limits of human attention. An unfortunate decision has been made to restrict the technomimetic story only to well-developed, matured areas, such as molecular tweezers, tubes, gears,

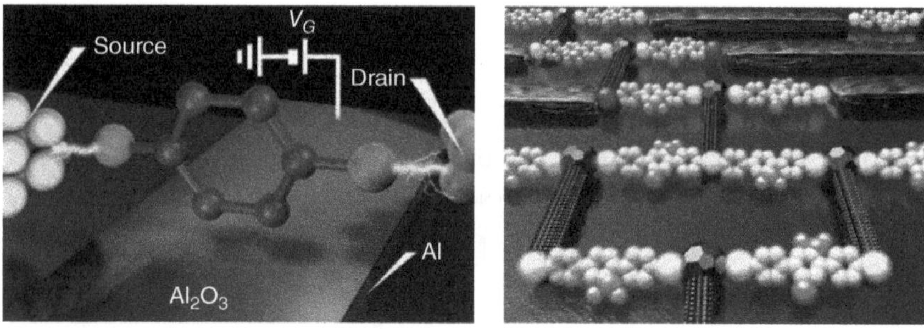

Figure 1.21 Examples of the proposed molecular electronic devices. *Source*: Tsutsui and Taniguchi 2012 [111]. http://www.mdpi.com/1424-8220/12/6/7259/htm. Licensed under CC BY 4.0. *Additional credits*: M. A. Reed/Yale University and P. Lugli/The Technical University of Munich.

and container compounds. In these areas, the critical development phase has already been achieved. Therefore, any further additions are expected in an incremental manner. This would ensure the lasting relevance of the book, with the acknowledgment of the inevitable loss of the in-depth coverage of the less developed, recently introduced areas of research.

Totally omitted were also technomimetic molecules related to modern electronic and computer devices, such as molecular wires, transistors, diodes, and other components (Figure 1.21) [109–112]. The current volume of the literature in this field is actually approaching the number of the references in "mechanical" technomimetic areas [113, 114]. In addition, modes of action and structures of these molecular electronic components are quite different compared to mechanical technomimetics [115]. This implies that the overview of the developments in molecular electronics should be best presented as a separate, stand-alone book.

References

1 Gakh, A.A., Sachleben, R.A., and Bryan, J.C. (1997) Molecular gearing systems. *Chemtech.*, **27**, 26–33.
2 Chen, C.W. and Whitlock, H.W. (1978) Molecular tweezers: a simple model of bifunctional intercalation. *J. Am. Chem. Soc.*, **100**, 4921–4922.
3 Rapenne, G. (2005) Synthesis of technomimetic molecules: towards rotation control in single-molecular machines and motors. *Org. Biomol. Chem.*, **3**, 1165–1169.
4 Balzani, V., Credi, A., and Venturi, M. (2008) *Molecular Devices and Machines: Concepts and Perspectives for the Nanoworld*, John Wiley & Sons.
5 Cram, D.J. (1992) Molecular container compounds. *Nature*, **356**, 29–36.

6 Rudkevich, D.M. (2002) Nanoscale molecular containers. *Bull. Chem. Soc. Jpn.*, **75**, 393–413.
7 Iwamura, H. (1985) Molecular design of correlated internal rotation. *J. Mol. Struct.*, **126**, 401–412.
8 Iwamura, H. and Mislow, K. (1988) Stereochemical consequences of dynamic gearing. *Acc. Chem. Res.*, **21**, 175–182.
9 Segawa, Y., Yagi, A., Matsui, K., and Itami, K. (2016) Design and synthesis of carbon nanotube segments. *Angew. Chem. Int. Ed.*, **55**, 5136–5158.
10 Hardouin-Lerouge, M., Hudhomme, P., and Sallé, M. (2010) Molecular clips and tweezers hosting neutral guests. *Chem. Soc. Rev.*, **40**, 30–43.
11 Leblond, J. and Petitjean, A. (2011) Molecular tweezers: concepts and applications. *ChemPhysChem*, **12**, 1043–1051.
12 Kelly, T.R., Bowyer, M.C., Bhaskar, K.V. *et al.* (1994) A molecular brake. *J. Am. Chem. Soc.*, **116**, 3657–3658.
13 Kelly, T.R., De Silva, H., and Silva, R.A. (1999) Unidirectional rotary motion in a molecular system. *Nature*, **401**, 150–152.
14 Jimenez-Bueno, G. and Rapenne, G. (2003) Technomimetic molecules: synthesis of a molecular wheelbarrow. *Tetrahedron Lett.*, **44**, 6261–6263.
15 Shirai, Y., Morin, J.-F., Sasaki, T. *et al.* (2006) Recent progress on nanovehicles. *Chem. Soc. Rev.*, **35**, 1043–1055.
16 Kinbara, K., Muraoka, T., and Aida, T. (2008) Chiral ferrocenes as novel rotary modules for molecular machines. *Org. Biomol. Chem.*, **6**, 1871–1876.
17 Sherman, J.C., Knobler, C.B., and Cram, D.J. (1991) Host–guest complexation. 56. Syntheses and properties of soluble carceplexes. *J. Am. Chem. Soc.*, **113**, 2194–2204.
18 Warmuth, R. (2000) The inner phase of molecular container compounds as a novel reaction environment. *J. Incl. Phenom. Macrocycl. Chem.*, **37**, 1–37.
19 Warmuth, R. (2010) Reactions inside carcerands, in *Molecular Encapsulation: Organic Reactions in Constrained Systems* (eds U.H. Brinker and J.-L. Mieusset), John Wiley & Sons, pp. 227–268.
20 Galan, A. and Ballester, P. (2016) Stabilization of reactive species by supramolecular encapsulation. *Chem. Soc. Rev.*, **45**, 1720–1737.
21 Warmuth, R., Maverick, E.F., Knobler, C.B., and Cram, D.J. (2003) Through-shell alkyllithium additions and borane reductions. *J. Org. Chem.*, **68**, 2077–2088.
22 Wong, T.-H., Chang, J.-C., Lai, C.-C. *et al.* (2014) Hemicarceplexes modify the solubility and reduction potentials of C_{60}. *J. Org. Chem.*, **79**, 3581–3586.
23 Popov, A.A., Yang, S., and Dunsch, L. (2013) Endohedral fullerenes. *Chem. Rev.*, **113**, 5989–6113.

24 Yang, S. and Wang, C.-R. (2014) *Endohedral Fullerenes: From Fundamentals to Applications*, World Scientific.

25 Lu, X., Echegoyen, L., Balch, A.L. *et al.* (eds) (2014) *Endohedral Metallofullerenes: Basics and Applications*, CRC Press.

26 Shinohara, S.H. and Tagmatarchis, S.N. (2015) *Endohedral Metallofullerenes: Fullerenes with Metal Inside*, John Wiley & Sons.

27 Burn-Callander, R. (2015) This is the Most Expensive Material in the World. The Telegraph.

28 Morton, J.J.L., Tyryshkin, A.M., Ardavan, A. *et al.* (2006) Electron spin relaxation of N@C_{60} in CS_2. *J. Chem. Phys.*, **124**, 14508.

29 Tang, Q., Abella, L., Hao, Y. *et al.* (2015) $Sc_2O@C_{2v}(5)$-C_{80}: dimetallic oxide cluster inside a C_{80} fullerene cage. *Inorg. Chem.*, **54**, 9845–9852.

30 Xu, W., Feng, L., Calvaresi, M. *et al.* (2013) An experimentally observed trimetallofullerene $Sm_3@I_h$-C_{80}: encapsulation of three metal atoms in a cage without a nonmetallic mediator. *J. Am. Chem. Soc.*, **135**, 4187–4190.

31 Wang, T.-S., Feng, L., Wu, J.-Y. *et al.* (2010) Planar quinary cluster inside a fullerene cage: synthesis and structural characterizations of $Sc_3NC@C_{80}$-I_h. *J. Am. Chem. Soc.*, **132**, 16362–16364.

32 Sanada, K., Ube, H., and Shionoya, M. (2016) Rotational control of a dirhodium-centered supramolecular four-gear system by ligand exchange. *J. Am. Chem. Soc.*, **138**, 2945–2948.

33 Iwamura, H., Ito, T., Ito, H. *et al.* (1984) Crystal and molecular structure of bis(9-triptycyl) ether. *J. Am. Chem. Soc.*, **106**, 4712–4717.

34 Behrend, R., Meyer, E., and Rusche, F. (1905) Ueber condensationsproducte aus glycoluril und formaldehyd. *Justus Liebigs Ann. Chem.*, **339**, 1–37.

35 Liu, Y.-Y., Lin, J.-Y., Bo, Y.-F. *et al.* (2016) Synthesis and crystal structure of highly strained [4]cyclofluorene: green-emitting fluorophore. *Org. Lett.*, **18**, 172–175.

36 Tong, L.H., Wietor, J.-L., Clegg, W. *et al.* (2008) Supramolecular assemblies of tripodal porphyrin hosts and C_{60}. *Chem. Eur. J.*, **14**, 3035–3044.

37 Muraoka, T., Kinbara, K., Kobayashi, Y., and Aida, T. (2003) Light-driven open-close motion of chiral molecular scissors. *J. Am. Chem. Soc.*, **125**, 5612–5613.

38 Muraoka, T., Kinbara, K., and Aida, T. (2006) Mechanical twisting of a guest by a photoresponsive host. *Nature*, **440**, 512–515.

39 Vives, G. and Tour, J.M. (2009) Synthesis of single-molecule nanocars. *Acc. Chem. Res.*, **42**, 473–487.

40 Joachim, C. and Rapenne, G. (2013) Molecule concept nanocars: chassis, wheels, and motors? *ACS Nano*, **7**, 11–14.

41 Akimov, A.V. and Kolomeisky, A.B. (2012) Unidirectional rolling motion of nanocars induced by electric field. *J. Phys. Chem. C*, **116**, 22595–22601.

42 Shirai, Y., Osgood, A.J., Zhao, Y. *et al.* (2005) Directional control in thermally driven single-molecule nanocars. *Nano Lett.*, **5**, 2330–2334.

43 Chu, P.-L.E., Wang, L.-Y., Khatua, S. *et al.* (2013) Synthesis and single-molecule imaging of highly mobile adamantane-wheeled nanocars. *ACS Nano*, **7**, 35–41.

44 Morin, J.-F., Sasaki, T., Shirai, Y. *et al.* (2007) Synthetic routes toward carborane-wheeled nanocars. *J. Org. Chem.*, **72**, 9481–9490.

45 Sasaki, T. and Tour, J.M. (2007) Synthesis of a dipolar nanocar. *Tetrahedron Lett.*, **48**, 5821–5824.

46 Khatua, S., Godoy, J., Tour, J.M., and Link, S. (2010) Influence of the substrate on the mobility of individual nanocars. *J. Phys. Chem. Lett.*, **1**, 3288–3291.

47 Vives, G. and Tour, J.M. (2009) Synthesis of a nanocar with organometallic wheels. *Tetrahedron Lett.*, **50**, 1427–1430.

48 Morin, J.-F., Shirai, Y., and Tour, J.M. (2006) En route to a motorized nanocar. *Org. Lett.*, **8**, 1713–1716.

49 Kudernac, T., Ruangsupapichat, N., Parschau, M. *et al.* (2011) Electrically driven directional motion of a four-wheeled molecule on a metal surface. *Nature*, **479**, 208–211.

50 Sasaki, T., Morin, J.-F., Lu, M., and Tour, J.M. (2007) Synthesis of a single-molecule nanotruck. *Tetrahedron Lett.*, **48**, 5817–5820.

51 Wang, L., Hampel, F., and Gladysz, J.A. (2006) "Giant" gyroscope-like molecules consisting of dipolar Cl-Rh-CO rotators encased in three-spoke stators that define 25–27-membered macrocycles. *Angew. Chem. Int. Ed.*, **45**, 4372–4375.

52 Nawara, A.J., Shima, T., Hampel, F., and Gladysz, J.A. (2006) Gyroscope-like molecules consisting of PdX_2/PtX_2 rotators encased in three-spoke stators: synthesis via alkene metathesis, and facile substitution and demetalation. *J. Am. Chem. Soc.*, **128**, 4962–4963.

53 Shima, T., Hampel, F., and Gladysz, J.A. (2004) Molecular gyroscopes: $\{Fe(CO)_3\}$ and $\{Fe(CO)_2(NO)\}^+$ rotators encased in three-spoke stators; facile assembly by alkene metatheses. *Angew. Chem. Int. Ed.*, **43**, 5537–5540.

54 Hess, G.D., Hampel, F., and Gladysz, J.A. (2007) Octahedral gyroscope-like molecules with $M(CO)_3(X)$ rotators encased in three-spoked diphosphine stators: syntheses by alkene metathesis/hydrogenation sequences, structures, dynamic properties, and reactivities. *Organometallics*, **26**, 5129–5131.

55 Commins, P. and Garcia-Garibay, M.A. (2014) Photochromic molecular gyroscope with solid state rotational states determined by an azobenzene bridge. *J. Org. Chem.*, **79**, 1611–1619.
56 Commins, P., Nuñez, J.E., and Garcia-Garibay, M.A. (2011) Synthesis of bridged molecular gyroscopes with closed topologies: triple one-pot macrocyclization. *J. Org. Chem.*, **76**, 8355–8363.
57 Inagaki, Y., Yamaguchi, K., and Setaka, W. (2014) A crystalline molecular gyrotop with germanium junctions between a phenylene rotor and alkyl spokes. *RSC Adv.*, **4**, 58624–58630.
58 Zigon, N., Guenet, A., Graf, E. *et al.* (2013) A platinum turnstile with a palladium lock. *Dalton Trans.*, **42**, 9740–9745.
59 Bedard, T.C. and Moore, J.S. (1995) Design and synthesis of molecular turnstiles. *J. Am. Chem. Soc.*, **117**, 10662–10671.
60 Zigon, N., Larpent, P., Jouaiti, A. *et al.* (2014) A luminescent molecular turnstile. *Dalton Trans.*, **43**, 15779–15784.
61 Wang, G., Xiao, H., He, J. *et al.* (2016) Molecular turnstiles regulated by metal ions. *J. Org. Chem.*, **81**, 3364–3371.
62 Zigon, N., Kyritsakas, N., and Hosseini, M.W. (2013) Organometallic turnstiles: acid and base locking and unlocking. *Dalton Trans.*, **43**, 152–157.
63 Garcia-Garibay, M.A. (2005) Crystalline molecular machines: encoding supramolecular dynamics into molecular structure. *Proc. Natl. Acad. Sci. U.S.A.*, **102**, 10771–10776.
64 Vogelsberg, C.S. and Garcia-Garibay, M.A. (2012) Crystalline molecular machines: function, phase order, dimensionality, and composition. *Chem. Soc. Rev.*, **41**, 1892–1910.
65 Kottas, G.S., Clarke, L.I., Horinek, D., and Michl, J. (2005) Artificial molecular rotors. *Chem. Rev.*, **105**, 1281–1376.
66 Schalley, C.A. (2002) Of molecular gyroscopes, matroshka dolls, and other "nano"-toys. *Angew. Chem. Int. Ed.*, **41**, 1513–1515.
67 Day, A.I., Blanch, R.J., Arnold, A.P. *et al.* (2002) A cucurbituril-based gyroscane: a new supramolecular form. *Angew. Chem. Int. Ed.*, **41**, 275–277.
68 Isobe, H., Hitosugi, S., Yamasaki, T., and Iizuka, R. (2013) Molecular bearings of finite carbon nanotubes and fullerenes in ensemble rolling motion. *Chem. Sci.*, **4**, 1293–1297.
69 Isobe, H., Nakamura, K., Hitosugi, S. *et al.* (2015) Theoretical studies on a carbonaceous molecular bearing: association thermodynamics and dual-mode rolling dynamics. *Chem. Sci.*, **6**, 2746–2753.
70 Coulston, R.J., Onagi, H., Lincoln, S.F., and Easton, C.J. (2006) Harnessing the energy of molecular recognition in a nanomachine having a photochemical on/off switch. *J. Am. Chem. Soc.*, **128**, 14750–14751.

71 Mössinger, D., Hornung, J., Lei, S. *et al.* (2007) Molecularly defined shape-persistent 2D oligomers: the covalent-template approach to molecular spoked wheels. *Angew. Chem. Int. Ed.*, **46**, 6802–6806.
72 Mössinger, D., Chaudhuri, D., Kudernac, T. *et al.* (2010) Large all-hydrocarbon spoked wheels of high symmetry: modular synthesis, photophysical properties, and surface assembly. *J. Am. Chem. Soc.*, **132**, 1410–1423.
73 Aggarwal, A.V., Jester, S.-S., Taheri, S.M. *et al.* (2013) Molecular spoked wheels: synthesis and self-assembly studies on rigid nanoscale 2D objects. *Chem. Eur. J.*, **19**, 4480–4495.
74 May, R., Jester, S.-S., and Höger, S. (2014) A giant molecular spoked wheel. *J. Am. Chem. Soc.*, **136**, 16732–16735.
75 Vaughan, O.P.H., Williams, F.J., Bampos, N., and Lambert, R.M. (2006) A chemically switchable molecular pinwheel. *Angew. Chem. Int. Ed.*, **45**, 3779–3781.
76 Norikane, Y. and Tamaoki, N. (2004) Light-driven molecular hinge: a new molecular machine showing a light-intensity-dependent photoresponse that utilizes the trans–cis isomerization of azobenzene. *Org. Lett.*, **6**, 2595–2598.
77 Liu, S., Kondratuk, D.V., Rousseaux, S.A.L. *et al.* (2015) Caterpillar track complexes in template-directed synthesis and correlated molecular motion. *Angew. Chem. Int. Ed.*, **54**, 5355–5359.
78 Badjić, J.D., Balzani, V., Credi, A. *et al.* (2004) A molecular elevator. *Science*, **303**, 1845–1849.
79 Erbas-Cakmak, S., Leigh, D.A., McTernan, C.T., and Nussbaumer, A.L. (2015) Artificial molecular machines. *Chem. Rev.*, **115**, 10081–10206.
80 Kleppe, K., Ohtsuka, E., Kleppe, R. *et al.* (1971) Studies on polynucleotides: XCVI. Repair replication of short synthetic DNA's as catalyzed by DNA polymerases. *J. Mol. Biol.*, **56**, 341–361.
81 Saiki, R.K., Scharf, S., Faloona, F. *et al.* (1985) Enzymatic amplification of beta-globin genomic sequences and restriction site analysis for diagnosis of sickle cell anemia. *Science*, **230**, 1350–1354.
82 Feuerbacher, N. and Vögtle, F. (1998) Iterative synthesis in organic chemistry, in *Dendrimers*, Topics in Current Chemistry, vol. **197** (ed. F. Vögtle), Springer, Berlin/Heidelberg, pp. 1–18.
83 Breidenbach, S., Ohren, S., Nieger, M., and Vögtle, F. (1995) An iterative synthetic approach to nanometre-scale molecular ribbons. *J. Chem. Soc. Chem. Commun.*, 1237–1238.
84 Breidenbach, S., Ohren, S., and Vögtle, F. (1996) Nanometre-scale molecular ribbons. *Chem. Eur. J.*, **2**, 832–837.
85 Diederich, F. and Stang, P. (eds) (2000) *Templated Organic Synthesis*, John Wiley & Sons, Weinheim.

86 Gerbeleu, N.V., Arion, V.B., and Burgess, J.P. (2008) *Template Synthesis of Macrocyclic Compounds*, John Wiley & Sons.
87 Wulff, G. (2002) Enzyme-like catalysis by molecularly imprinted polymers. *Chem. Rev.*, **102**, 1–28.
88 Haupt, K. and Mosbach, K. (2000) Molecularly imprinted polymers and their use in biomimetic sensors. *Chem. Rev.*, **100**, 2495–2504.
89 Vasapollo, G., Sole, R.D., Mergola, L. *et al.* (2011) Molecularly imprinted polymers: present and future prospective. *Int. J. Mol. Sci.*, **12**, 5908–5945.
90 Jasat, A. and Sherman, J.C. (1999) Carceplexes and hemicarceplexes. *Chem. Rev.*, **99**, 931–968.
91 Chapman, R.G., Chopra, N., Cochien, E.D., and Sherman, J.C. (1994) Carceplex formation: scope of a remarkably efficient encapsulation reaction. *J. Am. Chem. Soc.*, **116**, 369–370.
92 Robbins, T.A., Knobler, C.B., Bellew, D.R., and Cram, D.J. (1994) Host–guest complexation. 67. A highly adaptive and strongly binding hemicarcerand. *J. Am. Chem. Soc.*, **116**, 111–122.
93 Neuhaus, P., Cnossen, A., Gong, J.Q. *et al.* (2015) A molecular nanotube with three-dimensional π-conjugation. *Angew. Chem. Int. Ed.*, **54**, 7344–7348.
94 Yan, Y., Miao, J., Yang, Z. *et al.* (2015) Carbon nanotube catalysts: recent advances in synthesis, characterization and applications. *Chem. Soc. Rev.*, **44**, 3295–3346.
95 Kumar, M. (2011) Carbon nanotube synthesis and growth mechanism, in *Carbon Nanotubes – Synthesis, Characterization, Applications* (ed. S. Yellampalli), InTech, pp. 147–170.
96 Sanchez-Valencia, J.R., Dienel, T., Gröning, O. *et al.* (2014) Controlled synthesis of single-chirality carbon nanotubes. *Nature*, **512**, 61–64.
97 Halford, B. (2014) Seeds sprout select nanotubes. *Chem. Eng. News*, **92**, 5.
98 Gakh, A.A. (1999) Self-replicating peptides. *Chem. Eng. News*, **77**, 6.
99 Brown, T. (2011) *Introduction to Genetics: A Molecular Approach*, Garland Science.
100 Lewandowski, B., Bo, G.D., Ward, J.W. *et al.* (2013) Sequence-specific peptide synthesis by an artificial small-molecule machine. *Science*, **339**, 189–193.
101 Huang, W.-H., Zavalij, P.Y., and Isaacs, L. (2008) Cucurbit[*n*]uril formation proceeds by step-growth cyclo-oligomerization. *J. Am. Chem. Soc.*, **130**, 8446–8454.
102 Vidonne, A. and Philp, D. (2009) Making molecules make themselves – the chemistry of artificial replicators. *Eur. J. Org. Chem.*, **2009**, 593–610.
103 Terfort, A. and von Kiedrowski, G. (1992) Self-replication by condensation of 3-aminobenzamidines and 2-formylphenoxyacetic acids. *Angew. Chem. Int. Ed. Engl.*, **31**, 654–656.

104 Whitesides, G.M., Mathias, J.P., and Seto, C.T. (1991) Molecular self-assembly and nanochemistry: a chemical strategy for the synthesis of nanostructures. *Science*, **254**, 1312–1319.

105 Rebek, J. (2000) Host–guest chemistry of calixarene capsules. *Chem. Commun.*, 637–643.

106 Kobayashi, K. and Yamanaka, M. (2014) Self-assembled capsules based on tetrafunctionalized calix[4]resorcinarene cavitands. *Chem. Soc. Rev.*, **44**, 449–466.

107 Hof, F., Craig, S.L., Nuckolls, C., and Rebek, J. (2002) Molecular encapsulation. *Angew. Chem. Int. Ed.*, **41**, 1488–1508.

108 Choi, H.-J., Park, Y.S., Cho, C.S. *et al.* (2004) Unusually stable molecular capsule formation of a tetraphenyleneurea cavitand. *Org. Lett.*, **6**, 4431–4433.

109 Joachim, C., Gimzewski, J.K., and Aviram, A. (2000) Electronics using hybrid-molecular and mono-molecular devices. *Nature*, **408**, 541–548.

110 Heath, J.R. and Ratner, M.A. (2003) Molecular electronics. *Phys. Today*, **56**, 43–49.

111 Tsutsui, M. and Taniguchi, M. (2012) Single molecule electronics and devices. *Sensors*, **12**, 7259–7298.

112 Xiang, D., Wang, X., Jia, C. *et al.* (2016) Molecular-scale electronics: from concept to function. *Chem. Rev.*, **116**, 4318–4440.

113 Moth-Poulsen, K. (ed.) (2016) *Handbook of Single-Molecule Electronics*, CRC Press.

114 Kiguchi, M. (ed.) (2016) *Single-Molecule Electronics: An Introduction to Synthesis, Measurement and Theory*, Springer, Singapore.

115 Cuevas, J.C. and Scheer, E. (2010) *Molecular Electronics: An Introduction to Theory and Experiment*, World Scientific.

2

Container Compounds

CHAPTER MENU

2.1 Introduction

Containers are one of the simplest technological devices that were invented and used throughout the human history. In essence, containers serve to separate a limited space from the environment for preservation, transportation, and transformation of the content. Many primitive containers do not have lids and rely on gravity to achieve storage and release of the content (such as vases and mugs), or require a physical breakup to release the content (tin cans). The most prominent feature of advanced containers is the controlled release mechanism. This controlled release mechanism can be as simple as a wine bottle cork, based on friction and requiring substantial external force to open, or can be as complex as a computer-controlled valve found in modern cars, refrigerators, and washing machines (Figure 2.1). Most modern chemical engineering designs also employ various containers with controlled release mechanisms.

The construction of conventional containers usually entails the joining of several curved or tubular components, with lids or valves being manufactured separately and then assembled during the last stage of the production sequence.

Molecular Devices: An Introduction to Technomimetics and Its Biological Applications,
First Edition. Andrei A. Gakh.

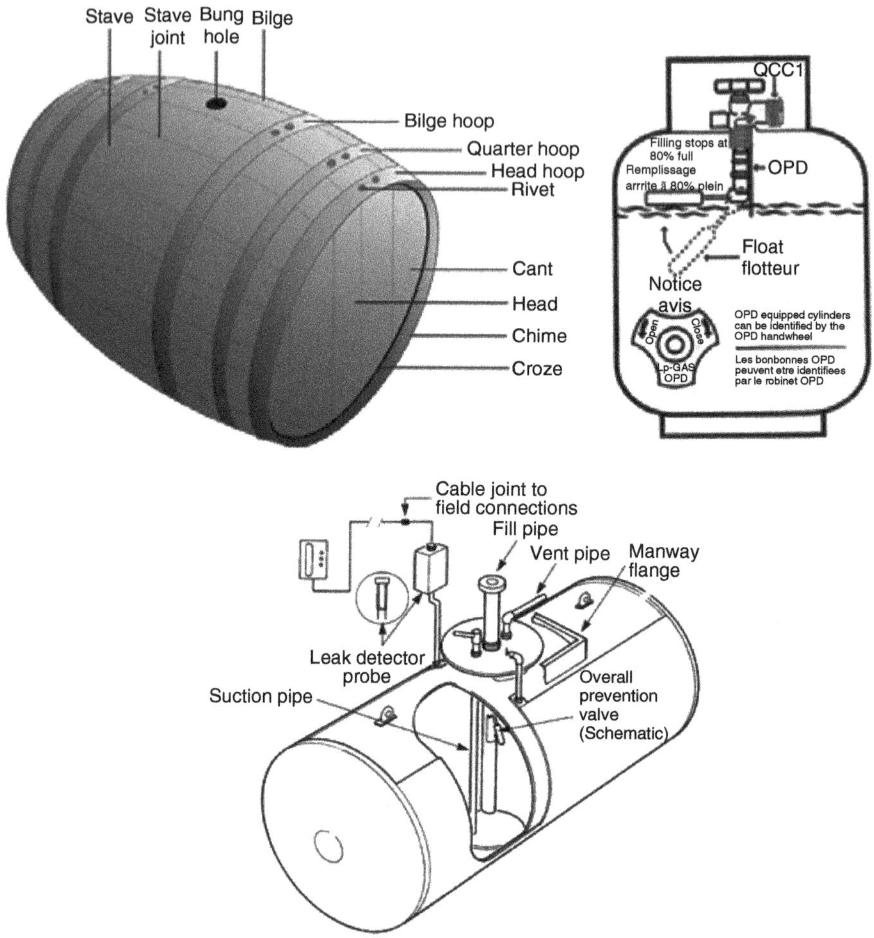

Figure 2.1 Conventional containers. *Credits*: Salut! Wine Co, National Fire Protection Association, JA Envirotanks.

Some notable examples include traditional wooden barrels, gas tanks, and large portable water reservoirs found in many provincial towns. In the molecular domain, template synthesis is similar to this conventional container construction technology. In this case, an appropriate guest molecule is being used as a template to properly align the container building blocks (Figure 2.2) [1]. The yield of the container compound is strongly influenced by the effectiveness of the template used, which can dramatically improve both the kinetic and the thermodynamic parameters of this reaction [2].

There are also conventional technologies that allow for the assembly of a container in one step from the melt (such as glass and plastic bottles), but these technologies proved to be difficult to replicate at the molecular level using a rational synthetic approach. The only notable example of a similar construction

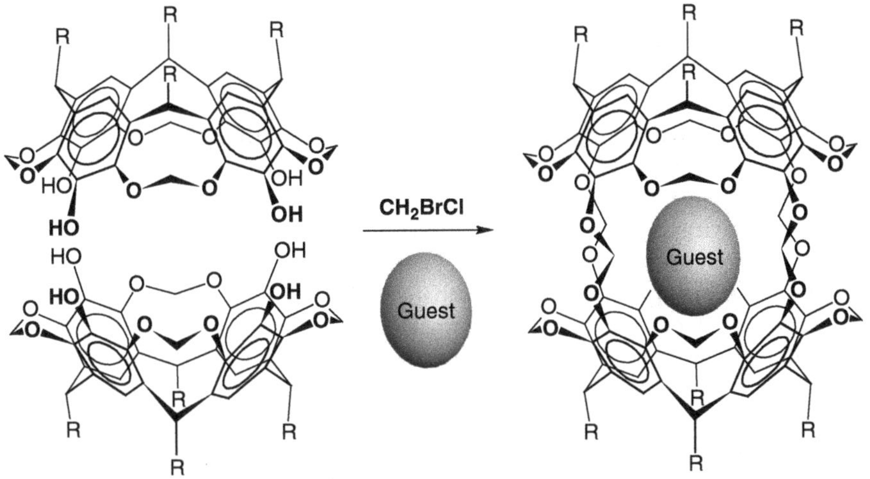

Figure 2.2 Template-directed synthesis of a pyrogallol[4]arene-based molecular container compound ($R = CH_2CH_2Ph$).

technology within the molecular domain is the preparation of fullerenes from graphite in arc-discharge conditions. Unfortunately, the exact mechanism of this remarkable reaction remains elusive, even after a quarter of century from the initial discovery [3].

Container compounds [4–6], as implied by their name, are typically three-dimensional molecular frameworks having a closed shape and composed of relatively inflexible fragments, which are conformationally rigid to prevent collapse. These molecular frameworks have an internal cavity capable of hosting an ion, an atom, or a small molecule. Container hosts usually have closed-surface spherical shells, but can also contain portals in their shells, with or without covering lids. In the case of closed shell container molecules, guest molecules can be permanently incarcerated during the synthesis. In the case of shells with portals, the molecular containers can be filled with complementary guest molecules after synthesis via a guest exchange process. A hybrid synthetic approach, when one guest molecule is being used as a template for the synthesis and then replaced by an appropriate guest, is also frequently employed.

Molecular containers are unique compounds because they can provide a potentially perfect chemical isolation of the appropriate guest from the environment. Some molecular containers with tight molecular mesh walls (such as fullerenes) are essentially nonpermeable at room temperature, but the preponderance of molecular containers allows certain permeability. In these semipermeable structures, the guest molecules are held by weak interactions with the interior via hydrogen bonding; ion–ion, ion–dipole, CH–π, and π–π interactions; van der Waals attractive forces; metal–ligand attractions; or solvophobic interactions [7]. It should be noted that complexes of carcerands, hemicarcerands, and self-assembling capsules also employ the so-called constrictive binding: the stability of these complexes is predominantly attained not through the host–guest attraction but through steric inhibition of the decomplexation process [8].

Depending on the construction and building parts involved, the extended family of molecular container compounds can be further classified as cavitands and carcerands. Cavitands (bowl-shaped molecules capable of holding various small guest molecules inside the cavity) can also be further subdivided into deep cavitands, bis-cavitands, and clamp-shell molecular containers (Figure 2.3). The major characteristic feature of cavitands is an open orifice, which provides easy access to the internal content. They are molecular analogs of conventional cups, buckets, and vases, which use gravity to retain the internal content. Since gravity cannot be effectively used in the molecular environment, small guests are retained in the molecular cavity by the combination of molecular attraction forces described earlier [7].

Carcerands and hemi-carcerands are fully closed molecular systems and could potentially keep guest molecules indefinitely without specific binding of these guest molecules to the interior (Figure 2.3). In practice, however, the preponderance of carcerands is filled with molecules that have at least some attraction to the inside environment. Fullerenes filled with inert gases, such as He, represent rare examples of carcerands where guest molecules are retained by pure mechanical forces, since guest escape would require a temporary rearrangement of the fullerene carbon framework (see also Section 2.11).

Similar to conventional containers, molecular container compounds have substantial application potential. These compounds can be used for selective binding, separation and sensing of small molecules and ions, molecular transport and delivery, stabilization of reactive intermediates, and catalysis via encapsulation (see also Section 2.12) [5–7, 9].

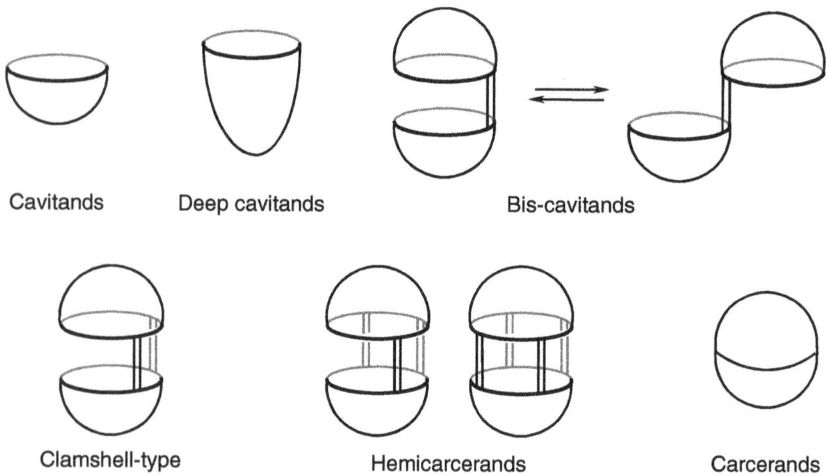

Figure 2.3 Typical structures of molecular container compounds.

2.2 Cavitands

Cavitands (see Figure 2.3) are bowl-shaped, open-ended molecules with reinforced internal cavities capable of binding complementary guests [4–7]. They are usually constructed from several aromatic or heteroaromatic units interconnected with short linking elements (usually CH_2 or CH(Alk) groups). The shape of these cavitands is determined by the number of aromatic units (usually three, four, or five) and relative link positions (*meta* or *ortho*).

Four important bowl-shaped macrocycles yielding a number of cavitands are cyclotriveratrylenes, calix[4]arenes, resorcin[4]arenes, and pyrogallol[4]arenes (Figure 2.4). The name *calixarenes* was derived from a type of Greek mixing bowl called “calyx krater” (Figure 2.4). Conformationally flexible calix[4]arenes, resorcin[4]arenes, and pyrogallol[4]arenes exist in several conformations. The

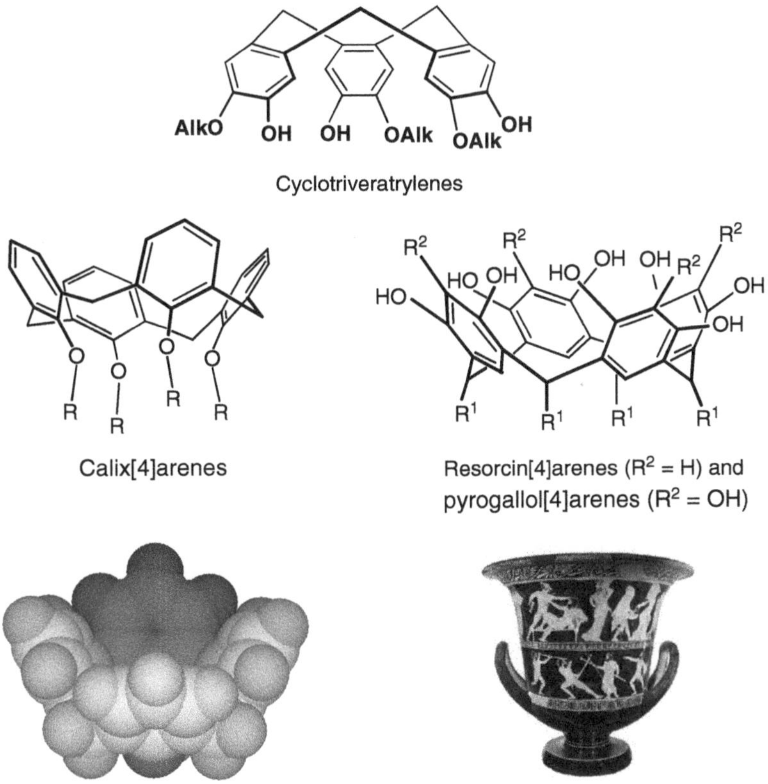

Figure 2.4 Structures of some bowl-shaped macrocycles frequently used for the construction of cavitands: cyclotriveratrylenes (top), calix[4]arenes (middle left, R = Alk), resorcin[4]arenes (middle right, R^1 = H, Alk; R^2 = H), and pyrogallol[4]arenes (middle right, R^1 = H, Alk; R^2 = OH), together with the energy-minimized space fill model of the simplest calix[4]arene in a cone confirmation (R = H, bottom left) and a picture of a calyx krater (bottom right). *Source*: Courtesy of The Classical Art Research Centre, University of Oxford.

desired bowl-shaped cone conformation can be fixed through appropriate derivatization [10–14].

Cyclotriveratrylenes are usually produced by acid-catalyzed trimerization of veratryl alcohol analogs in acidic conditions [10, 15]. Calixarenes can be synthesized by either acid- or base-catalyzed condensation of phenols and aldehydes [12, 13]. Similarly, resorcinarenes and pyrogallolarenes can be prepared from resorcinol, pyrogallol, and their derivatives by condensation with aldehydes under acidic or basic conditions [13, 14].

The bowl rim of all these macrocycles can be easily modified by the addition of functional substituents and bridging groups. For example, resorcin[4]arenes were successfully used in the preparation of alkylenedioxy- and dialkylsilicon-bridged cavitands [14] as presented in Figure 2.5. Compared to the parent resorcin[4]arenes, these simple cavitands have rigid structures and exist predominantly in the desired cone conformation. Although bridged resorcin[4]arenes have relatively narrow and shallow cavities, they can accommodate small linear molecules, such as CS_2 (Figure 2.5) [16].

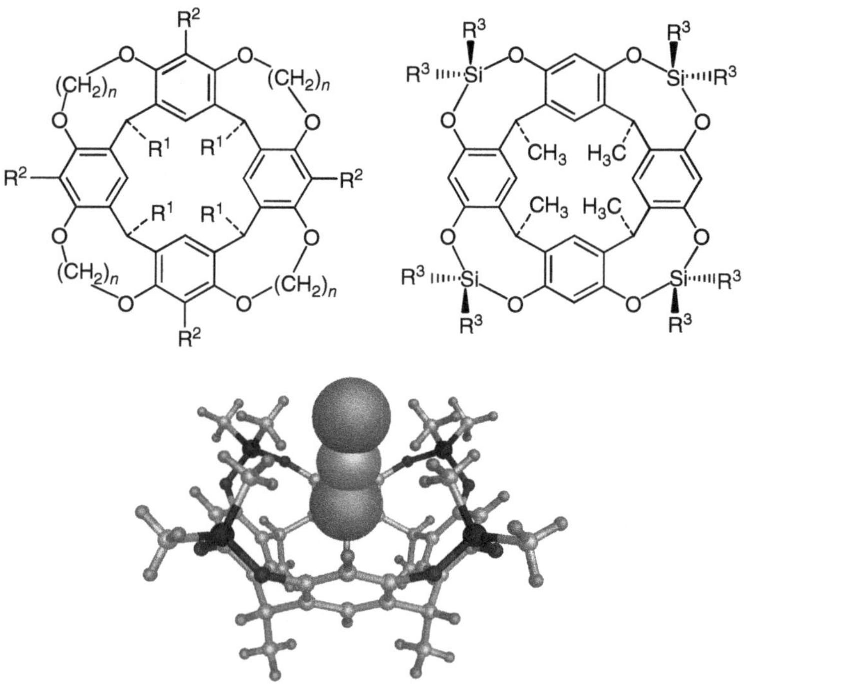

Figure 2.5 Examples of simple cavitands prepared from resorcin[4]arenes (top left, $n = 1–3$; $R^1 = H$, Alk; $R^2 = H$, Alk, OH, Hal, CH_2Hal; top right, $R^3 = Alk$) and the X-ray structure [16] of the dialkylsilicon-bridged cavitand complex with CS_2 (bottom, $R^3 = Me$).

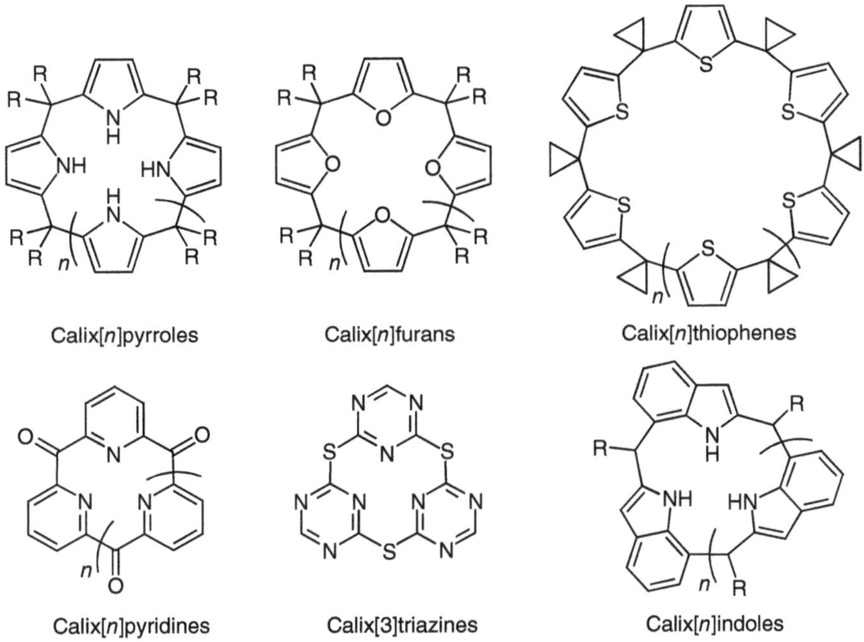

Figure 2.6 Examples of heterocalix[*n*]arenes [17].

Heterocalixarenes [17] (polycyclic molecules composed of various heterocycles – pyrroles, furans, thiophenes, pyridines, triazines, indoles, etc.) can also be used as starting materials for the preparation of cavitands with enhanced features attributed to the presence of heterocyclic moieties in their frameworks. Some examples of heterocalixarenes are presented in Figure 2.6. Mixed heterocalixarenes, such as multifarenes, are also known [18].

While many simple cavitands have been successfully employed as vase-type container compounds for various types of guest molecules, poorly developed cavities limit their ability for selective and strong entrapment of the internal content. These structural limitations can be overcome with the construction of deeper cavities by the expansion of aromatic walls of these cavitands.

2.3 Deep Cavitands

Previously described simple cavitands have rigid, but shallow cavities. In the preponderance of cavitand complexes, guest molecules are located only partially inside the cavitands, usually centering above the upper rim of the macrocyclic host [5, 7]. Several approaches have been developed to enlarge the cavity and to achieve a better guest entrapment. The schematic structure of *deep cavitands* is presented in Figure 2.3.

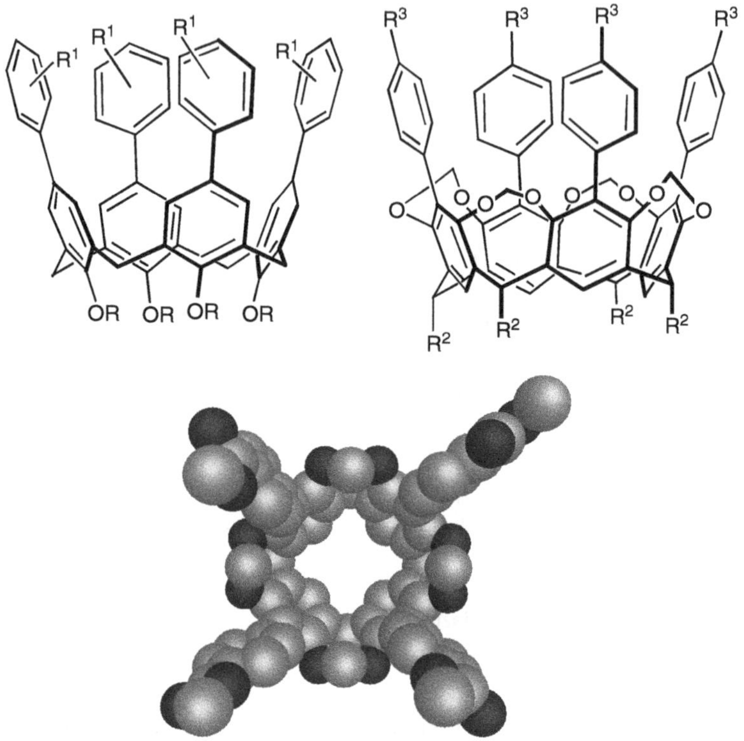

Figure 2.7 Deep cavitands prepared by direct C-arylation of calixarenes and resorcinarenes (top left, R = H, Alk; R^1 = H, OH, OAlk, NO_2, NH_2, NHC(O)Alk, NHC(O)NHAlk, Ph, *p*-C_6H_4OPh; top right, R^2 = Alk; R^3 = Br, CN, OH, NO_2, NH_2, NHC(O)Alk, NHC(O)NHAlk, COOMe) and the X-ray structure of one of these deep cavitands (bottom, $R^2 = C_5H_{11}$, R^3 = COOMe, hydrogen atoms and C_5H_{11} groups were removed for clarity).

The extension of the upper rim of calixarenes and resorcinarenes is frequently used for the construction of larger and deeper molecular containers. One popular approach toward larger (>10 Å) hydrophobic cavities entails direct C-arylation of the aromatic rings in calixarenes and resorcinarenes. For example, additional aryl fragments can be attached to the upper rim of calixarenes via palladium/boron-mediated (Suzuki) synthesis [19–21]. Similarly, resorcinarene-based cavitands can also be expanded via C-arylation by employing Suzuki coupling reactions (Figure 2.7) [22, 23].

Another strategy to prepare deep cavitands is based on proper bridging of residual resorcinarene hydroxyl groups [24–30]. For example, a family of deep cavitands was obtained by the reaction of resorcin[4]arenes with various benzylidene dibromides (Figure 2.8) [26, 27]. The condensation of resorcin[4]arenes with 2,3-dichloroquinoxaline resulted in the formation of

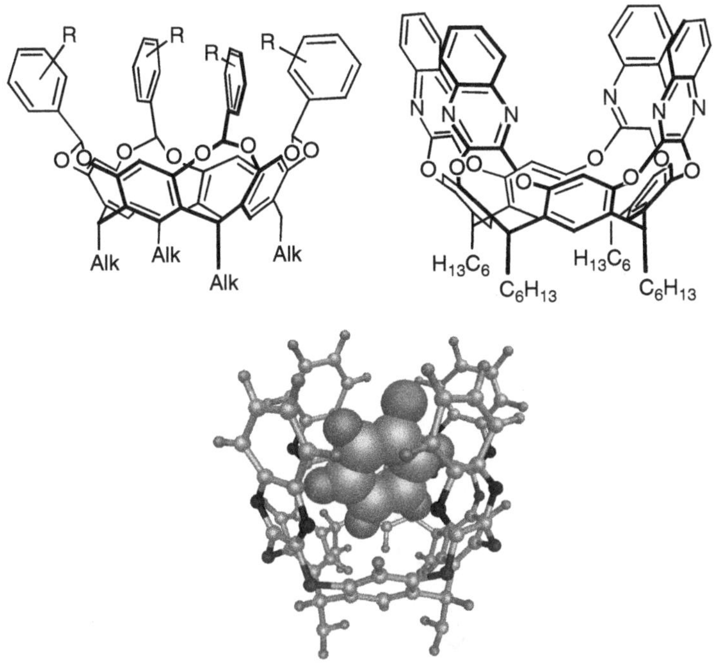

Figure 2.8 Deep cavitands: extended cavitands prepared by the reaction of resorcin[4]arenes with benzylidene dibromides (top left, Alk = CH_2CH_2Ph; R=H, OH, Br, I, Me, NO_2, CN, Ph, OCH_2OEt), 2,3-dichloroquinoxaline-based deep cavitand (top right), and the X-ray structure of its complex with fluorobenzene (bottom, C_5H_{11} alkyl chains were removed for clarity).

deep cavitands with shielded, well-defined inner cavities [28–30] (Figure 2.8). According to X-ray data, one of these deep cavitands has an internal cavity of about 7 Å wide and 8 Å deep. This cavity is large enough to accommodate small aromatic molecules, such as fluorobenzene (Figure 2.8) [28].

An advanced approach for producing deep cavitands involves further linking of additional aromatic units attached to already bridged resorcinarenes (Scheme 2.1). The size of the second rim in these double-bridged deep cavitands can be controlled by the proper selection of diamine units X (ethylenediamine, 1,3-diaminopropane, and *m*-xylylene diamine). Double bridging produces a large internal cavity capable of capturing large guest molecules [31].

A deep cavitand with acenaphtho(1,2-*b*)quinoxaline bridges (Figure 2.9) has a better organized, elongated internal cavity due to its extended heteroaromatic walls. The cavity of this container compound is about 14 Å deep and 12 Å wide. This deep cavitand is among the largest cavitands prepared from resorcin[4]arenes. According to molecular modeling calculations, a cavity of this size can accommodate several medium-sized guest molecules. However, the

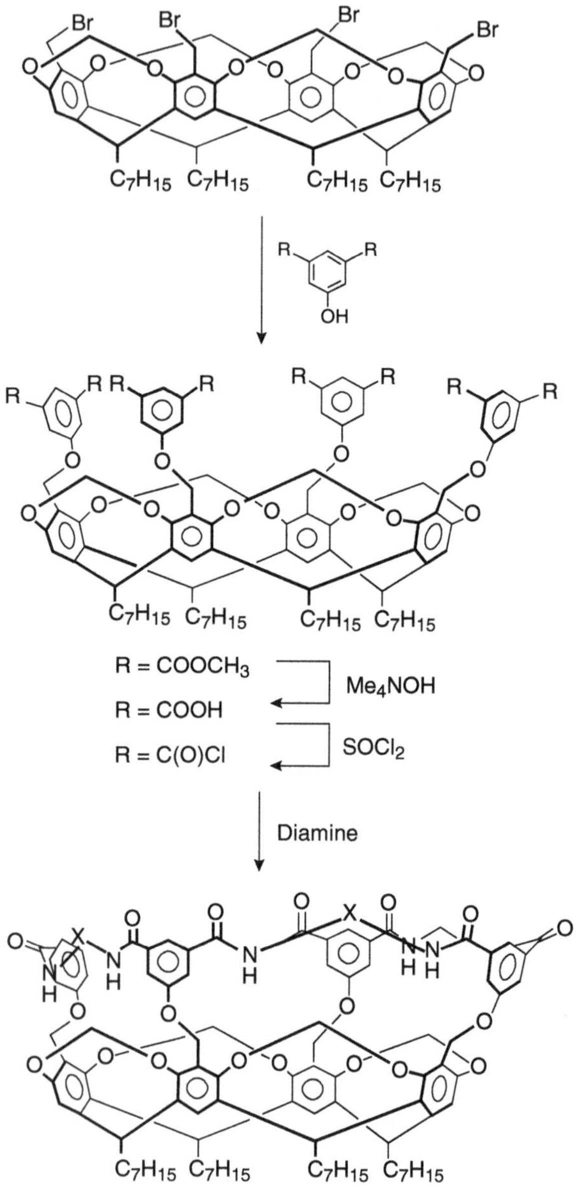

Scheme 2.1 The synthesis of double-bridged deep cavitands (X = CH_2CH_2, $CH_2CH_2CH_2$, m-$CH_2C_6H_4CH_2$).

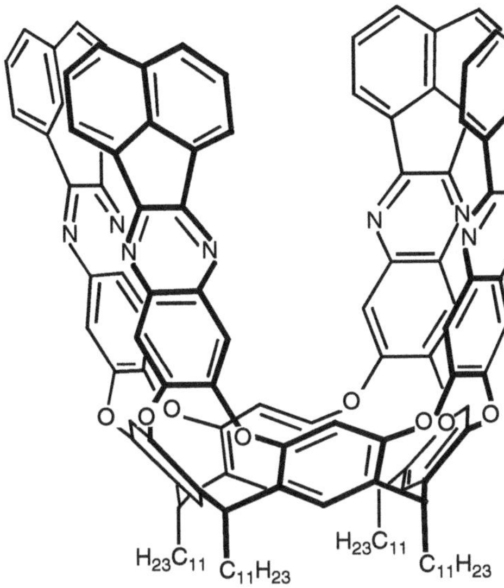

Figure 2.9 The deep cavitand with acenaphtho(1,2-*b*)quinoxaline bridges.

ability of this cavitand to form complexes with small organic guests appears to be poor. At the same time, the aromatic interiors of deep cavities interact well with π-rich surface of fullerene C_{60}. The association constant (K_a) of the fullerene C_{60} complex is $900 \pm 250\ M^{-1}$ in toluene [32].

In general, deep cavitands provide good guest selectivity and enhanced stability of host–guest complexes. Even better results can be achieved using semiclosed molecular architectures of bis-cavitands.

2.4 Bis-cavitands

One of the simplest approaches to improve the enclosure of inner space in cavitands involves the dimerization of two bowl-shaped cavitands. In these dimers, two cavitand hemispheres are usually linked together in a rim-to-rim manner via a flexible bridge, forming *bis-cavitand* structures (see Figure 2.3).

Small bis-cavitands presented in Figure 2.10 are synthesized by linking the upper rims of two rigid bis(crown-3) calix[4]arene units. These bis-cavitands bind strongly with heteroaromatic cations, such as *N*-methylpyridinium and *N*-methylpicolinium cations. The inner cavity of bis-cavitand with short CH_2 bridges is almost spherical. The use of a longer C≡C spacer between the calix[4]arene hemispheres results in an elliptical cavity (~7 × 11 Å) [33].

Also known are bis-cavitands containing larger calix[5]arene units. The directly linked bis-calix[5]arene forms 1:1 complexes with fullerenes C_{60} and

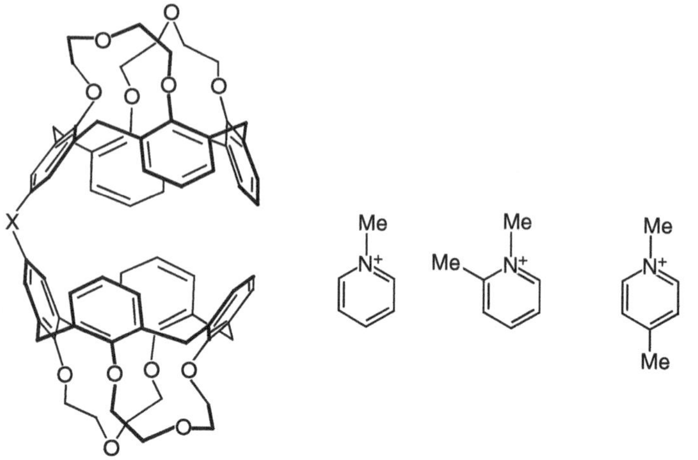

Figure 2.10 Small bis-cavitands based on calix[4]arenes (left, X = CH_2, C≡C) and their cationic guests (right).

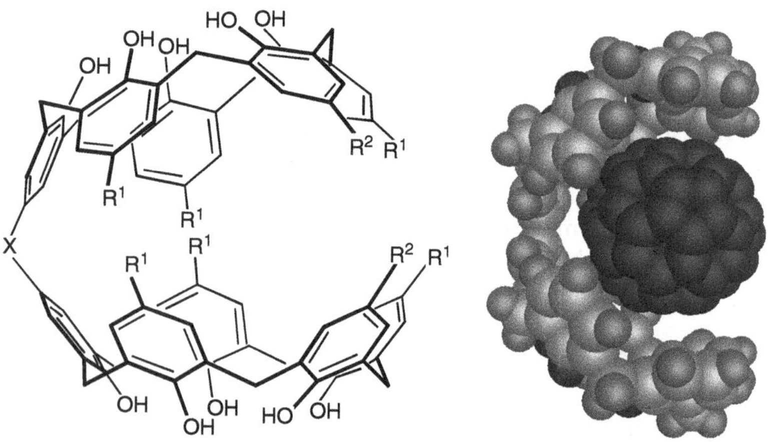

Figure 2.11 Large bis-cavitands based on calix[5]arenes (left, X = –, R^1, R^2 = H; X = $CH_2CH{=}CHCH_2$, R^1, R^2 = $CH_2CH{=}CH_2$; R^1 = Me, R^2 = I, X = 5-methyl-1,3-diethynyl benzene) and the X-ray structure of the fullerene C_{60} complex (right, X = –, R^1, R^2 = H; fullerene C_{60} guest molecule is partially disordered in the crystal).

C_{70} in CS_2 solution. The reported K_a values are 43 ± 5 and $233 \pm 30\,M^{-1}$ for C_{60} and C_{70}, respectively. The complex with·C_{60} was isolated and characterized by single-crystal X-ray diffraction. In this complex, the two calixarene units are in a *syn*-conformation, forming a jaw-shaped structure with C_{60} captured inside (Figure 2.11, X = –, R^1, R^2 = H) [34].

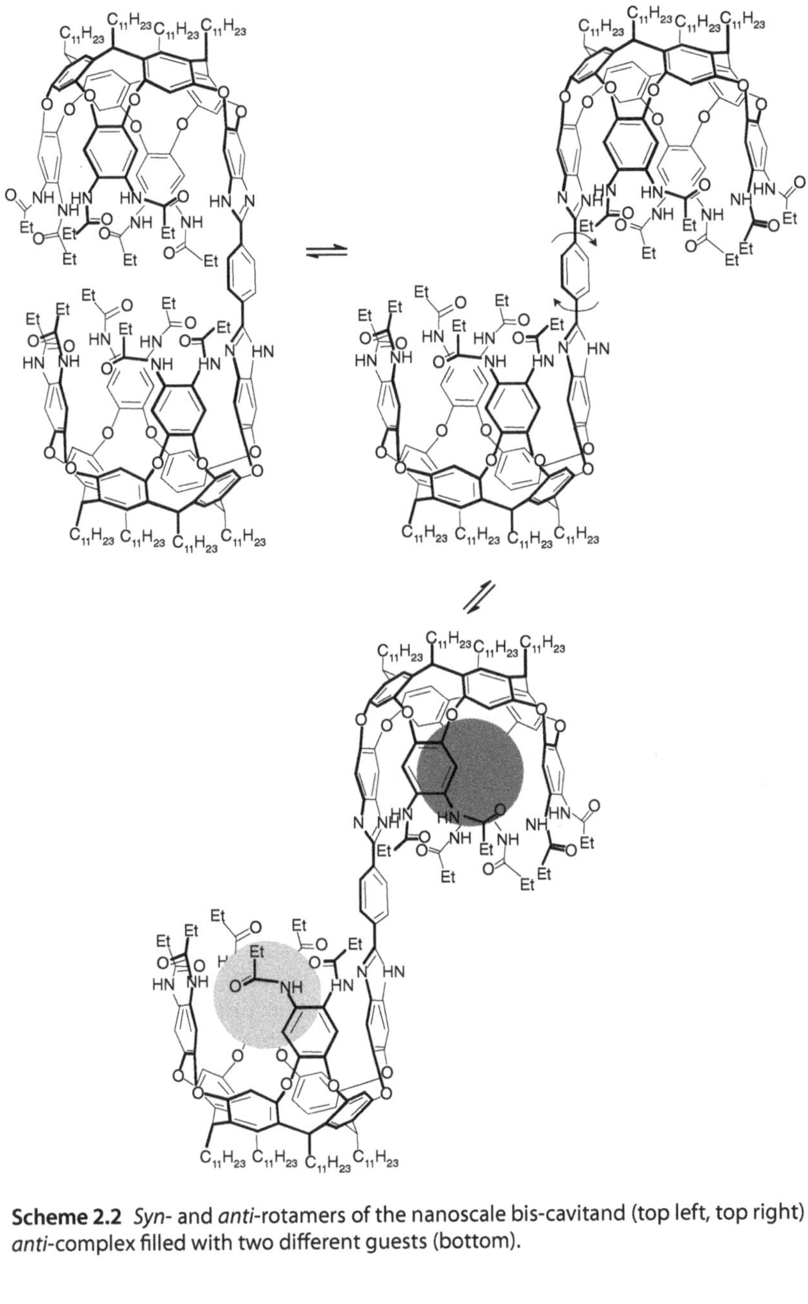

Scheme 2.2 *Syn-* and *anti-*rotamers of the nanoscale bis-cavitand (top left, top right) and *anti-*complex filled with two different guests (bottom).

Larger bis-calix[5]arene, which is linked at the upper rims with a 2-butenyl spacer and substituted with *p*-allyl fragments at the remaining aromatic rings (Figure 2.11, $X = CH_2CH{=}CHCH_2$, R^1, $R^2 = CH_2CH{=}CH_2$), shows stronger affinity toward C_{60} and C_{70} with K_a values in toluene of 1300 ± 65 and $625 \pm 32\,M^{-1}$, respectively. These K_a values are substantially higher than K_a values for the parent calix[5]arene mono-cavitand under the same conditions: $K_a = 30 \pm 2$ and $51 \pm 3\,M^{-1}$ for C_{60} and C_{70}, respectively [35].

The binding is even stronger when two calix[5]arene hemispheres are linked via a rigid and long 1,3-diethynylbenzene spacer (Figure 2.11, $R^1 = Me$, $R^2 = I$, X = 5-methyl-1,3-diethynylbenzene). This large bis-cavitand appears to be the best host for fullerenes C_{60} and C_{70} with K_a in toluene of $7.6 \pm 0.5 \times 10^4$ and $1.63 \pm 0.16 \times 10^5\,M^{-1}$, respectively [36].

A nanoscale bis-cavitand presented in Scheme 2.2 has two resorcinarene units interconnected through an extended rotationally flexible *p*-phenylene spacer. This container compound has a 10×23 Å cavity with an internal volume approaching 800 Å^3. The molecule is flexible: the *syn*- and *anti*-conformers interconvert fast on the NMR timescale. The *syn*-conformer of this container compound prefers to accommodate elongated guests, while the *anti*-conformer has two cavities, which can be filled independently with two spherical guest molecules. This S-shaped *anti*-conformer forms stable complexes with a series of adamantane- and cyclohexane-based guests, one per each cavity, with either two identical or different guest molecules [37].

Relatively flexible construction of bis-cavitands opens up an opportunity for fine-tuning of these host molecules for a particular guest. However, *syn–anti* rotation in bis-cavitands represents a major problem. To avoid this rotation, more bridges between the two cavitand hemispheres are required, thus leading to the design of clamshell-shaped molecular containers.

2.5 Clamshell-Shaped Molecular Containers

Clamshell-shaped molecular containers [5] (Figure 2.3) usually contain two links between the cavitand units. The shape of the resulting hosts resembles a clamshell. These hosts provide a better binding environment than cavitands and bis-cavitands. Adding more linking elements between the two cavitand units leads to the construction of hemicarcerands (Figure 2.3).

Clamshell-shaped molecular containers with resorcin[4]arene hemispheres interconnected via two short methylene links in *syn*-fashion are presented in Figure 2.12. These hosts formed kinetically stable 1:1 complexes with some heterocyclic guests such as pyrazine, dioxane, and pyridine. It should be noted that structurally similar resorcinarene-based bis-cavitands, interconnected with only one methylene linker, and the parent mono-cavitands do not form complexes under the same conditions [38].

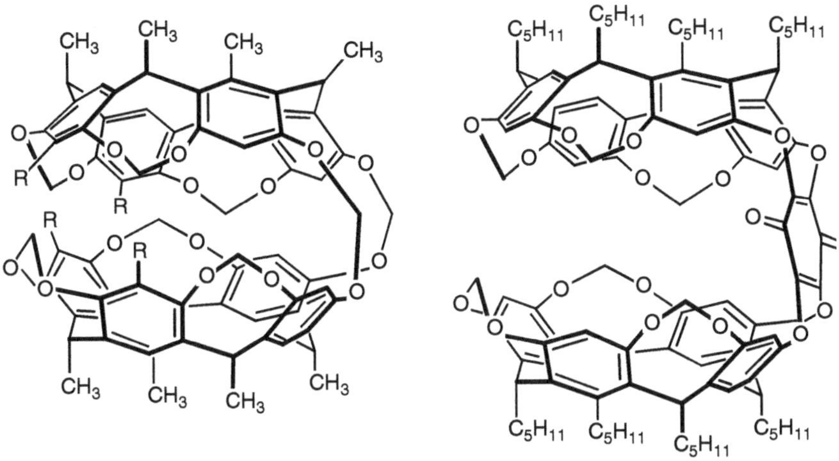

Figure 2.12 Resorcinarene-based clamshell-shaped molecular containers with two methylene bridges (left, R = H, OH, OMe) and with the *p*-benzoquinone spacer (right).

According to single-crystal X-ray analysis, the extended clamshell-shaped container molecule with the *p*-benzoquinone spacer, recrystallized from the CH_2Cl_2–CH_3CN mixture, has one CH_2Cl_2 molecule encapsulated in one cavitand hemisphere and one CH_3CN molecule situated in the other hemisphere (Figure 2.12). This host exhibits only very weak affinity toward aromatic guest molecules [39].

A very large, rigid clamshell-shaped molecular container presented in Figure 2.13 has two stable *syn*- and *anti*-isomers, which can be easily separated via liquid chromatography. This molecule features two cavities that are preorganized for cooperative binding with long and rigid guests (~17–18 Å), such as diaryl adamantanes. The complexation processes involving this host are slow on the NMR timescale, thus permitting a thorough evaluation of the kinetic and thermodynamic parameters for this reaction [37].

As a logical extension of this design, a cavitand–porphyrin clamshell-shaped molecular container was also synthesized (Figure 2.13). This host has a very large cavity containing two cavitand units bridged with the Zn-porphyrin spacer. The molecule is about 25 Å deep and 10 Å wide. The metalloporphyrin spacer provides an additional binding functionality for a number of heterocyclic guests, especially if a guest molecule is capable of interacting with the two cavitand units and the spacer simultaneously. The ability to host two guest molecules was also reported [40].

Clamshell-shaped container compounds bind guests stronger than most cavitands, but reversibly. Ultimate binding of the guest molecules can only be achieved in sealed molecular containers such as hemicarcerands, carcerands, and fullerenes.

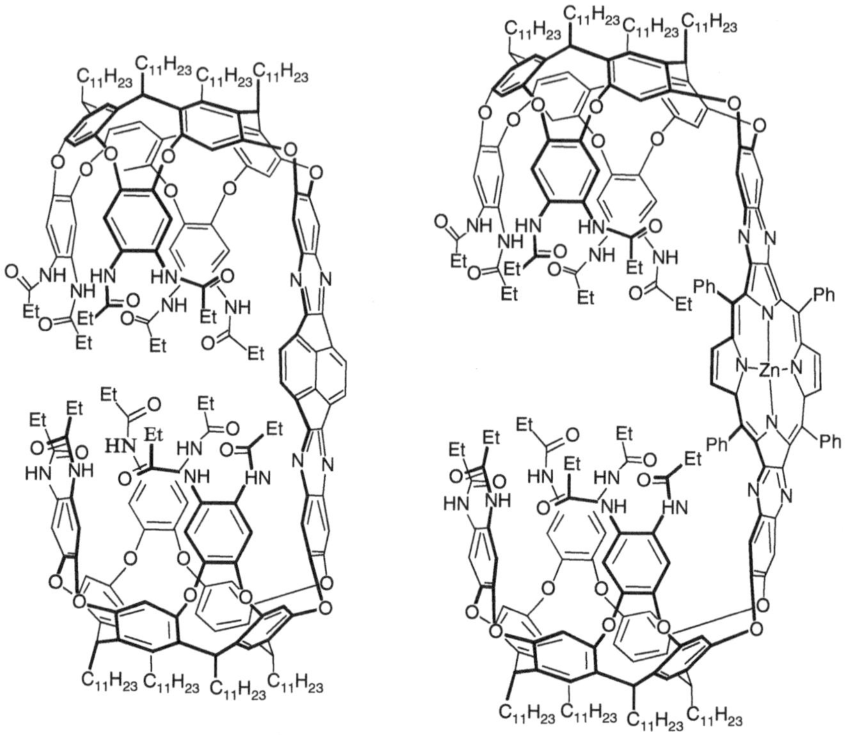

Figure 2.13 A very large clamshell-shaped molecular container (left) and its cavitand–porphyrin analog (right).

2.6 Hemicarcerands

Hemicarcerands [41, 42] are commonly constructed from two bowl-shaped cavitands interconnected with several bridge elements, providing a cage-like structure. Hemicarcerands usually allow guests to enter and exit the cavity at high temperatures, but form kinetically stable complexes at ambient temperatures. A typical hemicarcerand structure is presented in Figure 2.3.

Hemicarcerands are commonly divided into two structural types that have three or four bridges connecting two hemispheres, respectively (Figure 2.3). When a hemicarcerand has only three bridges, one wider portal is always available for guest molecule access. In hemicarcerand molecules with four bridges, conformationally flexible linkers are required to open portals for guest molecule access in a dynamic fashion. Typical hemicarcerand molecules prepared from resorcin[4]arenes with three and four bridges, as well as the X-ray structure of nitrobenzene complex of one of these compounds, are presented in Figure 2.14 [1, 8, 43, 44].

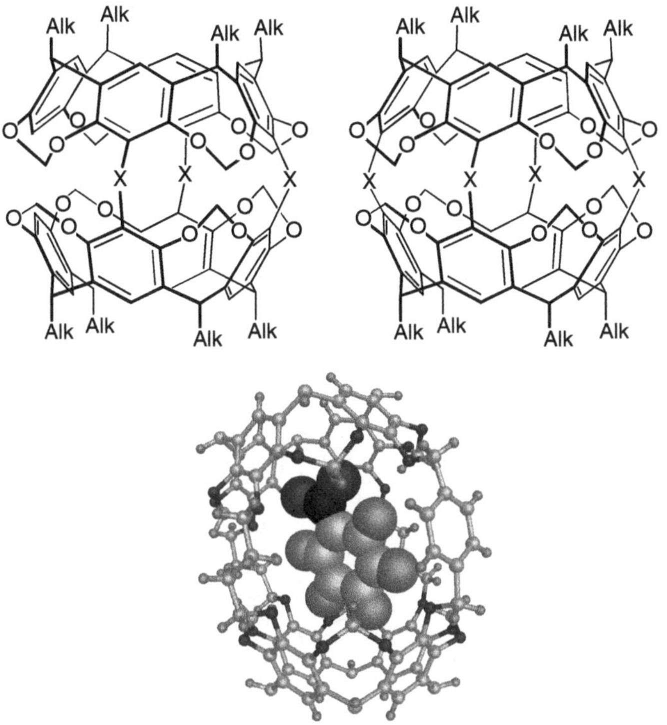

Figure 2.14 Typical structures of hemicarcerands prepared from resorcin[4]arenes with three (top left, X = OCH_2O, Alk = CH_2CH_2Ph) and four (top right, X = $O(CH_2)_4O$, *o*-$OCH_2C_6H_4CH_2O$, Alk = CH_2CH_2Ph) bridges and the X-ray structure [44] of the hemicarcerand complex with nitrobenzene (bottom, X = 3 × $O(CH_2)_4O$, 1 × *p*-$OCH_2C_6H_4CH_2O$, Alk = CH_2CH_2Ph; Alk fragments were removed for clarity).

Functionalized hemicarcerands can also be prepared from pyrogallol[4]arenes, including four-directionally functionalized hemicarcerands. These compounds are synthesized via one-pot reactions between pyrogallol[4]arenes and bridging reagents as shown in Scheme 2.3. Eight bonds are formed in one step, leading to the desired hemicarcerands in moderate yields (7–11%). One of these hemicarcerands (X = CN) forms well-defined complexes with mesitylene and *t*-butylbenzene. Reactive tetracyano- and tetrabromo groups of these hemicarcerands can be further transformed into various functional groups required for the construction of two-dimensional (2D) square grid networks [45].

Heterobridged hemicarcerands in which the fourth bridging link differs from the other three can also be synthesized. One of the examples is a hemicarcerand shown in Scheme 2.4. This hemicarcerand was prepared (as a $CHCl_3$ complex)

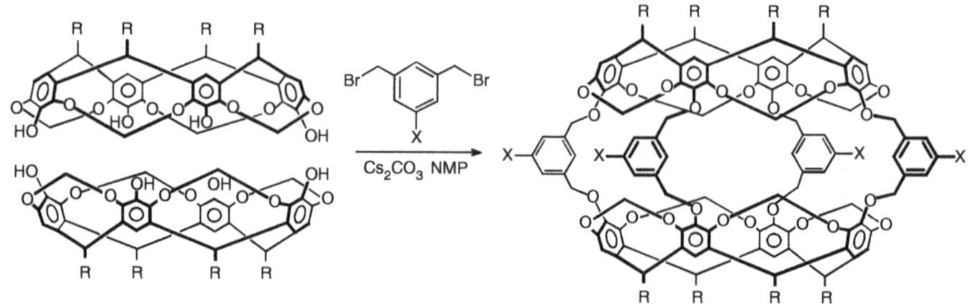

Scheme 2.3 The synthesis of four-directionally functionalized hemicarcerands (X = CN, Br; R = $CH_3(CH_2)_6$).

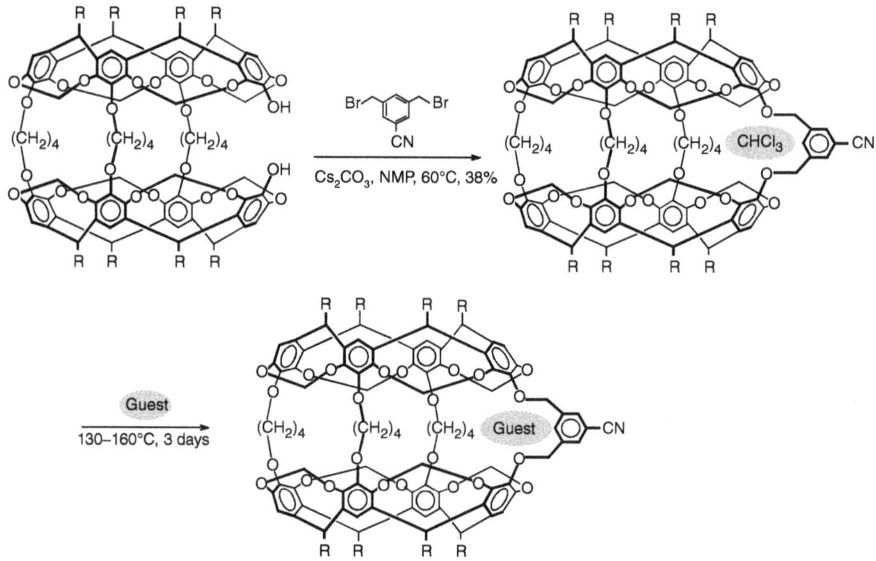

Scheme 2.4 The synthesis of the heterobridged hemicarcerand and its complexes (guest = $(MeO)_3C_6H_3$, $(MeO)_2C_6H_4$, $(Me)_2C_6H_4$; R = $(CH_2)_6CH_3$).

from the corresponding diol and 3,5-bis(bromomethyl)benzonitrile under the high dilution conditions in 38% yield. Other complexes of the abovementioned hemicarcerand can be obtained by heating this $CHCl_3$ complex with an excess of an appropriate guest (Scheme 2.4) [46].

The heterobridged hemicarcerand complex with $CHCl_3$ was also successfully used for the construction of dimeric molecular container systems through the metal coordination with $Pd(PhCN)_2Cl_2$ (Scheme 2.5). A self-organized 2D network of container molecules can be envisioned with the proper multi-fictionalization of this hemicarcerand [46].

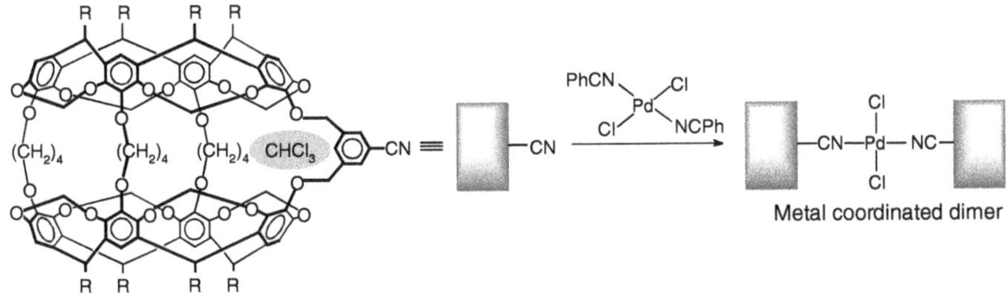

Scheme 2.5 The synthesis of the metal-coordinated dimeric molecular container system ($R = (CH_2)_6CH_3$).

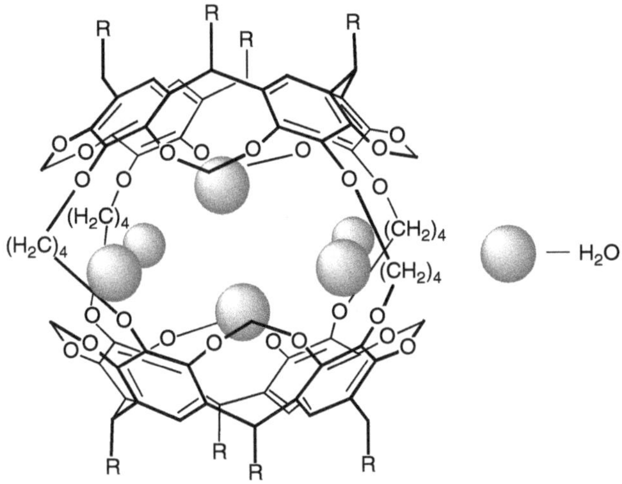

Figure 2.15 Encapsulation of six water molecules in the large hemicarcerand ($R = CH_2CH_2Ph$).

Large hemicarcerands are capable of binding two or more guest molecules. For example, an extended hemicarcerand having long $O(CH_2)_4O$ links crystallizes with six H_2O molecules encapsulated within the host (Figure 2.15). The whole structure is kept together by the network of hydrogen bonds between the water molecules and also with the hemicarcerand walls. This arrangement has high symmetry and matches the spatial characteristics of the inner cavity of the molecular container [47].

Large portals often increase the permeability of hemicarcerands for bulky organic guests, but this can in turn decrease the kinetic stability of the complexes [4]. To improve kinetic stability, one can prepare container molecules with rigidified portals. These hemicarcerands were constructed from two cavitand hemispheres and four relatively rigid *m*-$CH{=}NC_6H_4N{=}CH$ or

$OCH_2C{\equiv}C{-}C{\equiv}CCH_2O$ bridging units. These container compounds have large 28- and 30-membered portals, respectively (Figure 2.16), and form stable 1:1 complexes with a number of bulky organic molecules [48, 49].

Two resorcin[4]arene cavitands having $OCH(ArCH_2OH)O$ bridges can be linked via hydroxyl groups of CH_2OH units. The structure of a large hemicarcerand having this alternative linking pattern is presented in Figure 2.17.

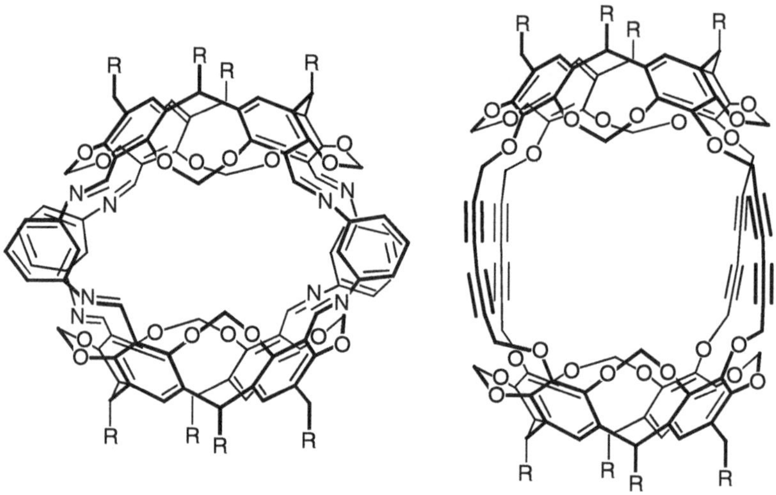

Figure 2.16 Hemicarcerands with large rigidified portals ($R = CH_2CH_2Ph$).

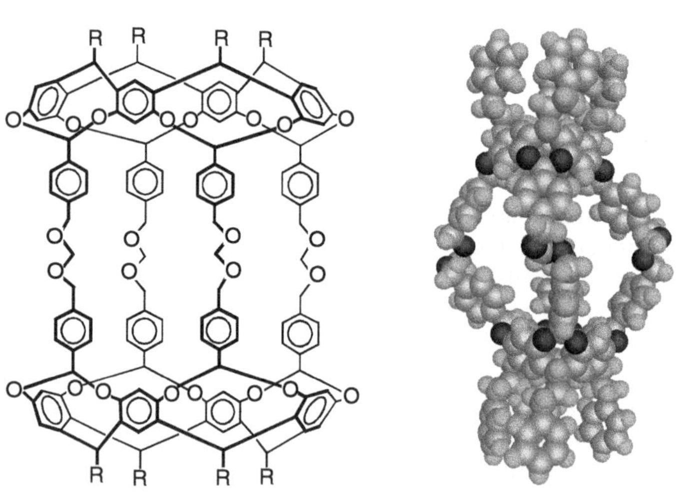

Figure 2.17 The hemicarcerand molecule with alternative linking of two resorcin[4]arene cavitands (left, $R = CH_2CH_2C_6H_5$) and its X-ray structure (right).

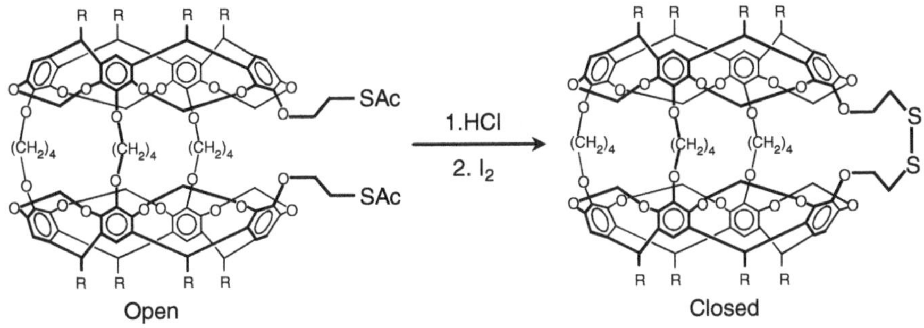

Scheme 2.6 The hemicarcerand with the redox-controlled gates ($R = (CH_2)_4CH_3$).

This container compound has an approximately 19 × 15 Å cavity, with the portals of about 9.5 × 11.5 Å. The yield of the hemicarcerand was relatively high in dimethyl sulfoxide (DMSO) (80%) due to a plausible self-assembly of the two cavitands via complementary hydrogen bonding, even without the use of a special molecular template [50].

Hemicarcerands with controllable gates are also known [51]. These gates can be controlled either chemically (reduction–oxidation, redox) [52] or photochemically [53]. An example of a hemicarcerand with redox-controlled gates is presented in Scheme 2.6. In the reduced form, an open bridge with two SAc terminal groups enables this hemicarcerand to accommodate guest molecules easily, but in the oxidized (closed) form, the S—S disulfide bridge hinders the access to the inner space, which can be filled with guest molecules only at elevated temperatures [52].

All hemicarcerands described earlier were constructed on the basis of pyrogallol[4]arenes, which can be classified as [1.1.1.1]*meta*cyclophanes. An additional structural diversity can be attained with the use of cyclotriveratrylenes, which are [1.1.1]*ortho*cyclophanes. However, the preparation of cyclotriveratrylene-derived container compounds, *cryptophanes*, requires somewhat different synthetic approaches.

2.7 Cryptophanes

Syn- and *anti-*cryptophanes represent a transitional class of compounds between hemicarcerands and carcerands. They are constructed from two cup-shaped, C_3-symmetric [1.1.1]*ortho*cyclophane units interconnected via three bridges. Several synthetic routes have been developed for the synthesis of these container compounds [10, 54–56]. An example of a two-step synthesis that entails trimerization of bis-veratryl alcohols is presented in Scheme 2.7 [57].

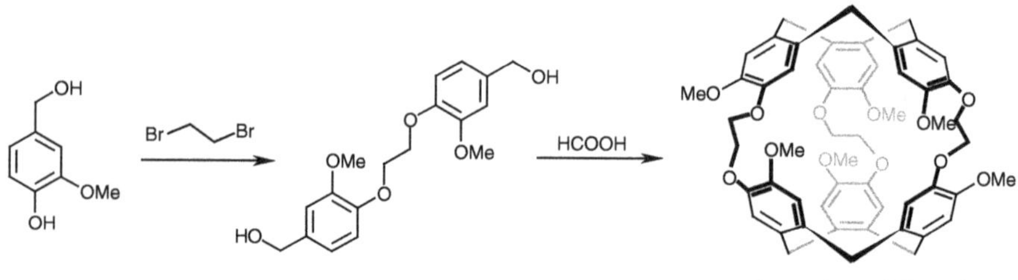

Scheme 2.7 The synthesis of cryptophane-A via trimerization of the corresponding bis-veratryl alcohol.

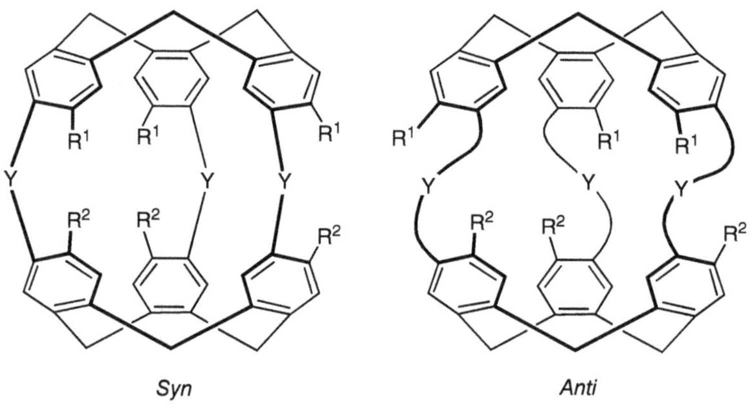

Figure 2.18 General structures of *syn*- and *anti*-cryptophanes.

In many cases, the synthesis of cryptophanes yields two geometrical isomers, *anti* and *syn* (Figure 2.18). Both isomers have preorganized, three-dimensional cavities suitable for accommodation of complementary guest molecules. A large number of cryptophanes with different cavity volumes, portal sizes, and periphery functional groups were prepared by varying the nature of bridges Y and peripheral substituents R [10, 54, 55, 58].

Cryptophanes form stable complexes with a variety of charged and neutral guests [59, 60]. The X-ray structure of one of these complexes with tetrahydrofuran (THF) is presented in Figure 2.19. This complex is quite strong and starts losing THF guest molecules from the internal cavity only at temperatures above 100 °C [60].

Perhaps the most valuable property of cryptophanes as container compounds is their ability to produce strong complexes with heavy noble gases, such as xenon and radon [61]. The X-ray structure of one of these complexes is presented in Figure 2.20 [62].

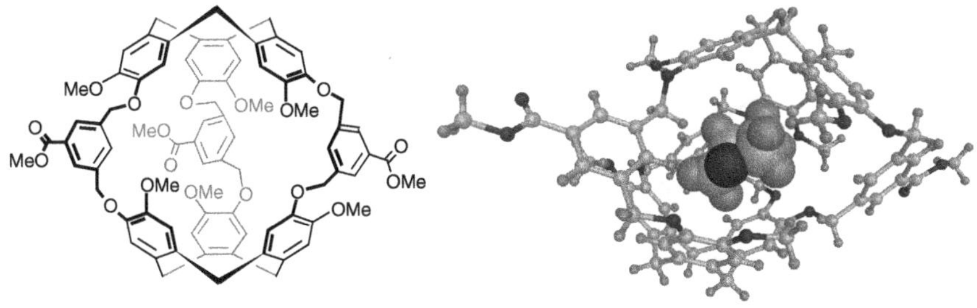

Figure 2.19 *Anti*-cryptophane with 3,5-$(CH_2)_2C_6H_3COOCH_3$ links (left) and the X-ray structure of its complex with THF (right).

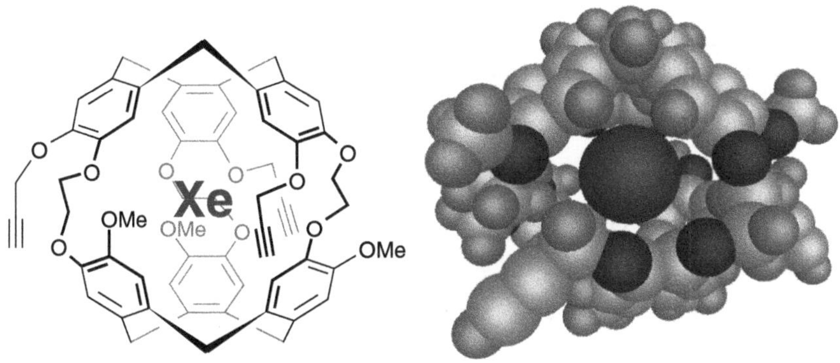

Figure 2.20 The X-ray structure of the Xe complex with the optimized cryptophane derivative having three propargyl groups.

The strength of these xenon complexes, in best cases approaching $5 \times 10^4\,M^{-1}$, is unexpected since internal volumes of even the smallest cryptophanes are larger than a xenon atom. This controversy was finally resolved with the discovery of "induce fit" effect [62]. In essence, relatively flexible cryptophanes are capable of adjusting their geometry to provide an optimal environment for a guest (Figure 2.20). This mode of action is a direct analog to conventional mesh bags, which can accommodate a variety of odd-shaped objects due to their supple design.

Similar to traditional cryptophanes, cyclotriveratrylene-based molecular containers with six links can be prepared via acid-catalyzed trimerization of properly functionalized dibenzo macrocyclic ethers [63–66]. A representative example of these container compounds with six long 3,6-dioxaoctane-1,8-diyl bridges is shown in Figure 2.21. This host molecule forms strong complexes with large aromatic dications, such as 4,4′-biphenylbisdiazonium [63].

The internal cavity of the cyclotriveratrylene-based container compounds with six decane-1,10-diyl links is large enough to accommodate fullerenes

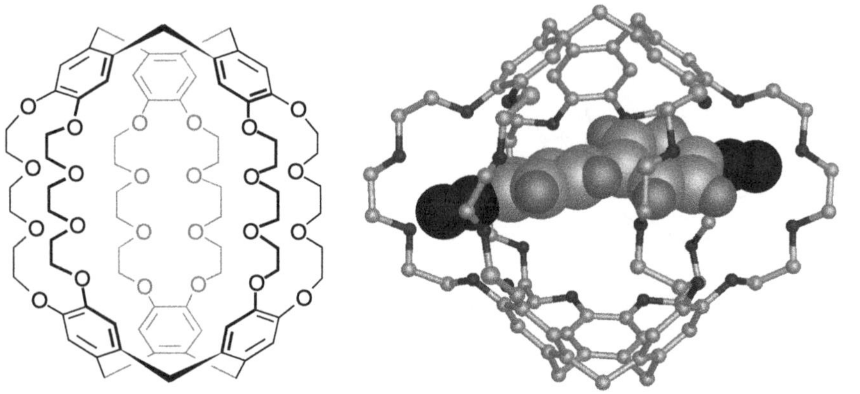

Figure 2.21 The cyclotriveratrylene-based molecular container with six 3,6-dioxaoctane-1,8-diyl links (left) and the X-ray structure of its complex with 4,4′-biphenylbisdiazonium dication (right, counteranions and some hydrogen atoms were removed for clarity) [63].

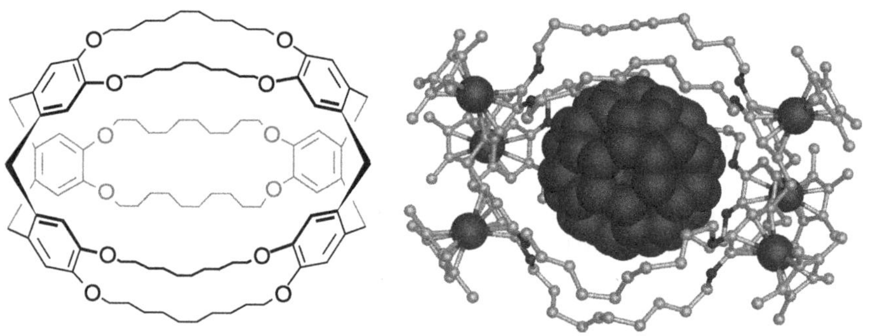

Figure 2.22 The cyclotriveratrylene-based container compound with six long links (left) and the X-ray structure of its Rh^+Cp^*-derivatized complex with fullerene C_{60} (right, counteranions and hydrogen atoms were removed for clarity) [66].

C_{60} and C_{70}. Encapsulation of fullerenes greatly improves their solubility in nonpolar solvents. Additional fictionalization with positively charged Rh^+Cp^* sandwich moieties can also improve the solubility in polar solvents. The X-ray structure of one of these fullerene C_{60} complexes is presented in Figure 2.22.

Hemicarcerands and cryptophanes, such as those described earlier, possess significant, but not unlimited, constrictive binding. In many cases, guest molecules in their cavities can be exchanged without cage rupture, especially at elevated temperatures. The ultimate constrictive binding can be achieved in *carcerands*, discussed in the following chapter.

2.8 Carcerands

Carcerands [4, 7] are sealed host molecules with inner cavities capable of holding guest atoms and small molecules (see Figure 2.3). The name carcerand refers to the capability of the host to imprison guest molecules and atoms as in a jail (Lat. "carcer"). Similar to hemicarcerands, carcerands can be prepared by tightly bonding two cavitands at their upper rims. The portals of carcerands are typically narrower than in hemicarcerands. These small portals forbid the guest molecules from escaping without breaking the covalent bonds. Stability-wise and synthesis-wise, complexes of carcerands represent a special case of host–guest interactions. Most carcerands include guests only during their synthesis. In many cases, these guest molecules serve as templates for the final bonding steps. These guests, and sometimes products of their subsequent transformations, cannot be removed or exchanged after the synthesis is completed. Due to irreversible inclusion, carcerands are suitable as molecular containers for guests of various sizes and shapes, from noble gases to fullerenes.

The early examples of carcerands were prepared by coupling two functionalized cavitands under high dilution conditions. For example, if one cavitand had four reactive chloromethyl groups and the other had four complementary thiomethyl groups, the resulting carcerand would have four dimethylenesulfide (CH_2SCH_2) links (Figure 2.23) [67]. Similarly, two identical cavitand units can be joined together by employing the appropriate bifunctional linking agents.

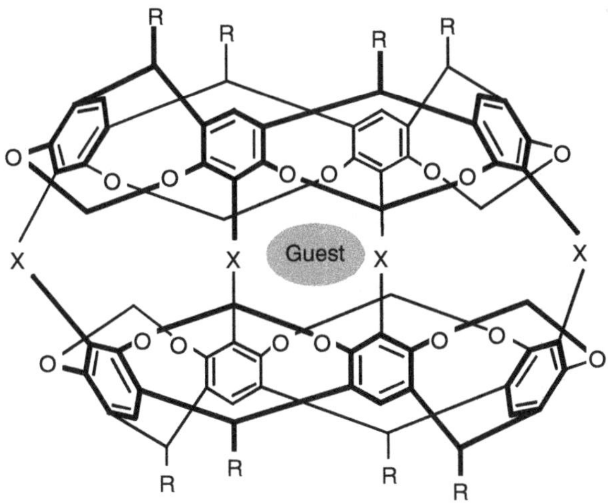

Figure 2.23 General structure of the early examples of carcerands ($X = CH_2SCH_2$; $O(CH_2)_nO$, $n = 1–3$; $R = CH_3$, C_5H_{11}, CH_2CH_2Ph).

Among other linking fragments, $O(CH_2)_nO$ was the most frequently used fragment for the construction of these carcerands. They were produced by joining two pyrogallol[4]arene units with appropriate bridging agents, such as $BrCH_2Cl$, in the presence of fitting templates. The X-ray structures of two carcerand complexes utilizing $O(CH_2)_nO$ bridges are presented in Figure 2.24. Typically, guest molecules in these complexes are solvents [68], or special templates, such as pyrazine [2].

Previously mentioned examples of carcerands (shown in Figure 2.24) are assembled from two bowl-shaped cavitands. Larger carcerands can be constructed from three cavitand units. For example, a big carcerand complex with three irreversibly retained molecules of dimethylformamide (DMF) was synthesized from the cavitand trimer and 1,3,5-tris(bromomethyl)-2,4,6-trimethyl benzene (Scheme 2.8). This carcerand complex holds the captured DMF molecules even upon heating in nitrobenzene at 160 °C for 6 h [69].

A very large molecular container with six cavitand subunits entraps seven guest DMSO molecules during the synthesis (Scheme 2.9). It can be prepared in three steps from the corresponding unsymmetrical dihydroxy cavitand building blocks. The key step of the synthesis is the cyclization of this dihydroxy

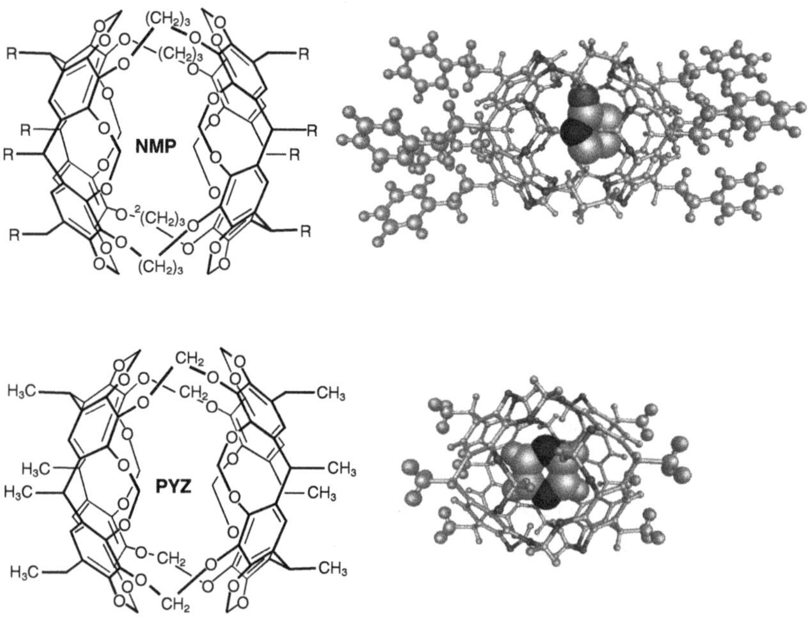

Figure 2.24 The X-ray structures of the early examples of carcerand complexes: with *N*-methylpyrrolidone (top, NMP – *N*-methylpyrrolidone, R = CH_2CH_2Ph, the disordered hydrogen atoms, and the methyl group of the guest molecule were removed for clarity in the X-ray structure) and with pyrazine (bottom, PYZ – pyrazine).

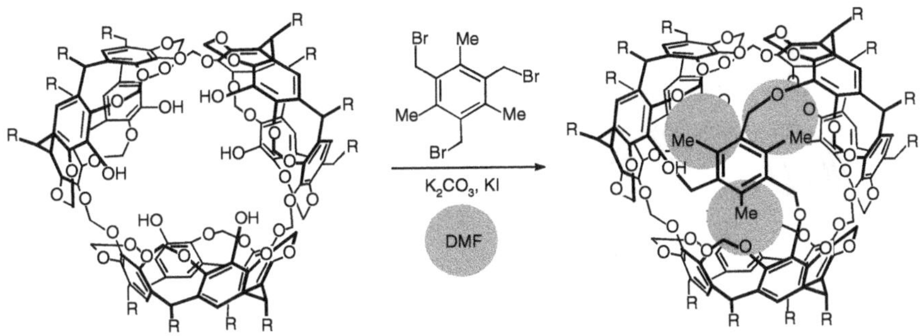

Scheme 2.8 The synthesis of the big carcerand complex ($R = CH_2CH_2Ph$) with three molecules of DMF.

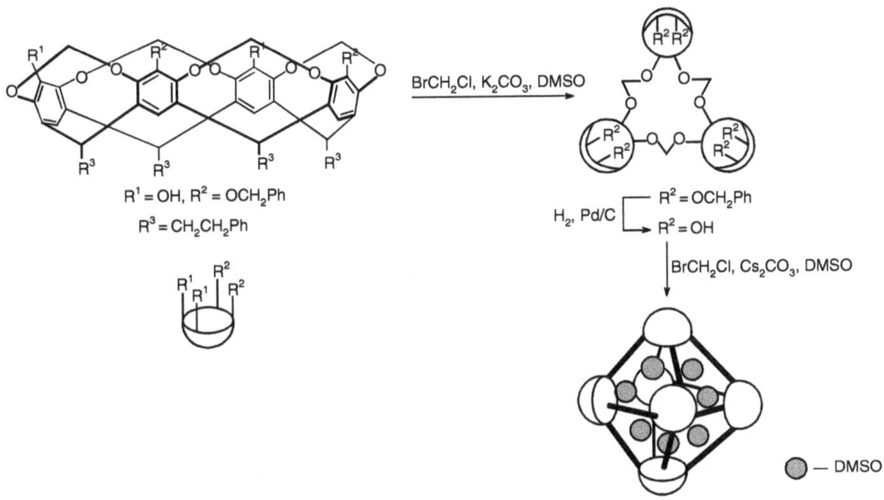

Scheme 2.9 The synthesis of the large carcerand complex with seven guest DMSO molecules.

cavitand with K_2CO_3 and CH_2BrCl in DMSO leading to a cyclic trimer (16%), which is subsequently debenzylated to give a hexahydroxy-trimer in 58% yield. In the final step of the synthesis, this hexahydroxy-trimer is successfully sealed using CH_2BrCl in DMSO to afford hexameric capsule incorporating seven DMSO molecules as guests in 35% yield [70].

Another interesting synthetic strategy entails the construction of hybrid carcerands from both calixarenes and resorcinarenes [71, 72]. For example, hybrid carcerand complexes presented in Figure 2.25 have a resorcin[4]arene unit and a calix[4]arene unit linked by four amide bridges. Complexes of these hybrid carcerands with various solvent molecules such as DMF, DMSO, and

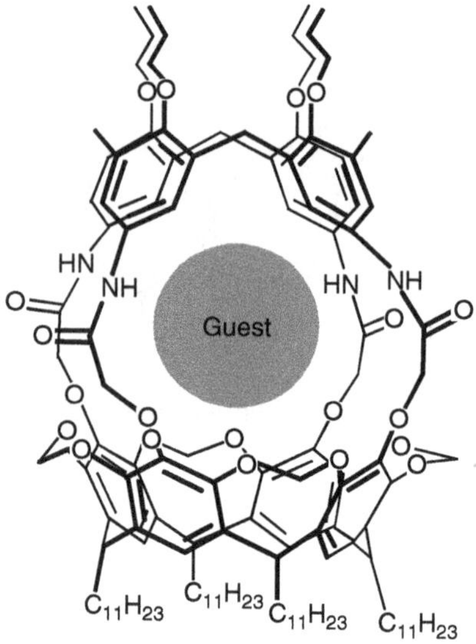

Figure 2.25 Hybrid carcerand complexes.

ethyl methyl sulfoxide can be prepared during the synthesis from the corresponding cavitands in almost quantitative yield. Hybrid carcerand complexes exhibit a new type of isomerism, called *carceroisomerism,* due to the inherent asymmetry of these host–guest assemblies. The guest molecules in these complexes can assume different orientations with respect to the calix[4]arene and the resorcin[4]arene halves of the molecular container [73].

All carcerands described in this section have cavitand subunits covalently bonded to each other, so the guest molecule cannot be removed without breaking at least one covalent bond of the container molecule. In other words, these carcerands do not have a controlled release mechanism, which is a valuable feature of nearly all storage containers used in conventional technological environments. The controlled release can easily be implemented in *reversible carcerands* presented in Section 2.9.

2.9 Reversible Carcerands

The reversible release on a molecular level is a permanent feature of carcerands employing noncovalent bonding of the cavitand subunits. Both hydrogen bonding and metal–ligand bonding were successfully employed for the construction of reversible carcerands using self-assembly methodology [74].

The hydrogen bonding approach was implemented using urea units attached to the upper rim of a calix[4]arene molecule (Scheme 2.10). Two self-complementary urea elements provide equatorial zipper-like bonding in the dimeric capsule via an array of 16 intermolecular hydrogen bonds [75]. The volume of the internal cavity of this dimeric carcerand is approximately 210 Å^3. This reversible molecular container forms complexes with a variety of aromatic, aliphatic, and cationic guests [76]. The terminal amino groups can be further modified with a number of functional groups, thus providing control functions for the self-assembly behavior of these calix[4]arene monomers [77]. Similar reversible capsules can be constructed from calix[4]pyrroles [78].

Two hydrogen-bonded dimeric capsules ("head-to-tail" and "head-to-head") are presented in Figure 2.26. The "head-to-tail" molecular container is formed when a calix[4]arene substituted with carboxylic groups at the upper rim interacts with a calix[4]arene substituted with 4-pyridyl fragments at the lower rim (Figure 2.26) [79]. A similar approach, but with both carboxylic groups and 4-pyridyl fragments attached to the upper rims of two calix[4]arene molecules, was used to obtain the "head-to-head" assembly (Figure 2.26) [80].

Besides dimeric capsules presented earlier, substituted resorcinarenes and pyrogallolarenes are also capable of self-organization in di-, tri-, tetra-, and hexameric reversible molecular containers, which are held together by weak noncovalent hydrogen bonding interactions [81, 82]. For example, alkyl-substituted resorcinarenes self-organize in container assemblies

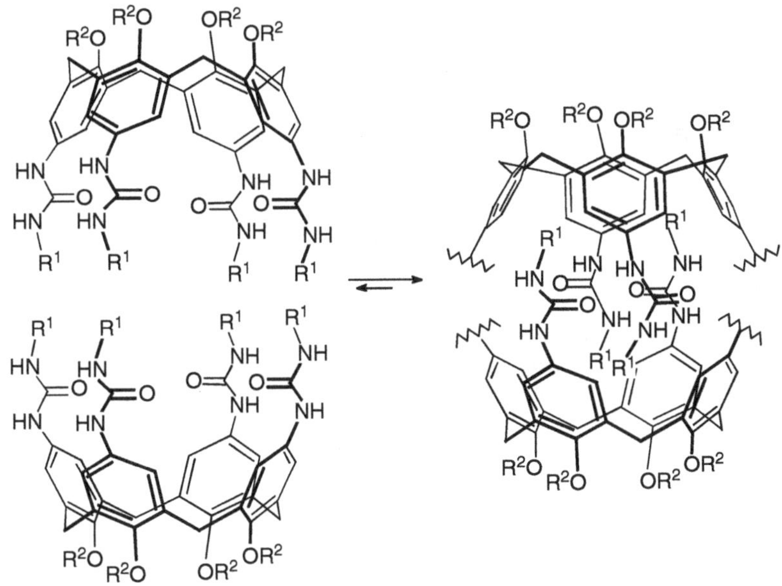

Scheme 2.10 The self-assembly of reversible cavitand-based capsules via the dimerization of urea-substituted calix[4]arenes ($R^1 = Ph$, p-FC_6H_4; $R^2 = CH_2Ph$).

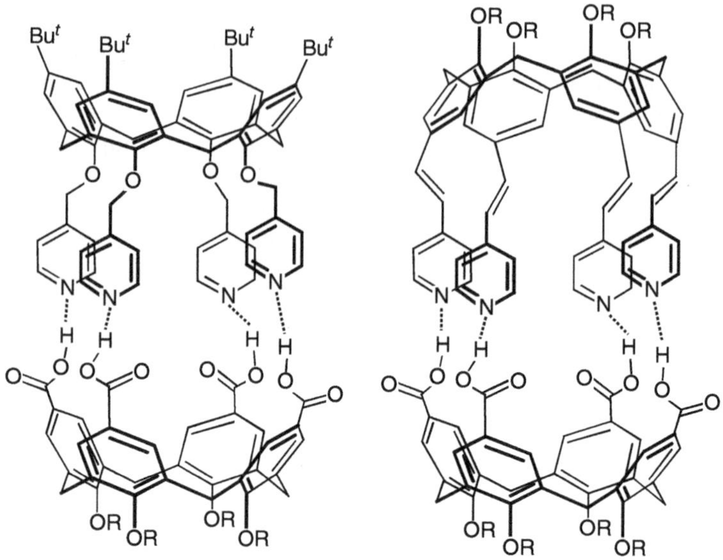

Figure 2.26 "Head-to-tail" (left, $R = C_8H_{17}$) and "head-to-head" (right, $R = C_3H_7$) hydrogen-bonded molecular capsules.

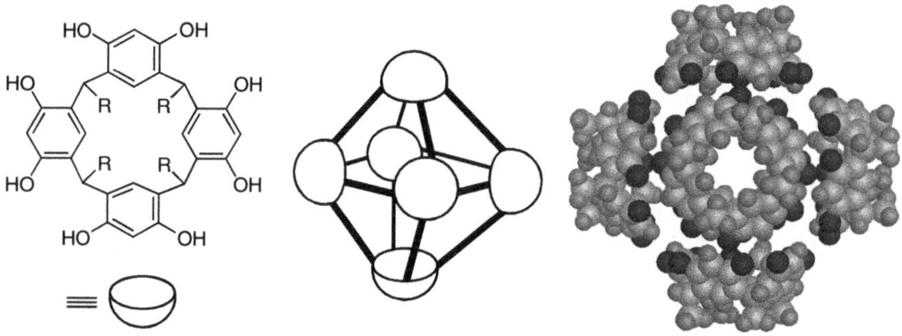

Figure 2.27 The hexameric capsule constructed from six resorcinarene molecules with six water molecules as additional linking elements (left) and its X-ray structure (right, $R = CH_3$, water molecules and poorly resolved hydrogen atoms of hydroxyl groups were removed for clarity) [83].

consisting of six resorcinarene units and six water molecules in the form of hexameric capsules. The structure of one of these assemblies was studied in detail by X-ray crystallography (Figure 2.27) [83]. The assembly is linked together by a total of 60 hydrogen bonds and has an internal cavity of about 1375 Å^3. This cavity is big enough to encapsulate large guest molecules. It is subsequently found that in the presence of tetraalkylammonium salts in water-saturated $CDCl_3$, this hexameric capsule incorporates only one tetraalkylammonium cation. The remaining space in the internal cavity is occupied by co-encapsulated solvent molecules. The number of these solvent molecules is decreasing with the increase in the tetraalkylammonium cation size. Guest exchange in this molecular container is facilitated by the reversible dissociation of one resorcinarene unit from the hexamer (similar to the lid on/lid off content release procedure of a conventional container) [84].

Unlike resorcin[4]arenes, pyrogallol[4]arenes, having four additional hydroxy groups, do not require linking water molecules to produce hexameric capsules. The X-ray structure of one of these hexameric capsules is presented in Figure 2.28 [85]. The assembly of six pyrogallol[4]arene cavitands is held together by a network of 72 hydrogen bonds. It is interesting to note that the solvent plays a key role in the formation of this hexamer. The hexamer self-assembly occurs spontaneously in ethyl acetate [85], whereas an acetonitrile–water solvent system favors the formation of dimers (Figure 2.28) [86].

Another class of weak interactions, metal–ligand coordination, was successfully used for the preparation of reversible molecular containers. A common synthetic strategy entails the self-assembly of two identical concave building blocks by linking of the upper rim substituents through metal atoms with the square-planar coordination geometry. For example, molecular containers shown in Figure 2.29 are typical examples of M–L assemblies where two tetracyano cavitand derivatives are connected through four Pd(II) or Pt(II) atoms. The spectral data (^{1}H NMR, electrospray ionization mass spectrometry) confirm that one triflate counteranion is included into the molecule interior of one of these container families during the synthesis, and it is too big to escape from the equatorial portals of the capsule (Figure 2.29). The addition of a competing ligand (such as triethylamine) collapses this molecular container. Subsequent addition of triflic acid restores the original complex. [87, 88]

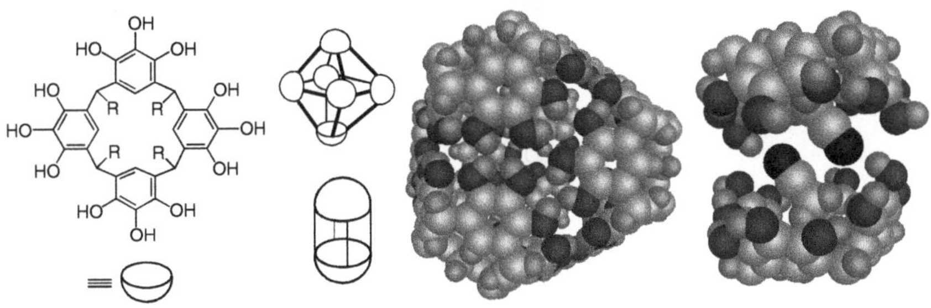

Figure 2.28 The pyrogallol[4]arene self-assemblies (left), the X-ray structure of the hexameric capsule (middle, $R = (CH_2)_5CH(CH_3)_2$ groups were removed for clarity), and the X-ray structure of the pyrogallol[4]arene dimer with two acetonitrile guest molecules (right, $R = (CH_2)_2CH_3$ groups were removed for clarity).

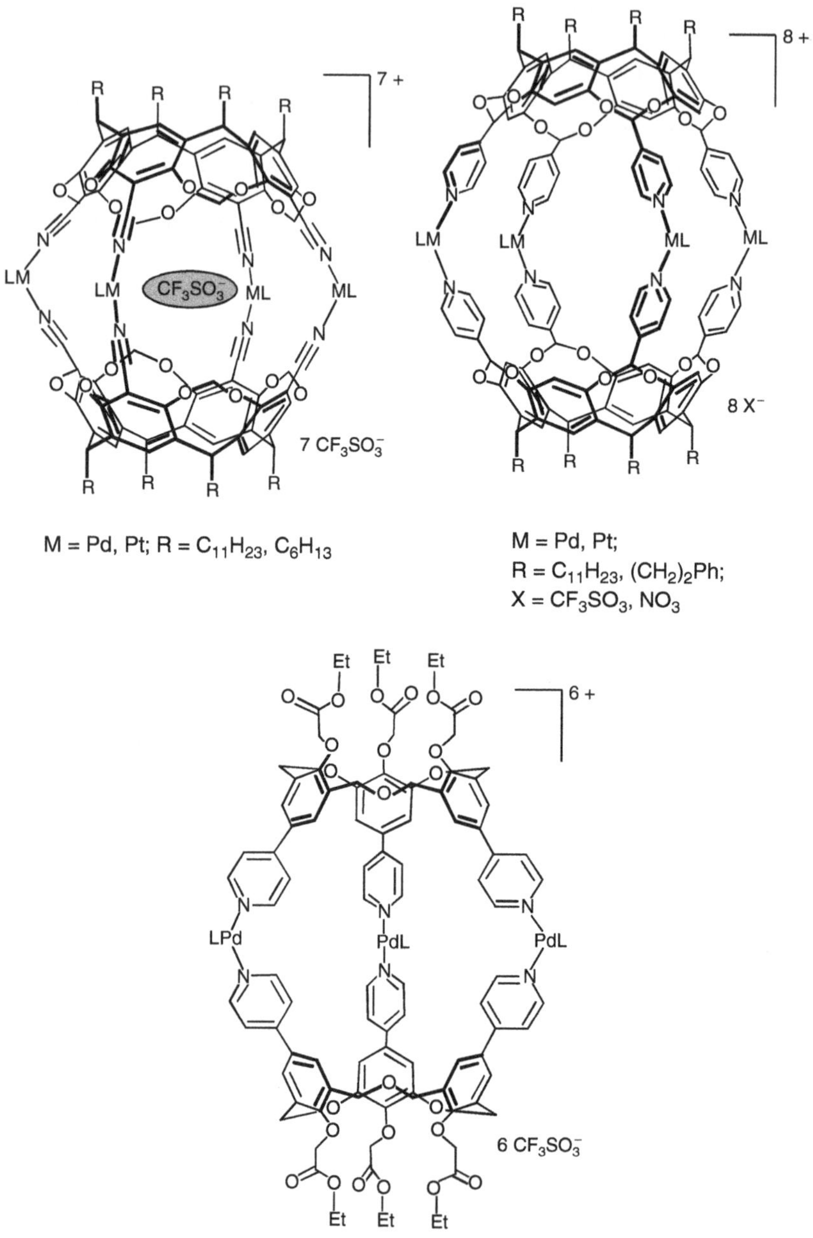

Figure 2.29 Reversible metal–ligand molecular containers (L = 1,3-bis(diphenylphosphino)propane).

The peripheral attachment of 4-pyridyl fragments to resorcin[4]arene cavitands facilitates the self-assembly of another family of metal–ligand reversible molecular container assemblies (Figure 2.29). These assemblies were prepared from two cavitand subunits and had four Pd(II) or Pt(II) metal–ligand links. The equatorial portals in these molecules are larger than in the previous examples of metal–ligand container complexes (Figure 2.29). They did not prevent counteranions from escaping the inner cavity [89].

Other cavitands containing 4-pyridyl units were successfully used for the synthesis of reversible metal–ligand capsules. One of these capsules (Figure 2.29) was prepared by the reaction of two 4-pyridyl-substituted homooxacalix[3]arenes with three equivalents of a *cis*-Pd(II) complex. This large container compound forms a kinetically stable 1:1 complex with fullerene C_{60}. The entrapment of C_{60} can be further improved by Li^+ coordination with the lower rims of the homooxacalix[3]arene subunits [90].

Better control of the capture-release processes in the bridged resorcinarene and calixarene assemblies can be achieved via a switchable molecular container design. An example of a switchable molecular "basket" is presented in Scheme 2.11. The compound was prepared by the oxidative coupling of an acetylenic precursor and isolated by high-performance gel-permeation chromatography. This host molecule forms stable complexes with size-matching cycloalkanes, such as cyclohexane [91].

The abovementioned basket molecular container can be switched into an open conformation upon the acidification with CF_3COOD, and this transformation turns off the complexation process without covalent bond breaking (Scheme 2.12). Storage capabilities of this molecular container can be restored by acid neutralization [91].

Similar gated molecular baskets [92] can be constructed on the basis of resorcin[4]arene [93], tris-norbornadiene [94, 95], and porphyrin frameworks [96].

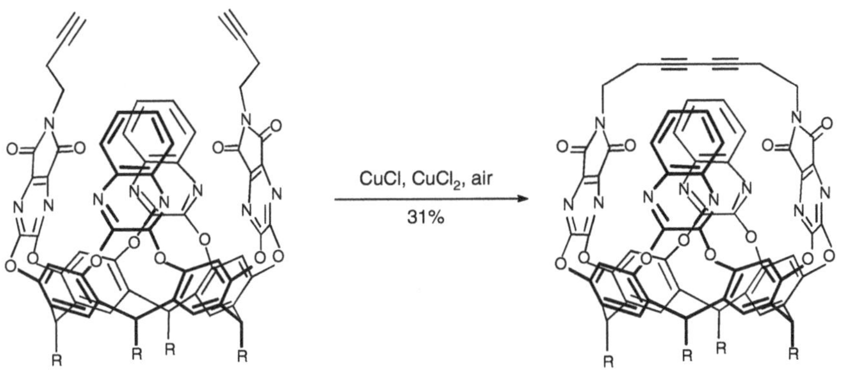

Scheme 2.11 The synthesis of the switchable "basket" molecular container ($R = (CH_2)_5CH_3$).

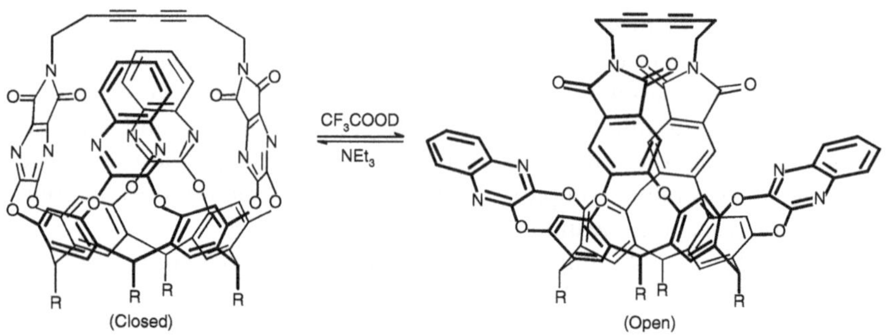

Scheme 2.12 The switchable molecular "basket" in open (left) and closed (right) forms ($R = (CH_2)_5CH_3$).

One of them with the porphyrin framework demonstrates stereoselective catalytic activity toward epoxidation of certain alkenes, such as *cis*-stilbene [96].

2.10 Spheriphanes

Another very interesting class of molecular containers is *spheriphanes*. These molecules are hydrocarbons made of interconnected benzene rings in the form of closed cages [97–100]. Many spheriphanes have rounded shapes with sizable intramolecular cavities and small portals. Some examples of spheriphanes are presented in Figure 2.30. These molecules are of interest as host compounds for hydrophobic guests and selected metal ions, which favor cation–π interactions, such as Ag^+ [98].

While spheriphanes are deceptively simple compounds, their synthesis is far more complicated than the synthesis of calixarene-, resorcinarene-, and pyrogallolarene-based container compounds. For example, a step-optimized synthetic approach toward spheriphane $C_{36}H_{36}$ involves the transformation

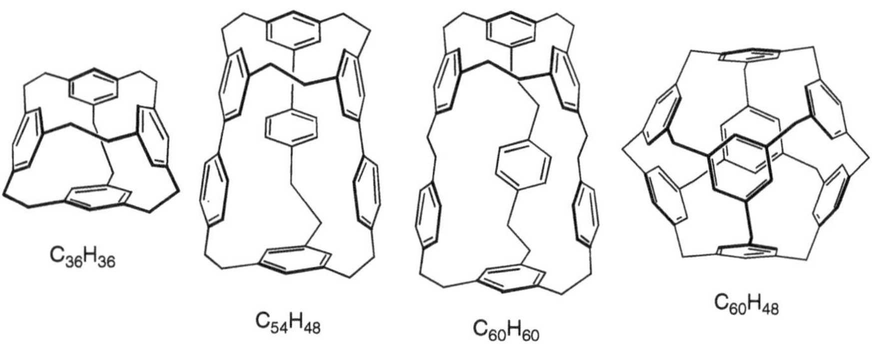

Figure 2.30 Spheriphanes $C_{36}H_{36}$, $C_{54}H_{48}$, $C_{60}H_{60}$, and $C_{60}H_{48}$.

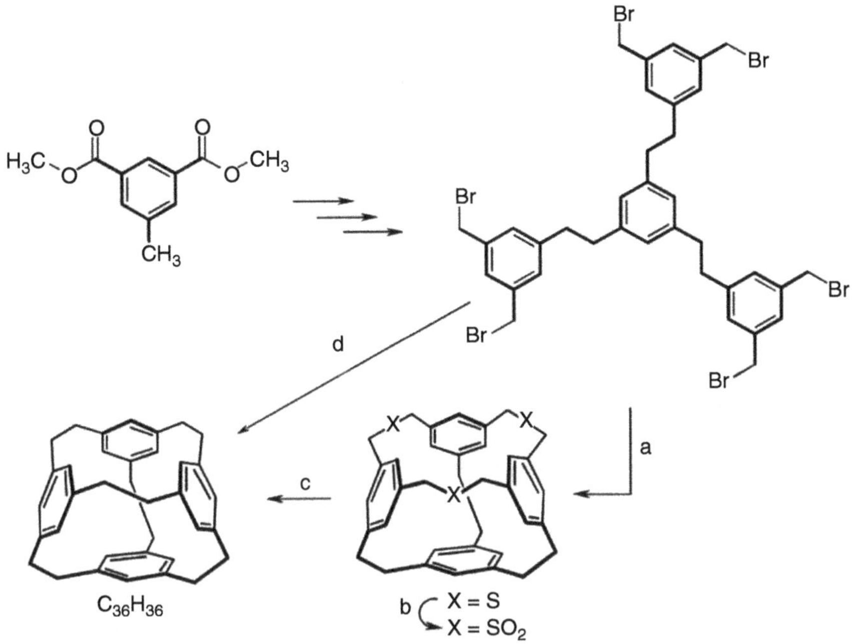

Scheme 2.13 The synthesis of the spheriphane $C_{36}H_{36}$: (a) Na_2S (18%); (b) *m*-ClC_6H_4COOOH (73%); (c) 10^{-2} Torr, 570 °C (45%); and (d) PhLi (11%).

of dimethyl 5-methylisophthalate via the hexabromide intermediate followed by cyclization leading to the desired compound in six steps. An alternative eight-step synthetic approach entails auxiliary cyclization of this key hexabromide with sodium sulfide, followed by the oxidation of the resulting tris-sulfide cage compound into the corresponding tris-sulfone. The pyrolytic desulfurization of this tris-sulfone yields the desired octacyclic hydrocarbon $C_{36}H_{36}$ (Scheme 2.13) [97].

The X-ray structure analysis of the spheriphane $C_{36}H_{36}$ (as a 1:1 adduct with chloroform) revealed the *P*-helical conformation of the carbon framework. The radius of the internal cavity in the spheriphane $C_{36}H_{36}$ is about 2.84 Å, and the size of the portals is about 3.86 Å (excluding van der Waals radii, Figure 2.31). Due to the relatively small cavity size, this compound produces only weak complexes with silver salts, which easily dissociate in methanol/dichloromethane mixtures [97].

Similar synthetic strategy was used for the preparation of concave hydrocarbons $C_{54}H_{48}$ and $C_{60}H_{60}$ with larger molecular frameworks. Unfortunately, this strategy cannot be used for the preparation of hitherto unknown spheriphane $C_{60}H_{48}$, which has 8 benzene rings interconnected with 12 methylene links. Spheriphanes $C_{36}H_{36}$, $C_{54}H_{48}$, and $C_{60}H_{60}$ encapsulate small metal ions and

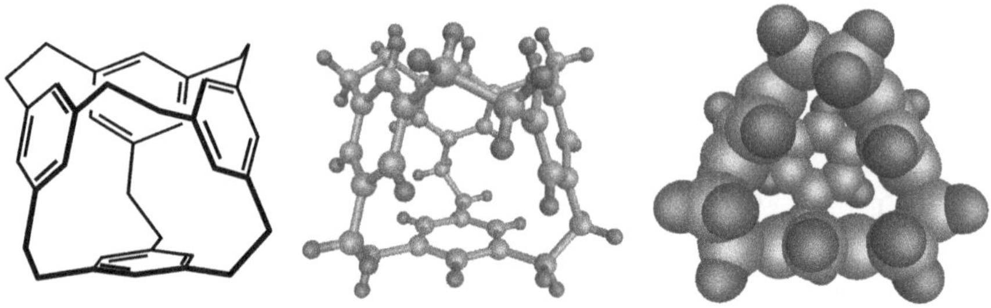

Figure 2.31 The X-ray structure of the spheriphane $C_{36}H_{36}$.

can be used for metal ion extraction from aqueous solution. Large spheriphane $C_{60}H_{60}$ shows enhanced selectivity toward silver and thallium cations [98, 99]. Even larger spheriphanes with acetylene links can be prepared via alkyne metathesis [101]. Despite the relatively large size of the cavity (about 14 Å in height), these compounds failed to produce container complexes with guest molecules with matching sizes, such as fullerene C_{60}.

Unlike all-hydrocarbon "classic" spheriphanes, structurally related spherical cage compounds containing heteroatoms are usually more affordable [102]. For example, a heterocyclic analog of the spheriphane $C_{36}H_{36}$, with six CH_2SSCH_2 links instead of six CH_2CH_2 links, can be prepared in one step from 1,3,5-tris(mercaptomethyl)benzene in the presence of $SbCl_3$. The main product of this reaction is actually the corresponding dimer of 1,3,5-tris(mercaptomethyl)benzene. It is plausible that $SbCl_3$ also serves as a template for this efficient one-step assembly of the spheriphane analog [103]. Similarly constructed spheriphane analogs with amide links [104, 105] and diimine links [106] are also known. The X-ray structure of a spheriphane analog with six methylamide links is presented in Figure 2.32.

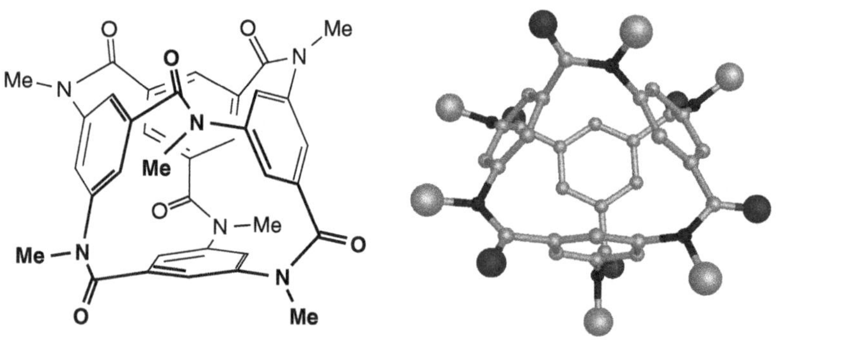

Figure 2.32 The spheriphane analog with six methylamide links (left) and its X-ray structure (right, hydrogen atoms were removed for clarity) [105].

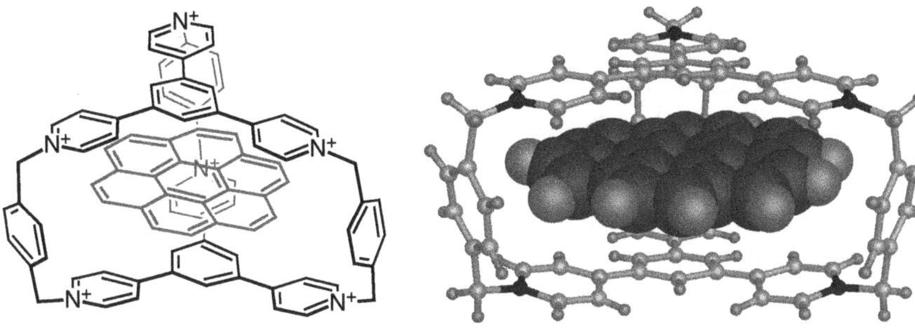

Figure 2.33 The complex of coronene with the positively charged heterocyclic spheriphane analog (left) and its X-ray structure (right, counteranions were removed for clarity).

Positively charged heterocyclic analogs of spheriphanes produce strong complexes with polycyclic aromatic compounds. The X-ray structure of one of these complexes with coronene is presented in Figure 2.33 [107]. The strength of this complex, with an estimated K_a above $3 \times 10^8\ M^{-1}$, can be attributed to a combination of face-to-face and edge-to-face π–π interactions between the electron-rich coronene guest molecule and electron-deficient, positively charged heteroaromatic walls of the molecular container.

As discussed earlier, the synthesis of molecular containers without heteroatoms is usually a formidable task, with one notable exception. This exception represents a unique class of spherical carbon allotropes called fullerenes, which can be prepared by a simple electric arc evaporation of various carbonaceous materials, such as graphite.

2.11 Fullerenes

Fullerenes are round container molecules consisting of only carbon atoms, typically assembled in a closed network of fused five- and six-membered carbon rings [108]. The first convincing evidence of the preferential formation of fullerene C_{60} during laser evaporation of graphite was obtained in 1985 [109]. Although these experiments proved the existence of this highly symmetric, all-carbon molecule, rapid progress in fullerene chemistry only started in 1990 with the advent of the arc-discharge evaporation method for the production of macroscopic quantities of fullerenes [110]. Given the high temperatures involved in the evaporation of graphite, this procedure yields predominantly the most thermodynamically favorable fullerenes, C_{60}-I_h (Figure 2.34) and C_{70}-D_{5h} (Figure 2.34), as the major components of "carbon soot". Although the arc-discharge process can be optimized for the production of other fullerenes, C_{60} and C_{70} retain their prominent roles in the fullerene science.

Less thermodynamically favorable higher fullerenes, such as C_{76}-C_{2v}, C_{78}-D_2, C_{84}-D_2, and larger, are also produced during the graphite evaporation, but in diminishingly lower yields. The most abundant higher fullerenes have the same isolated 12 pentagons as C_{60}-I_h, but with varied number of hexagons. It appears that there is no hard limit for the formation of large fullerenes, which can contain 500 or more carbon atoms. The variety of higher fullerenes is further extended by the ever-increasing number of available isomers, some of which were isolated in pure form by multistage HPLC [111]. Energy-minimized space-filling models of the major fullerene isomers are presented in Figure 2.35.

Small fullerenes (with less than 60 carbon atoms) can also be extracted from "carbon soot", but usually only as halogenated or hydrogenated derivatives. For example, fullerene C_{36} can be produced in the form of hydrides $C_{36}H_4$ and

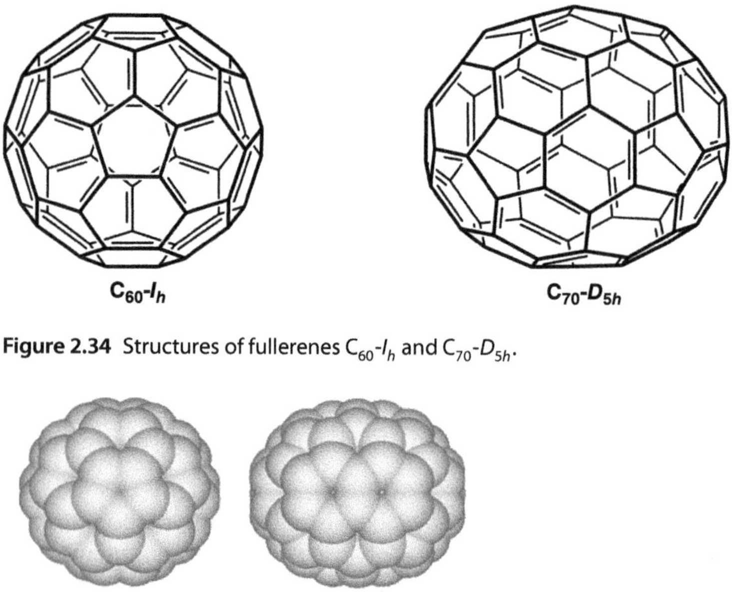

Figure 2.34 Structures of fullerenes C_{60}-I_h and C_{70}-D_{5h}.

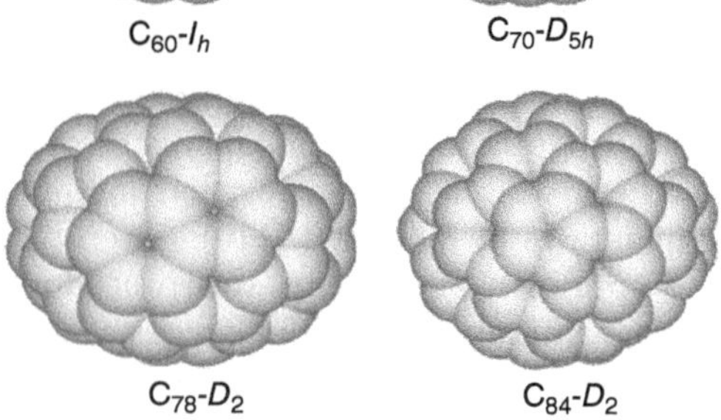

Figure 2.35 Space-filling models of the major fullerene isomers: C_{60}-I_h, C_{70}-D_2, C_{78}-D_{5h}, and C_{84}-D_2.

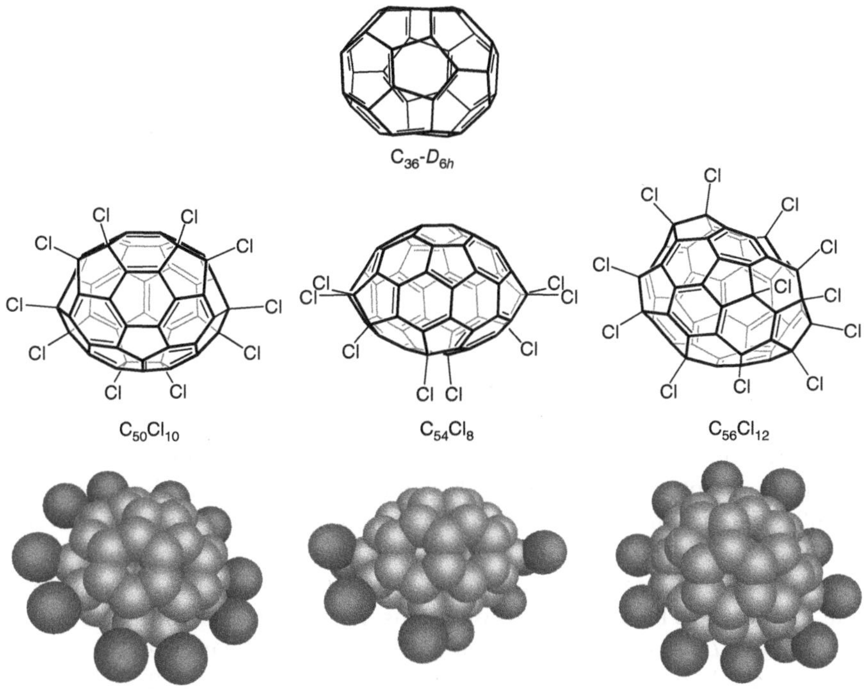

Figure 2.36 A proposed structure of the small fullerene C_{36}-D_{6h} (top) and the X-ray structures of the polychlorinated small fullerenes $C_{50}Cl_{10}$, $C_{54}Cl_8$, and $C_{56}Cl_{12}$ (middle and bottom rows).

$C_{36}H_6$ under certain conditions (Figure 2.36) [112, 113]. Fullerene C_{50}-D_{5h} was isolated in 0.002% yield as the corresponding decachloride $C_{50}Cl_{10}$ from "carbon soot" produced in the arc-discharge reactor retrofitted with a CCl_4-enhanced gas mixture [114]. Other small fullerene chlorides, such as $C_{54}Cl_8$ and $C_{56}Cl_{12}$, were also detected in the reaction mixture (Figure 2.36) [115, 116].

Excessive substitution of fullerenes usually leads to distortion of the carbon framework. For example, highly chlorinated fullerene $C_{60}Cl_{30}$ has a cylindrical "drum-like" carbon cage structure (Figure 2.37) [117]. Similar distortions were observed in highly fluorinated fullerene $C_{60}F_{48}$-D_3, which has a squashed prismatoid-like carbon cage structure, with little resemblance to the highly symmetrical ball-like structure of the parent fullerene C_{60} (Figure 2.37) [118, 119]. Major $C_{60}F_{36}$ isomers (as well as $C_{60}H_{36}$ isomers) [120] are also not spherical and have tetrahedron-like shape (Figure 2.37) [121, 122]. Extreme overcrowding conditions in highly fluorinated fullerenes lead to a unique phenomenon called "fluorine dance". Fluorine dance is an easy migration of fluorine atoms on the fullerene surface that leads predominantly to the most

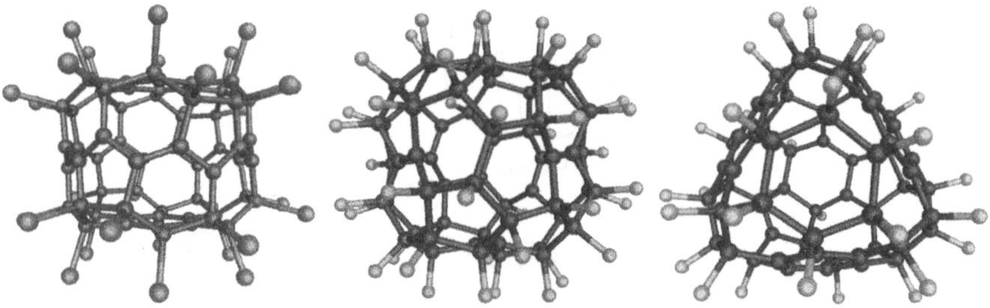

$C_{60}Cl_{30}$ $C_{60}F_{48}$-D_3 $C_{60}F_{36}$-C_1

Figure 2.37 The X-ray structures of the chlorinated fullerene $C_{60}Cl_{30}$ (left), the fluorinated fullerene $C_{60}F_{48}$-D_3 (middle), and the fluorinated fullerene $C_{60}F_{36}$-C_1 (right).

thermodynamically favorable isomers at elevated temperatures [123, 124]. Attempts to fluorinate fullerene C_{60} beyond $C_{60}F_{48}$ resulted in the rupture of the fullerene carbon network and the formation of "hyperfluorinated" species [125, 126].

Rearrangement of carbon bonds in the already formed fullerenes is also possible, but requires harsh reaction conditions. For example, chlorination of the fullerene C_{76}-D_2 in excess of $SbCl_5$ at 340 °C for several days yields flattened chlorinated fullerene $C_{76}Cl_{24}$-C_2 with "fused pentagons" (Figure 2.38) [127].

Fullerenes have tightly meshed "carbon wall" networks and relatively small inner cavities, although large enough to hold several atoms and even small molecules, thus forming *endohedral fullerenes* [128]. Since the carbon walls of fullerenes are generally not permeable at room temperature, the insertion of a guest in an empty fullerene molecule requires high pressures and temperatures, typically above 600 °C. Unfortunately, this "brute-force" approach is particularly effective only for noble gas atoms [129]. The X-ray structure of one of

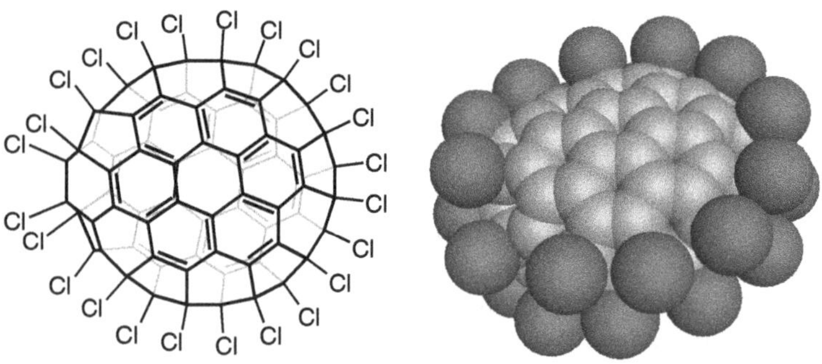

Figure 2.38 The chlorinated fullerene $C_{76}Cl_{24}$-C_2 with "fused pentagons" (left) and its crystal structure (right).

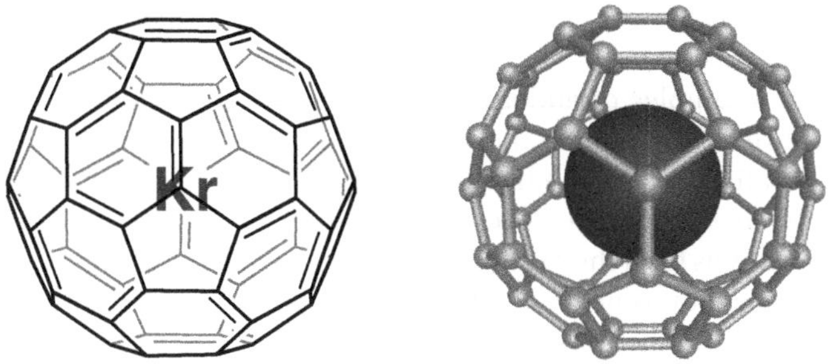

Figure 2.39 The endohedral fullerene Kr@C_{60} (left) and its X-ray structure (right).

these noble gas container compounds, Kr@C_{60}, is presented in Figure 2.39. This endohedral fullerene was prepared via a high-temperature (600–650 °C) and high-pressure (3.2 kbar of Kr) "brute-force" procedure with potassium cyanide as a catalyst in less than 0.3% yield. Enrichment of the sample to about 12% Kr@C_{60} content was performed by multistage high-pressure liquid chromatography. It is interesting to note that the introduction of the Kr atom in the center of the internal cavity does not induce any significant changes in the fullerene C_{60} carbon framework [130].

A variation of this "brute-force" synthesis of endohedral fullerenes is based on ion beam (or plasma) technology. It allows for the preparation of various endohedral fullerenes with a variety of atoms entrapped inside the carbon cage, from lithium to polonium, albeit in a very low yield. Of particular interest is endohedral fullerene with one nitrogen atom inside, N@C_{60}, in which the nitrogen atom retains its unpaired spin [131] and therefore can be used for spin labeling applications [132]. The crystal structure of another notable endohedral fullerene, Li^+@C_{60} (as 1,1′-bis(diphenylphosphino)ferrocene platinum complex), is presented in Figure 2.40 [133]. Li^+@C_{60} is now commercially available as a hexafluorophosphate salt [134], with potential applications in

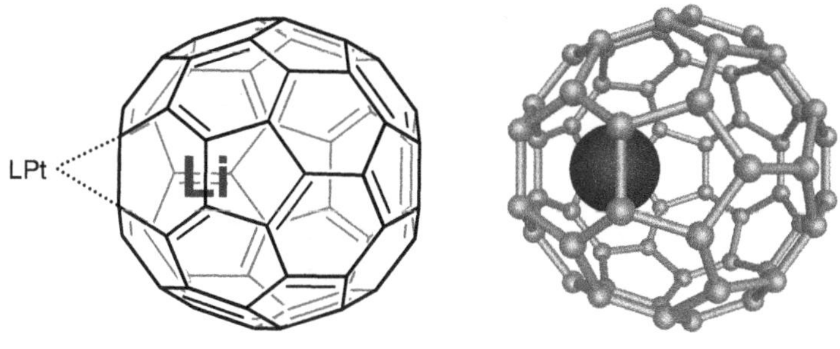

Figure 2.40 The endohedral fullerene Li^+@C_{60} as a complex with LPt (left, LPt = 1,1′-bis(diphenylphosphino)ferrocene platinum) and its X-ray structure (right, the LPt platinum moiety was removed for clarity).

electrochemical and photovoltaic areas due to its high ionic conductivity and attractive redox properties [135].

The abovementioned examples of endohedral fullerenes were produced by employing a forceful "breakthrough" in fullerenes. A more intelligent synthetic strategy for the preparation of endohedral fullerenes is "molecular surgery", in which the cage is opened and then closed in a series of site-selective reactions. Although more tedious than the one-step "brute-force" methodology, the molecular surgery approach in some cases allows for the preparation of almost 100% enriched samples of endohedral fullerenes [136–140].

The first attempts to perform molecular surgery on fullerene C_{60} were only partially successful. One of them was based on an efficient one-pot ring expansion reaction, which opened up a 14-membered ring portal in fullerene C_{60} (Scheme 2.14). Two small guests, a helium atom and molecular hydrogen, were inserted through the 14-membered ring orifice into the internal cavity of this bislactam derivative. However, even in high pressure and temperature conditions, the insertion of molecular hydrogen was achieved only in 5% yield. Maximum helium enrichment was substantially lower (1.5%). No attempts were made to close the portal, which would lead to the desired $H_2@C_{60}$ or $He@C_{60}$ [141].

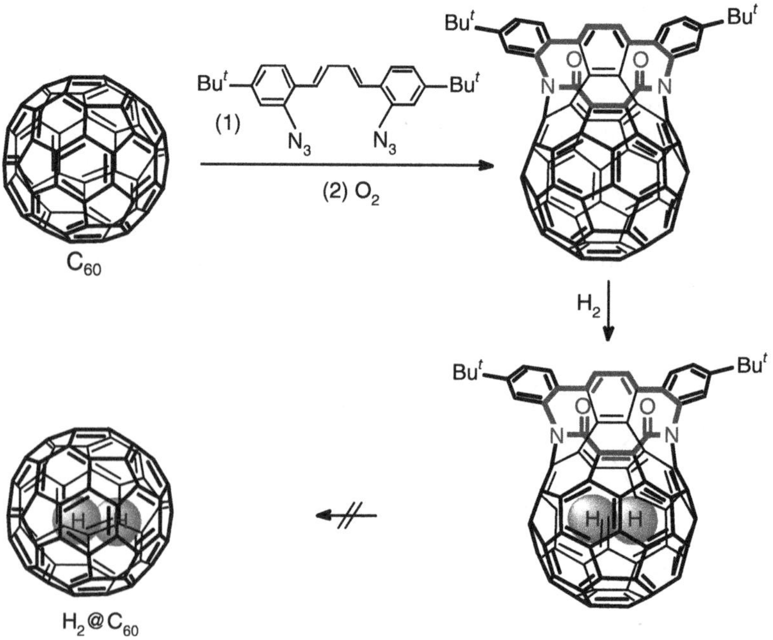

Scheme 2.14 An attempted synthesis of $H_2@C_{60}$.

Subsequently, a successful strategy was developed to close the orifice of an open-cage fullerene with a guest molecule inside [142]. In this case, a more permeable 13-membered ring orifice with a sulfur atom was used [143], which allowed for an almost quantitative encapsulation of molecular hydrogen [144]. A successful four-step orifice closure reaction sequence finally yielded the desired $H_2@C_{60}$ (Scheme 2.15). ^{13}C NMR studies show that the electronic configuration and the structure of the fullerene C_{60} framework are not materially affected by the presence of an H_2 molecule in the internal cavity. Endohedral fullerene $H_2@C_{60}$ demonstrates excellent thermal stability, comparable to the stability of intact fullerene C_{60}. For example, this filled molecular container compound easily withstands heating at 500 °C for 10 min without the loss of the encapsulated hydrogen molecule. This observation allows considering $H_2@C_{60}$ as a new type of hydrocarbon molecules, which are held together by constrictive bonding rather than C—H covalent interactions [142].

Further development of this molecular surgery strategy yields many important endohedral fullerenes C_{60} and C_{70} with various guest atoms and even small molecules trapped inside. The X-ray structure of the C_{60} container compound filled with a single molecule of water is presented in Figure 2.41. This compound is a unique example of a water molecule entrapped in a completely hydrophobic, all-carbon inner environment of fullerene C_{60} [145].

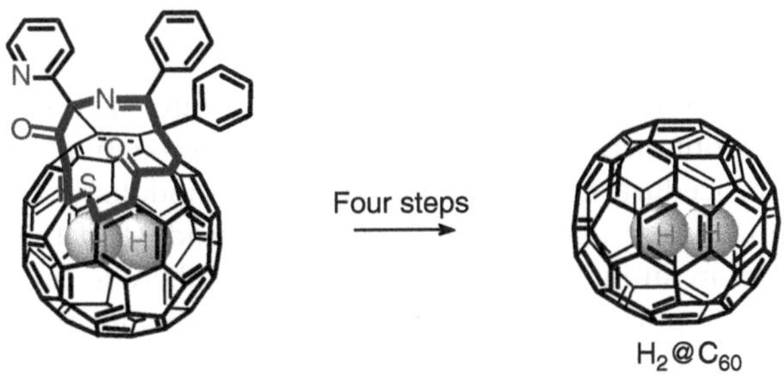

Scheme 2.15 The successful synthesis of $H_2@C_{60}$.

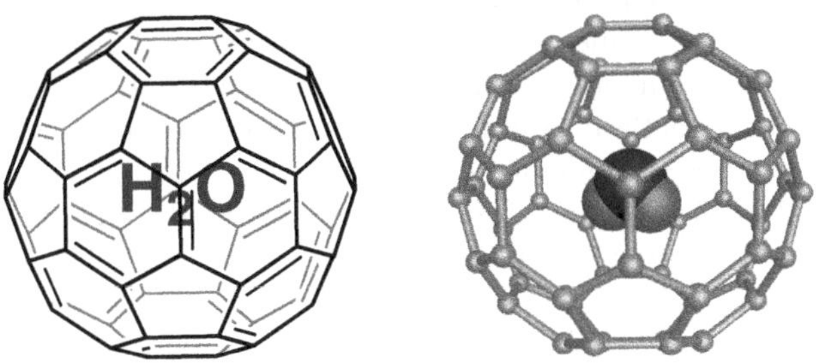

Figure 2.41 The endohedral fullerene $H_2O@C_{60}$ (left) and its X-ray structure (right) [145].

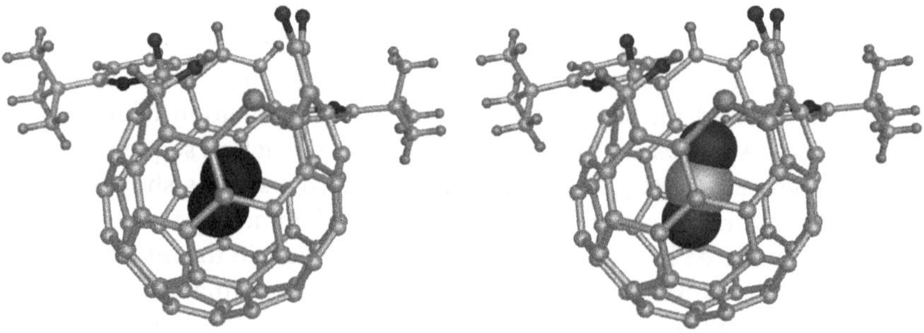

Figure 2.42 The X-ray structures of N_2 (left) and CO_2 (right) encapsulated in the open-cage fullerene container compound [146].

Encapsulation of other molecules, such as N_2 [146], CO_2 [146], HCN [147], CH_2O [147], NH_3 [148], and CH_4 [149], in the cavities of open-cage fullerenes is possible, but without the subsequent restoration of the original fullerene C_{60} carbon framework. The X-ray structures of N_2 and CO_2 container compounds are presented in Figure 2.42. Gated open-cage fullerenes, capable of controlled capture of guest molecules, are also known [150–152].

A large fullerene C_{70} molecular container has adequate cavity size for the entrapment of two small molecules [153, 154]. Examples of endohedral fullerenes C_{70} filled with one and two H_2O molecules are presented in Scheme 2.16.

In some rare cases, the introduction of a guest molecule into an open fullerene portal proceeds spontaneously. During a recent investigation of a new fullerene opening procedure employing oxidation methodology, an accidental discovery was made that one CO fragment, a by-product of this oxidation reaction, ended up in an internal cavity of the open fullerene cage (Scheme 2.17) [155]. Since the portal in this open fullerene is smaller than CO, this guest molecule is permanently stuck inside the cavity and can only be removed by further cutting of the fullerene carbon framework.

The preparation of endohedral metallofullerenes filled with some metals (usually rare earth metals) and their clusters proved to be a much easier task than the synthesis of endohedral fullerenes with other atoms or molecules mentioned earlier. These metallofullerenes can be produced by simply adding the appropriate ingredients directly into graphite rods before arc-discharge evaporation. The first fullerenes with metal atoms inside were detected along with C_{60} back in 1985 [156]. Subsequent research demonstrated that some metal atoms and clusters can serve as effective templates for the fullerene cage self-organization, sometimes producing almost exclusively endohedral fullerenes in high yield and purity [128, 157–161]. Metal-templating shifts thermally equilibrated pools of fullerenes toward different structures, including small endohedral metallofullerenes (such as $U@C_{28}$, Figure 2.43) [162] as

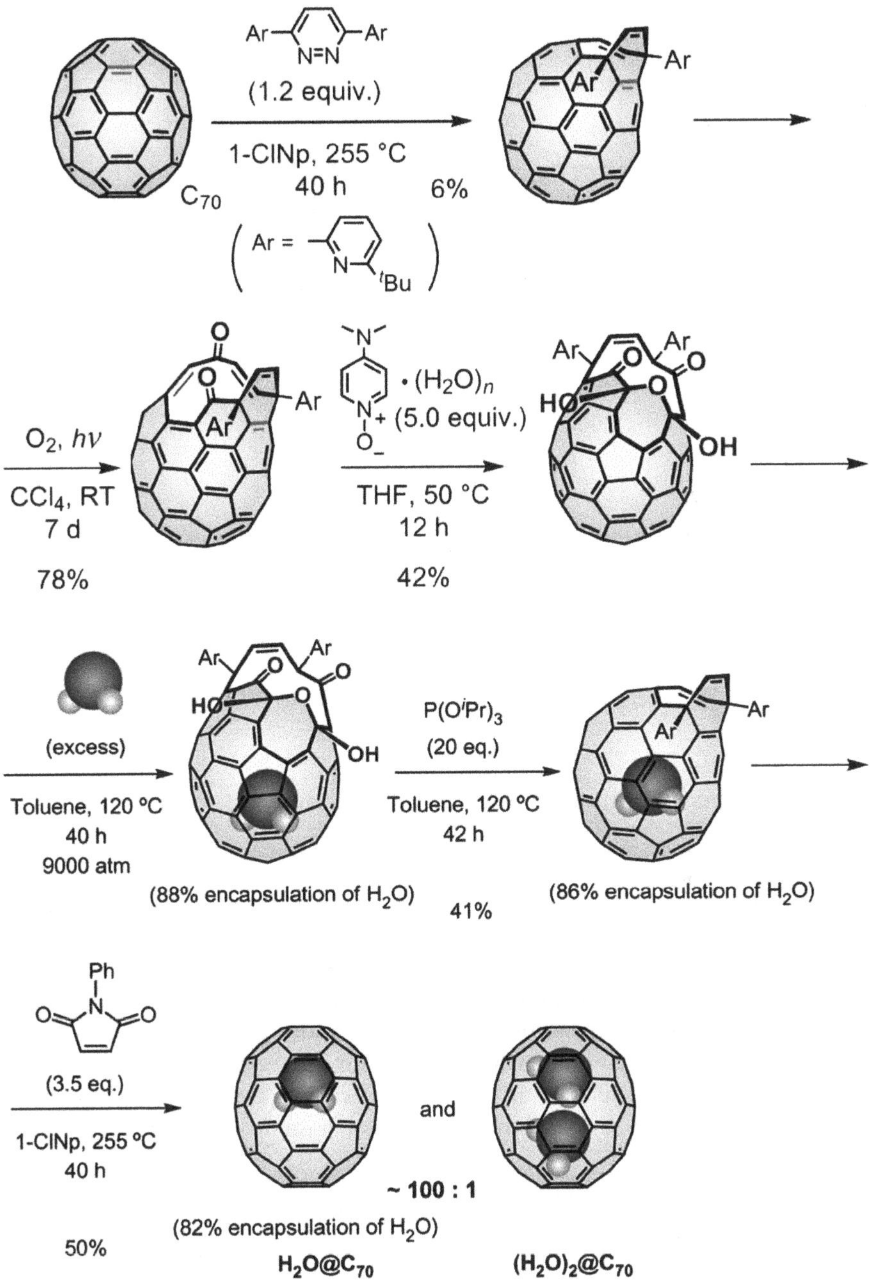

Scheme 2.16 The synthesis of the endohedral fullerenes $H_2O@C_{70}$ and $(H_2O)_2@C_{70}$ (1-ClNp = 1-chloronaphthalene). *Source*: Zhang *et al.* 2016 [154]. Adapted with permission of Nature Publishing Group.

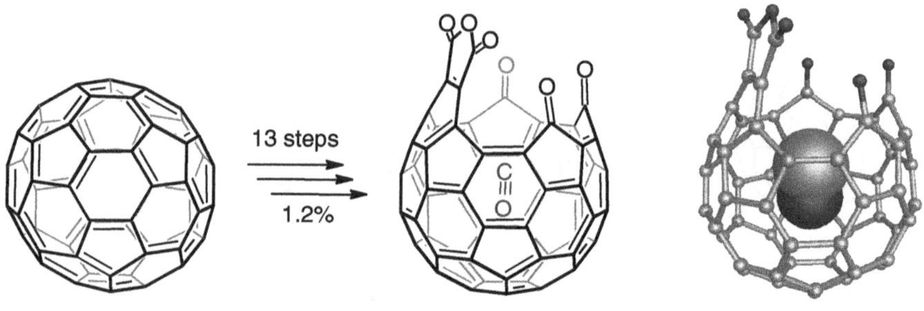

Scheme 2.17 The synthesis (left) and the X-ray structure (right) of the open endohedral fullerene $CO@C_{59}O_6$.

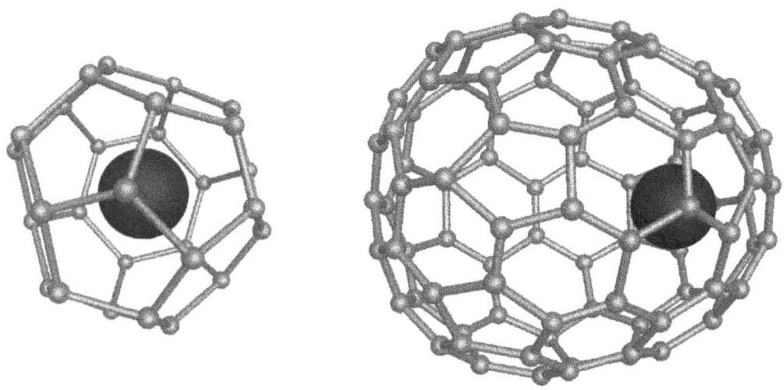

Figure 2.43 A proposed structure of small metallofullerene $U@C_{28}$ (left) [162] and the X-ray structure of large metallofullerene $Ca@C_{94}$-C_{3v} (right) [163].

well as large metallofullerenes (such as $Ca@C_{94}$-C_{3v}, Figure 2.43) [163], which are not very common in the absence of the metals.

The positions of the metal atoms in the preponderance of endohedral metallofullerenes are not center symmetric and are usually fixed within a relatively narrow region. This asymmetric metal position behavior is contrary to encapsulated noble gas atoms, which are usually centered in the middle of the cage and produce little effects on the fullerene carbon framework [130]. Some representative examples of endohedral rare-earth mono-metallofullerenes are presented in Figure 2.44.

Subsequent derivatization of endohedral mono-metallofullerenes provides an easy route for the construction of various container compounds, since the "payload" is usually retained during transformations that do not violate the integrity of the fullerene cage. In some cases, this derivatization allows for position control of the metal atom due to the alteration of the smooth and almost chemically indistinguishable all-carbon fullerene cage interior [167].

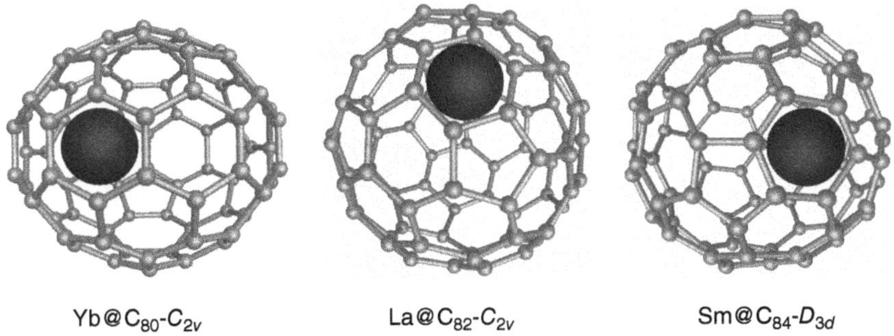

Figure 2.44 The X-ray structures of some typical rare-earth mono-metallofullerenes, $Yb@C_{80}$-C_{2v} [164], $La@C_{82}$-C_{2v} [165], and $Sm@C_{84}$-D_{3d} [166].

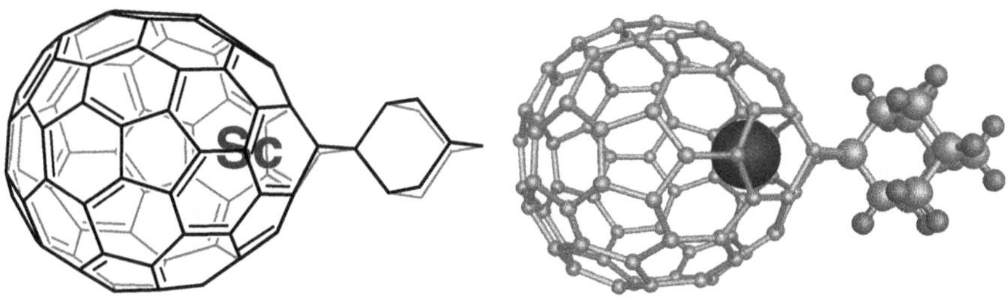

Figure 2.45 The adamantylidenyl derivative of endohedral fullerene $Sc@C_{82}$-C_{2v} (left) and its X-ray structure (right).

A notable example of this approach is the successful derivatization of endohedral fullerene $Sc@C_{82}$-C_{2v} via reaction with adamantylidene carbene (Figure 2.45) [168].

Elongated and large fullerenes provide adequate space for two metal atoms. These two metal atoms have a distinct tendency to occupy the opposite corners of the molecular containers due to unfavorable Coulomb interactions between the positively charged moieties. This tendency can be exemplified by observing X-ray structures of two endohedral di-metallofullerenes, $Sc_2@C_{66}$-C_{2v} (Figure 2.46) [169] and $Sm_2@C_{104}$-D_{3d} (Figure 2.46) [170].

While di-metallofullerenes are rare examples of simple poly-metallofullerenes, trimetallic nitride endohedral fullerenes are quite common and can sometimes be produced in yields comparable to the most abundant C_{60} by simply adding a source of nitrogen and metal during the arc-discharge evaporation of graphite. In fact, fullerenes containing a trimetallic nitride template (TNT) within the cage represent the most abundant class of endohedral metallofullerenes [171]. They have the general formula $A_xB_{3-x}N@C_{2n}$ (A, B = metal, $x = 0$–3) [172]. One of the prominent members of this class of

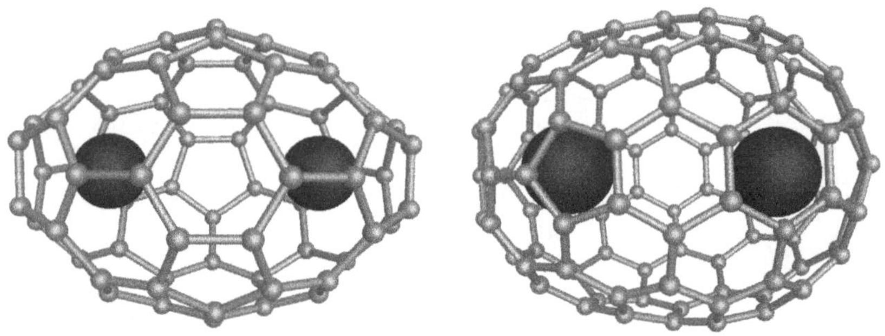

Figure 2.46 The X-ray structures of di-metallofullerenes $Sc_2@C_{66}$-C_{2v} and $Sm_2@C_{104}$-D_{3d}.

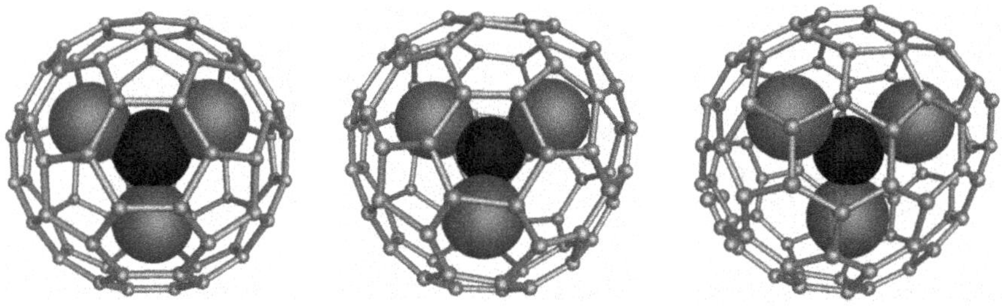

Figure 2.47 The X-ray structures of some typical TNT-fullerenes: $Sc_3N@C_{78}$-D_{3h} [175], $Du_3N@C_{80}$-I_h [176], and $Tb_3N@C_{84}$-C_s [177].

compounds, $Sc_3N@C_{80}$-I_h, can be prepared in 3–5% yields by employing a nitrogen–helium gas mixture in a conventional fullerene generator during the vaporization of graphite rods containing metal oxides [173]. The synthesis of nitride cluster fullerenes can be further enhanced by the addition of 5–10% of NH_3 as the "reactive gas". In this case, endohedral fullerenes are formed as the major products of the "carbon soot" mixture [174]. Some typical examples of TNT-fullerenes are presented in Figure 2.47.

Examples of endohedral clusterfullerenes [178, 179] also include carbides [180, 181], oxides [182, 183], sulfides [184, 185], and other [186, 187] container compounds. They can be prepared with the use of appropriate reactive gases, such as carbon dioxide [182, 183], sulfur dioxide [184, 185], and methane [188]. Some of these compounds are of interest as single-molecule magnets [188]. Other potential applications of endohedral fullerenes are discussed in Section 2.12.

2.12 Applications

Container compounds have numerous applications, particularly in material sciences and biomedical fields, as well as in sensors [189]. Here we present only one aspect of emerging applied chemistry of molecular containers related to the development of ^{129}Xe and Gd carriers as biological probes for magnetic resonance imaging (MRI).

Xenon-129 is actively explored as an imaging agent due to its unique ability to produce long-lasting hyperpolarized atoms via spin-exchange optical pumping technology, which improves the signal-to-noise MRI ratio by four orders of magnitude [190]. Because ^{129}Xe is a chemically inert noble gas, development of biologically relevant carriers is a complicated task. Container compounds appear to be a perfect choice of carrier vehicles for ^{129}Xe because they can entrap the noble gas atoms by means other than chemical bonding. One of the most promising molecular container carriers for ^{129}Xe is cryptophane-111. Cryptophane-111 forms a very strong complex with Xe [191]. In the solid state, this complex starts to lose Xe only above 290 °C (Figure 2.48) [192].

MRI probes based on ^{129}Xe–cryptophane complexes usually contain additional functional groups to ensure bioavailability, proper target delivery, and sensor functionality. Some examples of these MRI probes, which can be used as pH-dependent cell labeling, or biotin–avidin binding biosensors, are presented in Figure 2.49.

An advanced biological functionality can be attained through the modification of ^{129}Xe–cryptophane complexes with drug-like ligand fragments capable of selective binding to active sites of protein molecules. An example of this approach is shown in Figure 2.50. In this particular case, the cryptophane container was appended with a sulfonamide moiety as a site-selective ligand for human carbonic anhydrase (CA). Despite the presence of the bulky molecular container unit, this ^{129}Xe biosensor has surprisingly high binding affinity toward CA (K_d = 20–110 nM) and can differentiate two isoforms of this enzyme, CAI and CAII [195].

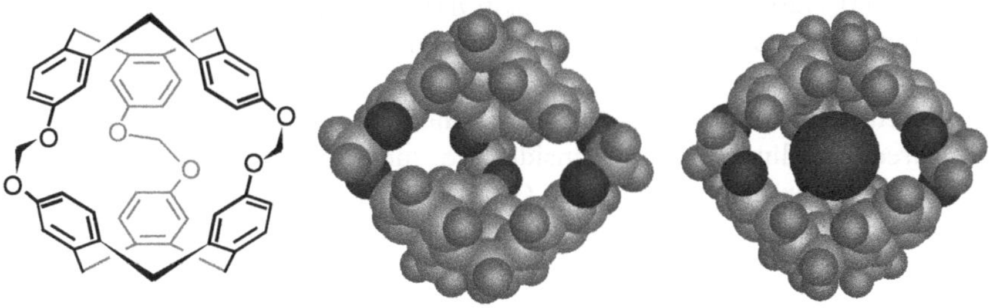

Figure 2.48 Cryptophane-111 (left), its X-ray structure (middle), and the X-ray structure of cryptophane-111 complex with Xe (right).

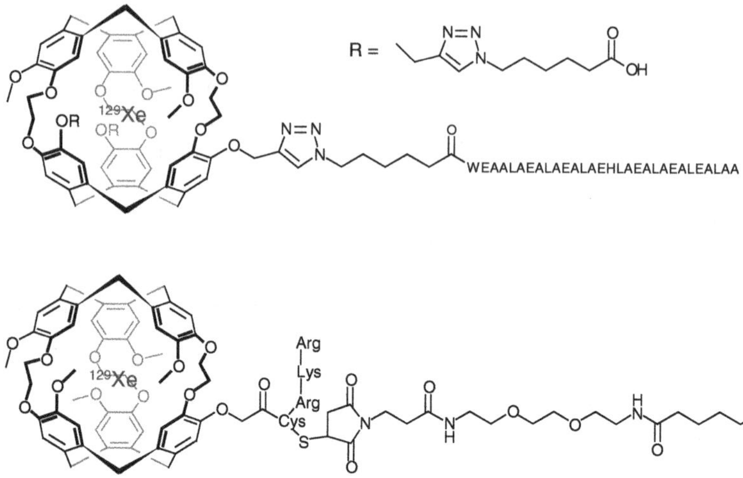

Figure 2.49 MRI probes based on ^{129}Xe–cryptophane complexes: the biosensor for pH-dependent cell labeling (top, WEAA(LAEA)$_2$LAEH(LAEA)$_2$LEALAA = GALA, pore – forming peptide) [193] and the biotin–avidin binding sensor (bottom, Arg = arginine, Cys = cysteine, Lys = lysine) [194].

Another emerging field of container compound applications is related to gadolinium endohedral fullerenes as MRI contrast agents [197]. These contrast agents are superior to commercial contrast agents by an order of magnitude [198]. The toxicity of these new MRI agents related to gadolinium "leak" should be minimal due to the total enclosure of gadolinium atoms inside the impermeable fullerene cages. The X-ray structures of some gadolinium endohedral metallofullerenes are presented in Figure 2.51.

To achieve bioavailability, gadolinium metallofullerenes are usually surface modified. One of the simplest modifications is the attachment of hydroxy groups on the fullerene surface. This modification can be implemented by the synthesis of various polyhydroxy derivatives of metallofullerenes, such as $Gd@C_{60}(OH)_n$ [202, 203] and $Gd@C_{82}(OH)_m$ [198] (Figure 2.52). Both the compounds demonstrate very good MRI relaxivities, in many cases exceeding $80\,mM^{-1}\,s^{-1}$.

Even better MRI relaxivities can be attained by the use of TNT fullerenes with three gadolinium atoms inside the fullerene molecular container. For example, some derivatives of $Gd_3N@C_{80}$ and $Gd_3N@C_{84}$ can reach relaxivities well above $100\,mM^{-1}\,s^{-1}$ [204]. The exceptional contrast properties of TNT fullerenes could substantially improve the efficiency of MRI imaging applications. Proper derivatization allows for the construction of practical gadolinium molecular container contrast agents with good bioavailability and target delivery while retaining excellent MRI relaxivity parameters. An

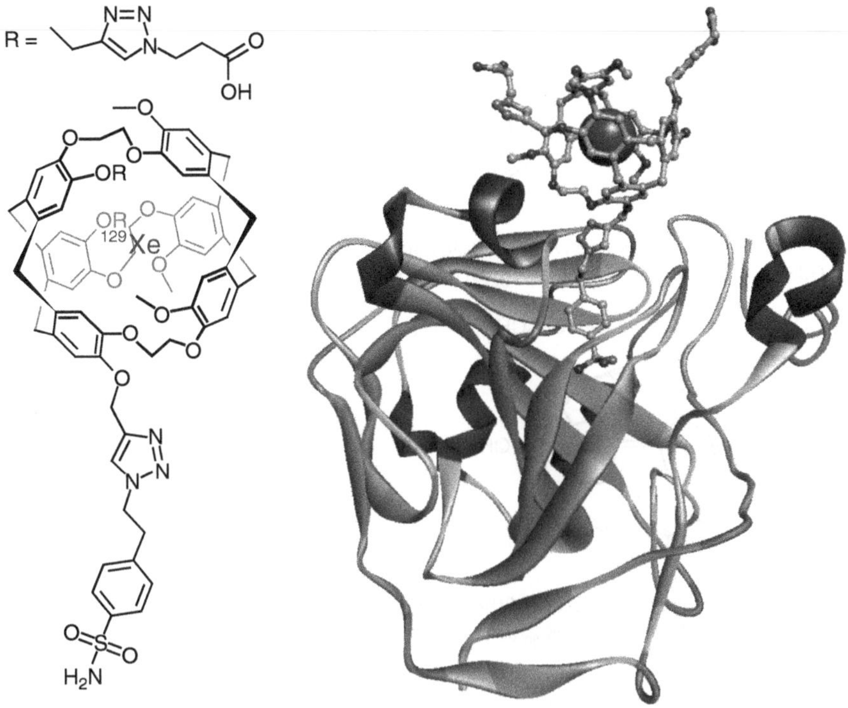

Figure 2.50 The structure of the drug-like cryptophane container compound as a ^{129}Xe carrier (left) and the X-ray structure of its complex with human carbonic anhydrase II (right) [196].

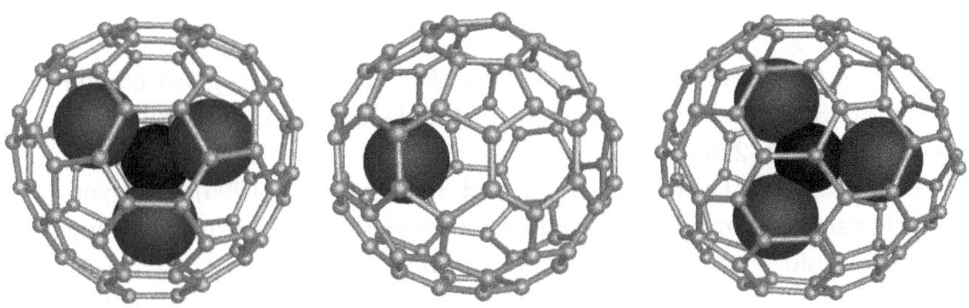

Figure 2.51 The X-ray structures of some gadolinium metallofullerenes: $Gd_3N@C_{80}$-I_h [199], $Gd@C_{82}$-C_{2v} [200], and $Gd_3N@C_{84}$-C_s [201].

example of an optimized water-soluble polyethyleneglycol derivative of TNT fullerene $Gd_3N@C_{80}$ is presented in Figure 2.53 [205].

Other potential applications of container compounds include single-molecule magnets [206], cancer therapy [207], chemical separation [208], solar cell components [209], atomic clock [210], and many others. While only

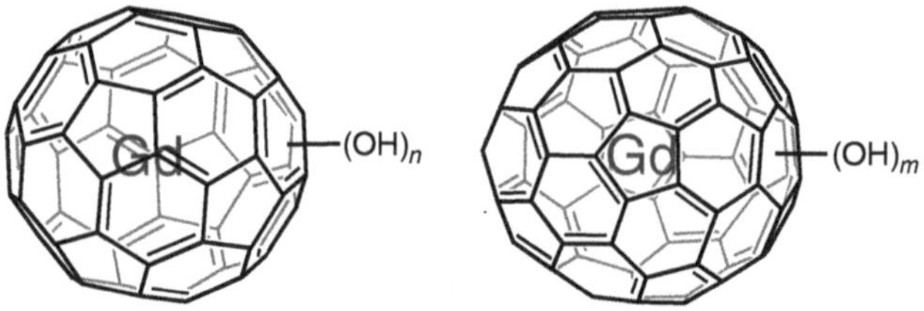

Figure 2.52 Structures of some water-soluble gadolinium polyhydroxy metallofullerenes (left, $Gd@C_{60}(OH)_n$, $n = 27$; right, $Gd@C_{82}(OH)_m$, $m = 40$).

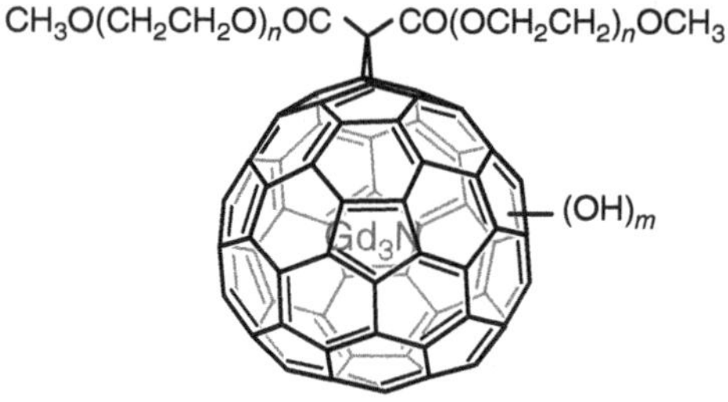

Figure 2.53 The proposed structure of the optimized water-soluble polyethyleneglycol derivative of TNT fullerene $Gd_3N@C_{80}$ ($n = 7$).

a few molecular containers are currently commercialized, the wide range of application fields makes it almost certain that we will witness a substantial number of commercial products in the near future.

References

1 Sherman, J.C., Knobler, C.B., and Cram, D.J. (1991) Host–guest complexation. 56. Syntheses and properties of soluble carceplexes. *J. Am. Chem. Soc.*, **113**, 2194–2204.
2 Fraser, J.R., Borecka, B., Trotter, J., and Sherman, J.C. (1995) An asymmetric carceplex and new crystal structure yield information regarding a 1 million-fold template effect. *J. Org. Chem.*, **60**, 1207–1213.
3 Dunk, P.W., Kaiser, N.K., Hendrickson, C.L. *et al.* (2012) Closed network growth of fullerenes. *Nat. Commun.*, **3**, 855.
4 Cram, D.J. (1992) Molecular container compounds. *Nature*, **356**, 29–36.
5 Rudkevich, D.M. (2002) Nanoscale molecular containers. *Bull. Chem. Soc. Jpn.*, **75**, 393–413.
6 Turner, D.R., Pastor, A., Alajarin, M., and Steed, J.W. (2004) Molecular containers: design approaches and applications, in *Supramolecular Assembly via Hydrogen Bonds I*, Structure and Bonding (ed. D.M.P. Mingos), Springer, Berlin/Heidelberg, pp. 97–168.

7 Cram, D.J. and Cram, J.M. (1997) *Container Molecules and Their Guests*, Monographs in Supramolecular Chemistry (ed. J.F. Stoddart), Royal Society of Chemistry, Cambridge.
8 Cram, D.J., Blanda, M.T., Paek, K., and Knobler, C.B. (1992) Host–guest complexation. 63. Constrictive and intrinsic binding in a hemicarcerand containing four portals. *J. Am. Chem. Soc.*, **114**, 7765–7773.
9 Galan, A. and Ballester, P. (2016) Stabilization of reactive species by supramolecular encapsulation. *Chem. Soc. Rev.*, **45**, 1720–1737.
10 André, C. (1987) Cyclotriveratrylenes and cryptophanes. *Tetrahedron*, **43**, 5725–5759.
11 Vicens, J. and Böhmer, V. (eds) (1991) *Calixarenes: A Versatile Class of Macrocyclic Compounds*, Topics in Inclusion Science, vol. **3**, Kluwer, Dordrecht.
12 Gutsche, C.D. (1998) *Calixarenes Revisited*, Royal Society of Chemistry.
13 Sliwa, W. and Kozlowski, C. (2009) *Calixarenes and Resorcinarenes. Synthesis, Properties and Applications*, 1st edn, Wiley-VCH.
14 Timmerman, P., Verboom, W., and Reinhoudt, D.N. (1996) Resorcinarenes. *Tetrahedron*, **52**, 2663–2704.
15 Hardie, M.J. (2010) Recent advances in the chemistry of cyclotriveratrylene. *Chem. Soc. Rev.*, **39**, 516–527.
16 Goldberg, I. (1986) Binding neutral guests to concave surfaces of molecular hosts. Rigid structural model for complexation between two lipophilic entities. *J. Incl. Phenom*, **4**, 191–198.
17 Kumar, S., Paul, D., and Singh, H. (2006) Syntheses, structures and interactions of heterocalixarenes. *ARKIVOC*, **9**, 17–25.
18 Parvari, G., Annamalai, S., Borovoi, I. *et al.* (2014) Multifarenes: new modular cavitands. *Chem. Commun.*, **50**, 2494–2497.
19 Juneja, R.K., Robinson, K.D., Johnson, C.P., and Atwood, J.L. (1993) Synthesis and characterization of rigid, deep-cavity calix[4]arenes. *J. Am. Chem. Soc.*, **115**, 3818–3819.
20 Gleave, C.A. and Sutherland, I.O. (1994) A new calix[4]arene binding site. Strong cooperativity in cation binding by a two site receptor. *J. Chem. Soc., Chem. Commun.*, 1873–1874.
21 Xu, B., Miao, Y.-J., and Swager, T.M. (1998) Palladium couplings on metallocalix[4]arenes: a efficient synthesis of new functionalized cavities. *J. Org. Chem.*, **63**, 8561–8564.
22 von dem Bussche-Hünnefeld, C., Helgeson, R.C., Bühring, D. *et al.* (1996) Bowl shaped cavitands dimerize and complex certain organic guests in organic solvents which themselves are poor guests. *Croat. Chem. Acta*, **69**, 447–458.
23 Ma, S., Rudkevich, D.M., and Rebek, J. (1998) "Deep-cavity" resorcinarenes dimerize through hydrogen bonding and self-inclusion. *J. Am. Chem. Soc.*, **120**, 4977–4981.

24 Biros, S.M. and Rebek, J. (2006) Structure and binding properties of water-soluble cavitands and capsules. *Chem. Soc. Rev.*, **36**, 93–104.

25 Mettry, M., Moehlig, M.P., and Hooley, R.J. (2015) Synthesis, guest binding, and metal coordination of functionalized self-folding deep cavitands. *Org. Lett.*, **17**, 1497–1500.

26 Xi, H., Gibb, C.L.D., and Gibb, B.C. (1999) Functionalized deep-cavity cavitands. *J. Org. Chem.*, **64**, 9286–9288.

27 Green, J.O., Baird, J.-H., and Gibb, B.C. (2000) Reduced-symmetry deep-cavity cavitands. *Org. Lett.*, **2**, 3845–3848.

28 Soncini, P., Bonsignore, S., Dalcanale, E., and Ugozzoli, F. (1992) Cavitands as versatile molecular receptors. *J. Org. Chem.*, **57**, 4608–4612.

29 Vincenti, M., Minero, C., Pelizzetti, E. *et al.* (1995) Host–guest chemistry in the gas phase and at the gas–solid interface: fundamental aspects and practical applications. *Pure Appl. Chem.*, **67**, 1075–1084.

30 Moran, J.R., Ericson, J.L., Dalcanale, E. *et al.* (1991) Vases and kites as cavitands. *J. Am. Chem. Soc.*, **113**, 5707–5714.

31 Kim, K. and Paek, K. (2003) Nano-scaled deep-cavity cavitands. *Bull. Korean Chem. Soc.*, **24**, 1374–1376.

32 Tucci, F.C., Rudkevich, D.M., and Rebek, J. (1999) Deeper cavitands. *J. Org. Chem.*, **64**, 4555–4559.

33 Arduini, A., Pochini, A., and Secchi, A. (2000) Rigid calix[4]arene as a building block for the synthesis of new quaternary ammonium cation receptors. *Eur. J. Org. Chem.*, **2000**, 2325–2334.

34 Wang, J., Bodige, S.G., Watson, W.H., and Gutsche, C.D. (2000) Complexation of fullerenes with 5,5′-biscalix[5]arene. *J. Org. Chem.*, **65**, 8260–8263.

35 Wang, J. and Gutsche, C.D. (1998) Complexation of fullerenes with bis-calix[*n*]arenes synthesized by tandem Claisen rearrangement. *J. Am. Chem. Soc.*, **120**, 12226–12231.

36 Haino, T., Yanase, M., and Fukazawa, Y. (1998) Fullerenes enclosed in bridged calix[5]arenes. *Angew. Chem. Int. Ed.*, **37**, 997–998.

37 Lücking, U., Tucci, F.C., Rudkevich, D.M., and Rebek, J. (2000) Self-folding cavitands of nanoscale dimensions. *J. Am. Chem. Soc.*, **122**, 8880–8889.

38 Chapman, R.G. and Sherman, J.C. (1998) Reversible molecular capsules composed of two cavitands linked via an assortment of charged-hydrogen bonds and covalent bonds. *J. Am. Chem. Soc.*, **120**, 9818–9826.

39 Cram, D.J., Tunstad, L.M., and Knobler, C.B. (1992) Host–guest complexation. 61. C- and Z-shaped ditopic cavitands, their binding characteristics, and monotopic relatives. *J. Org. Chem.*, **57**, 528–535.

40 Starnes, S.D., Rudkevich, D.M., and Rebek, J. (2001) Cavitand-porphyrins. *J. Am. Chem. Soc.*, **123**, 4659–4669.

41 Jasat, A. and Sherman, J.C. (1999) Carceplexes and hemicarceplexes. *Chem. Rev.*, **99**, 931–968.

42 Warmuth, R. and Yoon, J. (2001) Recent highlights in hemicarcerand chemistry. *Acc. Chem. Res.*, **34**, 95–105.
43 Cram, D.J., Tanner, M.E., and Thomas, R. (1991) The taming of cyclobutadiene. *Angew. Chem. Int. Ed.*, **30**, 1024–1027.
44 Yoon, J., Knobler, C.B., Maverick, E.F., and Cram, D.J. (1997) Dissymmetric new hemicarcerands containing four bridges of different lengths. *Chem. Commun.*, 1303–1304.
45 Chang, T.-Y. and Paek, K. (2008) Four-directionally functionalized hemicarcerands: intermediates for 2D square grid network of container molecules. *Bull. Korean Chem. Soc.*, **29**, 276–278.
46 Ye, B. and Pack, K. (2006) Molecular engineering. Part 12. Hemicarcerand having a metal coordinating ligand at a hetero-bridge. *Bull. Korean Chem. Soc.*, **27**, 305–308.
47 Robbins, T.A., Knobler, C.B., Bellew, D.R., and Cram, D.J. (1994) Host–guest complexation. 67. A highly adaptive and strongly binding hemicarcerand. *J. Am. Chem. Soc.*, **116**, 111–122.
48 Quan, M.L.C. and Cram, D.J. (1991) Constrictive binding of large guests by a hemicarcerand containing four portals. *J. Am. Chem. Soc.*, **113**, 2754–2755.
49 Cram, D.J., Jaeger, R., and Deshayes, K. (1993) Host–guest complexation. 65. Hemicarcerands that encapsulate hydrocarbons with molecular weights greater than two hundred. *J. Am. Chem. Soc.*, **115**, 10111–10116.
50 Gibb, C.L.D., Stevens, E.D., and Gibb, B.C. (2000) The self-assembly of benzyl alcohol derived deep-cavity cavitands: a new, highly efficient moiety for irreversible assemblies? *Chem. Commun.*, 363–364.
51 Liu, F., Helgeson, R.C., and Houk, K.N. (2014) Building on Cram's legacy: stimulated gating in hemicarcerands. *Acc. Chem. Res.*, **47**, 2168–2176.
52 Helgeson, R.C., Hayden, A.E., and Houk, K.N. (2010) Controlled gating of a hemicarcerand by disulfide–dithiol interchange. *J. Org. Chem.*, **75**, 570–575.
53 Wang, H., Liu, F., Helgeson, R.C., and Houk, K.N. (2013) Reversible photochemically gated transformation of a hemicarcerand to a carcerand. *Angew. Chem. Int. Ed.*, **52**, 655–659.
54 Collet, A. (1996) Cryptophanes, in *Molecular Recognition: Receptors for Molecular Guests*, Comprehensive Supramolecular Chemistry, vol. **2** (ed. F. Vögtle), Pergamon Press, Oxford, pp. 325–365.
55 Brotin, T. and Dutasta, J.-P. (2009) Cryptophanes and their complexes—present and future. *Chem. Rev.*, **109**, 88–130.
56 Hardie, M.J. (2012) Cyclotriveratrylene and cryptophanes, in *Molecular Recognition*, Supramolecular Chemistry: From Molecules to Nanomaterials, vol. **3** (eds J.W. Steed and P.A. Gale), John Wiley & Sons, pp. 895–916.
57 Canceill, J. and Collet, A. (1988) Two-step synthesis of D_3 and C_{3h} cryptophanes. *J. Chem. Soc., Chem. Commun.*, 582–584.

58 Collet, A., Dutasta, J.-P., Lozach, B., and Canceill, J. (1993) Cyclotriveratrylenes and cryptophanes: their synthesis and applications to host–guest chemistry and to the design of new materials, in *Supramolecular Chemistry I—Directed Synthesis and Molecular Recognition*, Topics in Current Chemistry, vol. **165**, Springer, Berlin/Heidelberg, pp. 103–129.

59 Garel, L., Dutasta, J., and Collet, A. (1993) Complexation of methane and chlorofluorocarbons by cryptophane-A in organic solution. *Angew. Chem. Int. Ed.*, **32**, 1169–1171.

60 Mough, S.T., Goeltz, J.C., and Holman, K.T. (2004) Isolation and structure of an "imploded" cryptophane. *Angew. Chem. Int. Ed.*, **43**, 5631–5635.

61 Jacobson, D.R., Khan, N.S., Collé, R. *et al.* (2011) Measurement of radon and xenon binding to a cryptophane molecular host. *Proc. Natl. Acad. Sci. U.S.A.*, **108**, 10969–10973.

62 Taratula, O., Hill, P.A., Khan, N.S. *et al.* (2010) Crystallographic observation of "induced fit" in a cryptophane host–guest model system. *Nat. Commun.*, **1**, 148.

63 Li, M.-J., Lai, C.-C., Liu, Y.-H. *et al.* (2009) Two guest complexation modes in a cyclotriveratrylene-based molecular container. *Chem. Commun.*, 5814–5816.

64 Li, M.-J., Huang, C.-H., Lai, C.-C., and Chiu, S.-H. (2012) Hemicarceplex formation with a cyclotriveratrylene-based molecular cage allows isolation of high-purity (≥99.0%) C_{70} directly from fullerene extracts. *Org. Lett.*, **14**, 6146–6149.

65 Ku, M.-Y., Huang, S.-J., Huang, S.-L. *et al.* (2014) Hemicarceplex formation allows ready identification of the isomers of the metallofullerene $Sc_3N@C_{80}$ using ^1H and ^{13}C NMR spectroscopy. *Chem. Commun.*, **50**, 11709–11712.

66 Wong, T.-H., Chang, J.-C., Lai, C.-C. *et al.* (2014) Hemicarceplexes modify the solubility and reduction potentials of C_{60}. *J. Org. Chem.*, **79**, 3581–3586.

67 Cram, D.J., Karbach, S., Kim, Y.H. *et al.* (1985) Shell closure of two cavitands forms carcerand complexes with components of the medium as permanent guests. *J. Am. Chem. Soc.*, **107**, 2575–2576.

68 Helgeson, R.C., Paek, K., Knobler, C.B. *et al.* (1996) Guest-assisted and guest-inhibited shell closures provide differently shaped carceplexes and hemicarceplexes. *J. Am. Chem. Soc.*, **118**, 5590–5604.

69 Chopra, N. and Sherman, J.C. (1999) A giant carceplex permanently entraps three organic molecules. *Angew. Chem. Int. Ed.*, **38**, 1955–1957.

70 Makeiff, D.A. and Sherman, J.C. (2005) A six-bowl carceplex that entraps seven guest molecules. *J. Am. Chem. Soc.*, **127**, 12363–12367.

71 van Wageningen, A.M.A., Timmerman, P., van Duynhoven, J.P.M. *et al.* (1997) Calix[4]arene-based (hemi)carcerands and carceplexes: synthesis, functionalization, and molecular modeling study. *Chem. Eur. J.*, **3**, 639–654.
72 Timmerman, P., Nierop, K.G.A., Brinks, E.A. *et al.* (1995) Hydrophobic concave surfaces and cavities by combination of calix[4]arenes and resorcin[4]arenes. *Chem. Eur. J.*, **1**, 132–143.
73 Timmerman, P., Verboom, W., van Veggel, F.C.J.M. *et al.* (1994) A novel type of stereoisomerism in calix[4]arene-based carceplexes. *Angew. Chem. Int. Ed.*, **33**, 2345–2348.
74 Kobayashi, K. and Yamanaka, M. (2014) Self-assembled capsules based on tetrafunctionalized calix[4]resorcinarene cavitands. *Chem. Soc. Rev.*, **44**, 449–466.
75 Shimizu, K.D. and Rebek, J. (1995) Synthesis and assembly of self-complementary calix[4]arenes. *Proc. Natl. Acad. Sci. U.S.A.*, **92**, 12403–12407.
76 Hamann, B.C., Shimizu, K.D., and Rebek, J. (1996) Reversible encapsulation of guest molecules in a calixarene dimer. *Angew. Chem. Int. Ed.*, **35**, 1326–1329.
77 Rebek, J. (2000) Host–guest chemistry of calixarene capsules. *Chem. Commun.*, 637–643.
78 Ballester, P. and Gil-Ramírez, G. (2009) Self-assembly of dimeric tetraurea calix[4]pyrrole capsules. *Proc. Natl. Acad. Sci. U.S.A.*, **106**, 10455–10459.
79 Vreekamp, R.H., Verboom, W., and Reinhoudt, D.N. (1996) Lower rim-upper rim hydrogen-bonded adducts of calix[4]arenes. *J. Org. Chem.*, **61**, 4282–4288.
80 Koh, K., Araki, K., and Shinkai, S. (1994) Self-assembled molecular capsule based on the hydrogen-bonding interaction between two different calix[4]arenes. *Tetrahedron Lett.*, **35**, 8255–8258.
81 Śliwa, W. (2006) Calixarene- and cavitand-based capsules. *ARKIVOC*, **5**, 137–159.
82 Śliwa, W. and Dondela, B. (2007) Trimeric and hexameric calixarene-based capsules. *ARKIVOC*, **2**, 201–223.
83 MacGillivray, L.R. and Atwood, J.L. (1997) A chiral spherical molecular assembly held together by 60 hydrogen bonds. *Nature*, **389**, 469–472.
84 Yamanaka, M., Shivanyuk, A., and Rebek, J. (2004) Kinetics and thermodynamics of hexameric capsule formation. *J. Am. Chem. Soc.*, **126**, 2939–2943.
85 Cave, G.W.V., Antesberger, J., Barbour, L.J. *et al.* (2004) Inner core structure responds to communication between nanocapsule walls. *Angew. Chem. Int. Ed.*, **43**, 5263–5266.
86 Shivanyuk, A., Friese, J.C., Döring, S., and Rebek, J. (2003) Solvent-stabilized molecular capsules. *J. Org. Chem.*, **68**, 6489–6496.

87 Jacopozzi, P. and Dalcanale, E. (1997) Metal-induced self-assembly of cavitand-based cage molecules. *Angew. Chem. Int. Ed.*, **36**, 613–615.
88 Fochi, F., Jacopozzi, P., Wegelius, E. *et al.* (2001) Self-assembly and anion encapsulation properties of cavitand-based coordination cages. *J. Am. Chem. Soc.*, **123**, 7539–7552.
89 Pirondini, L., Bertolini, F., Cantadori, B. *et al.* (2002) Design and self-assembly of wide and robust coordination cages. *Proc. Natl. Acad. Sci. U.S.A.*, **99**, 4911–4915.
90 Ikeda, A., Udzu, H., Yoshimura, M., and Shinkai, S. (2000) Inclusion of [60]fullerene in a self-assembled homooxacalix[3]arene-based dimeric capsule constructed by a Pd^{II}-pyridine interaction. The Li^+-binding to the lower rims can improve the inclusion ability. *Tetrahedron*, **56**, 1825–1832.
91 Gottschalk, T., Jaun, B., and Diederich, F. (2007) Container molecules with portals: reversibly switchable cycloalkane complexation. *Angew. Chem. Int. Ed.*, **46**, 260–264.
92 Hermann, K., Ruan, Y., Hardin, A.M. *et al.* (2014) Gated molecular baskets. *Chem. Soc. Rev.*, **44**, 500–514.
93 Fankhauser, D., Kolarski, D., Grüning, W.R., and Diederich, F. (2014) Resorcin[4]arene-based molecular baskets and water-soluble container molecules: synthesis and ^1H NMR host–guest complexation studies. *Eur. J. Org. Chem.*, **2014**, 3575–3583.
94 Wang, B.-Y., Bao, X., Yan, Z. *et al.* (2008) A 3-fold "butterfly valve" in command of the encapsulation's kinetic stability. Molecular baskets at work. *J. Am. Chem. Soc.*, **130**, 15127–15133.
95 Hu, L., Polen, S., Hardin, A.M. *et al.* (2015) On the transfer of chirality, thermodynamic stability, and folding characteristics of stereoisomeric gated baskets. *Eur. J. Org. Chem.*, **2015**, 6832–6840.
96 Wang, B.-Y., Žujović, T., Turner, D.A. *et al.* (2012) Design, preparation, and study of catalytic gated baskets. *J. Org. Chem.*, **77**, 2675–2688.
97 Vögtle, F., Gross, J., Seel, C., and Nieger, M. (1992) $C_{36}H_{36}$—tetrahedral clamping of four benzene rings in a spherical hydrocarbon framework. *Angew. Chem. Int. Ed.*, **31**, 1069–1071.
98 Gross, J., Harder, G., Vögtle, F. *et al.* (1995) $C_{60}H_{60}$ and $C_{54}H_{48}$: silver ion extraction with new concave hydrocarbons. *Angew. Chem. Int. Ed.*, **34**, 481–484.
99 Gross, J., Harder, G., Siepen, A. *et al.* (1996) Concave hydrocarbons. *Chem. Eur. J.*, **2**, 1585–1595.
100 Nierle, J. and Kuck, D. (2006) Synthesis of a benzoannelated spheriphane (globular cyclophane) on the way to potential precursors of C_{60}-fullerene. *Synlett*, **2006**, 2914–2920.
101 Lee, S., Yang, A., Moneypenny, T.P., and Moore, J.S. (2016) Kinetically trapped tetrahedral cages via alkyne metathesis. *J. Am. Chem. Soc.*, **138**, 2182–2185.

102 Zhang, G. and Mastalerz, M. (2014) Organic cage compounds – from shape-persistency to function. *Chem. Soc. Rev.*, **43**, 1934–1947.
103 Collins, M.S., Carnes, M.E., Nell, B.P. *et al.* (2016) A facile route to old and new cyclophanes via self-assembly and capture. *Nat. Commun.*, 7, 11052.
104 Masu, H., Katagiri, K., Kato, T. *et al.* (2008) Chiral spherical molecule constructed from aromatic amides: facile synthesis and highly ordered network structure in the crystal. *J. Org. Chem.*, **73**, 5143–5146.
105 Masu, H., Sagara, Y., Imabeppu, F. *et al.* (2010) Crystal structure of spherical aromatic amide: pseudopolymorphs and formation of infinite water cluster in the channel structure. *CrystEngComm*, **13**, 406–409.
106 Tozawa, T., Jones, J.T.A., Swamy, S.I. *et al.* (2009) Porous organic cages. *Nat. Mater.*, **8**, 973–978.
107 Dale, E.J., Vermeulen, N.A., Thomas, A.A. *et al.* (2014) ExCage. *J. Am. Chem. Soc.*, **136**, 10669–10682.
108 Kroto, H.W., Fischer, J.E., and Cox, D. (1993) *The Fullerenes*, Pergamon.
109 Kroto, H.W., Heath, J.R., O'Brien, S.C. *et al.* (1985) C_{60}: buckminsterfullerene. *Nature*, **318**, 162–163.
110 Kratschmer, W., Lamb, L.D., Fostiropoulos, K., and Huffman, D.R. (1990) Solid C_{60}: a new form of carbon. *Nature*, **347**, 354–358.
111 Diederich, F., Ettl, R., Rubin, Y. *et al.* (1991) The higher fullerenes: isolation and characterization of C_{76}, C_{84}, C_{90}, C_{94}, and $C_{70}O$, an oxide of D_{5h}-C_{70}. *Science*, **252**, 548–551.
112 Piskoti, C., Yarger, J., and Zettl, A. (1998) C_{36}, a new carbon solid. *Nature*, **393**, 771–774.
113 Koshio, A., Inakuma, M., Wang, Z.W. *et al.* (2000) In situ laser-furnace TOF mass spectrometry of C_{36} and the large-scale production by arc-discharge. *J. Phys. Chem. B*, **104**, 7908–7913.
114 Xie, S.-Y., Gao, F., Lu, X. *et al.* (2004) Capturing the labile fullerene[50] as $C_{50}Cl_{10}$. *Science*, **304**, 699.
115 Tan, Y.-Z., Li, J., Zhu, F. *et al.* (2010) Chlorofullerenes featuring triple sequentially fused pentagons. *Nat. Chem.*, **2**, 269–273.
116 Tan, Y.-Z., Han, X., Wu, X. *et al.* (2008) An entrant of smaller fullerene: C_{56} captured by chlorines and aligned in linear chains. *J. Am. Chem. Soc.*, **130**, 15240–15241.
117 Troshin, P.A., Lyubovskaya, R.N., Ioffe, I.N. *et al.* (2005) Synthesis and structure of the highly chlorinated [60]fullerene $C_{60}Cl_{30}$ with a drum-shaped carbon cage. *Angew. Chem. Int. Ed.*, **44**, 234–237.
118 Gakh, A.A., Tuinman, A.A., Adcock, J.L. *et al.* (1994) Selective synthesis and structure determination of $C_{60}F_{48}$. *J. Am. Chem. Soc.*, **116**, 819–820.
119 Troyanov, S.I., Troshin, P.A., Boltalina, O.V. *et al.* (2001) Two isomers of $C_{60}F_{48}$: an indented fullerene. *Angew. Chem. Int. Ed.*, **40**, 2285–2287.

120 Gakh, A.A., Romanovich, A.Y., and Bax, A. (2003) Thermodynamic rearrangement synthesis and NMR structures of C_1, C_3, and T isomers of $C_{60}H_{36}$. *J. Am. Chem. Soc.*, **125**, 7902–7906.
121 Gakh, A.A. and Tuinman, A.A. (2001) The structure of $C_{60}F_{36}$. *Tetrahedron Lett.*, **42**, 7133–7135.
122 Avent, A.G., Clare, B.W., Hitchcock, P.B. *et al.* (2002) $C_{60}F_{36}$: there is a third isomer and it has C_1 symmetry. *Chem. Commun.*, 2370–2371.
123 Gakh, A.A. and Tuinman, A.A. (2001) "Fluorine dance" on the fullerene surface. *Tetrahedron Lett.*, **42**, 7137–7139.
124 Avent, A.G. and Taylor, R. (2002) Fluorine takes a hike: remarkable room-temperature rearrangement of the C_1 isomer of $C_{60}F_{36}$ into the C_3 isomer via a 1,3-fluorine shift. *Chem. Commun.*, 2726–2727.
125 Tuinman, A.A., Gakh, A.A., Adcock, J.L., and Compton, R.N. (1993) Hyperfluorination of buckminsterfullerene: cracking the sphere. *J. Am. Chem. Soc.*, **115**, 5885–5886.
126 Gakh, A.A. and Tuinman, A.A. (2000) Chemical fragmentation of $C_{60}F_{48}$. *J. Phys. Chem. A*, **104**, 5888–5891.
127 Ioffe, I.N., Goryunkov, A.A., Tamm, N.B. *et al.* (2009) Fusing pentagons in a fullerene cage by chlorination: IPR D_2-C_{76} rearranges into non-IPR $C_{76}Cl_{24}$. *Angew. Chem. Int. Ed.*, **48**, 5904–5907.
128 Popov, A.A., Yang, S., and Dunsch, L. (2013) Endohedral fullerenes. *Chem. Rev.*, **113**, 5989–6113.
129 Saunders, M., Jiménez-Vázquez, H.A., Cross, R.J., and Poreda, R.J. (1993) Stable compounds of helium and neon: He@C_{60} and Ne@C_{60}. *Science*, **259**, 1428–1430.
130 Lee, H.M., Olmstead, M.M., Suetsuna, T. *et al.* (2002) Crystallographic characterization of Kr@C_{60} in (0.09Kr@C_{60}/0.91C_{60})·{Ni^{II}(OEP)}·2C_6H_6. *Chem. Commun.*, 1352–1353.
131 Weidinger, A., Waiblinger, M., Pietzak, B., and Almeida Murphy, T. (1998) Atomic nitrogen in C_{60}: N@C_{60}. *Appl. Phys. A – Mater.*, **66**, 287–292.
132 Plant, S.R. and Porfyrakis, K. (2014) Using electron paramagnetic resonance to map N@C_{60} during high throughput processing. *Analyst*, **139**, 4519–4524.
133 Watanabe, T., Itoh, M.F., Komuro, T. *et al.* (2014) Iridium and platinum complexes of Li^+@C_{60}. *Organometallics*, **33**, 608–611.
134 Okada, H., Komuro, T., Sakai, T. *et al.* (2012) Preparation of endohedral fullerene containing lithium (Li@C_{60}) and isolation as pure hexafluorophosphate salt ([Li^+@C_{60}][PF_6^-]). *RSC Adv.*, **2**, 10624–10631.
135 Ueno, H., Kokubo, K., Nakamura, Y. *et al.* (2013) Ionic conductivity of [Li^+@C_{60}](PF_6^-) in organic solvents and its electrochemical reduction to Li^+@C_{60}^-. *Chem. Commun.*, **49**, 7376–7378.
136 Murata, M., Murata, Y., and Komatsu, K. (2008) Surgery of fullerenes. *Chem. Commun.*, 6083–6094.

137 Vougioukalakis, G.C., Roubelakis, M.M., and Orfanopoulos, M. (2010) Open-cage fullerenes: towards the construction of nanosized molecular containers. *Chem. Soc. Rev.*, **39**, 817–844.

138 Gan, L., Yang, D., Zhang, Q., and Huang, H. (2010) Preparation of open-cage fullerenes and incorporation of small molecules through their orifices. *Adv. Mater.*, **22**, 1498–1507.

139 Murata, M., Murata, Y., and Komatsu, K. (2013) Organic synthesis of endohedral fullerenes encapsulating helium, dihydrogen, and water, in *Organic Nanomaterials: Synthesis, Characterization, and Device Applications* (eds T. Torres and G. Bottari), John Wiley & Sons, pp. 225–239.

140 Papastavrou, A.T., Pefkianakis, E.K., Roubelakis, M.M., and Vougioukalakis, G.C. (2016) Open-cage fullerenes, in *Fullerenes, Nanotubes, and Nanodiamonds*, Carbon Nanomaterials Sourcebook, vol. **1** (ed. K.D. Sattler), CRC Press, pp. 193–257.

141 Rubin, Y., Jarrosson, T., Wang, G. *et al.* (2001) Insertion of helium and molecular hydrogen through the orifice of an open fullerene. *Angew. Chem. Int. Ed.*, **40**, 1543–1546.

142 Komatsu, K., Murata, M., and Murata, Y. (2005) Encapsulation of molecular hydrogen in fullerene C_{60} by organic synthesis. *Science*, **307**, 238–240.

143 Murata, Y., Murata, M., and Komatsu, K. (2003) Synthesis, structure, and properties of novel open-cage fullerenes having heteroatom(s) on the rim of the orifice. *Chem. Eur. J.*, **9**, 1600–1609.

144 Murata, Y., Murata, M., and Komatsu, K. (2003) 100% Encapsulation of a hydrogen molecule into an open-cage fullerene derivative and gas-phase generation of $H_2@C_{60}$. *J. Am. Chem. Soc.*, **125**, 7152–7153.

145 Kurotobi, K. and Murata, Y. (2011) A single molecule of water encapsulated in fullerene C_{60}. *Science*, **333**, 613–616.

146 Futagoishi, T., Murata, M., Wakamiya, A., and Murata, Y. (2015) Trapping N_2 and CO_2 on the sub-nano scale in the confined internal spaces of open-cage C_{60} derivatives: isolation and structural characterization of the host–guest complexes. *Angew. Chem.*, **127**, 15004–15007.

147 Chen, C.-S., Kuo, T.-S., and Yeh, W.-Y. (2016) Encapsulation of formaldehyde and hydrogen cyanide in an open-cage fullerene. *Chem. Eur. J.*, **22**, 8773–8776.

148 Whitener, K.E., Frunzi, M., Iwamatsu, S. *et al.* (2008) Putting ammonia into a chemically opened fullerene. *J. Am. Chem. Soc.*, **130**, 13996–13999.

149 Whitener, K.E., Cross, R.J., Saunders, M. *et al.* (2009) Methane in an open-cage [60]fullerene. *J. Am. Chem. Soc.*, **131**, 6338–6339.

150 Zhang, Q., Pankewitz, T., Liu, S. *et al.* (2010) Switchable open-cage fullerene for water encapsulation. *Angew. Chem. Int. Ed.*, **49**, 9935–9938.

151 Xu, L., Liang, S., Sun, J., and Gan, L. (2015) Open-cage fullerene with a stopper acts as a molecular vial for a single water molecule. *Org. Chem. Front.*, **2**, 1500–1504.

152 Yu, Y., Shi, L., Yang, D., and Gan, L. (2013) Molecular containers with a dynamic orifice: open-cage fullerenes capable of encapsulating either H_2O or H_2 under mild conditions. *Chem. Sci.*, **4**, 814–818.
153 Murata, M., Maeda, S., Morinaka, Y. *et al.* (2008) Synthesis and reaction of fullerene C_{70} encapsulating two molecules of H_2. *J. Am. Chem. Soc.*, **130**, 15800–15801.
154 Zhang, R., Murata, M., Aharen, T. *et al.* (2016) Synthesis of a distinct water dimer inside fullerene C_{70}. *Nat. Chem.*, **8**, 435–441.
155 Shi, L., Yang, D., Colombo, F. *et al.* (2013) Punching a carbon atom of C_{60} into its own cavity to form an endohedral complex $CO@C_{59}O_6$ under mild conditions. *Chem. Eur. J.*, **19**, 16545–16549.
156 Heath, J.R., O'Brien, S.C., Zhang, Q. *et al.* (1985) Lanthanum complexes of spheroidal carbon shells. *J. Am. Chem. Soc.*, **107**, 7779–7780.
157 Cong, H., Yu, B., Akasaka, T., and Lu, X. (2013) Endohedral metallofullerenes: an unconventional core–shell coordination union. *Coord. Chem. Rev.*, **257**, 2880–2898.
158 Lu, X., Feng, L., Akasaka, T., and Nagase, S. (2012) Current status and future developments of endohedral metallofullerenes. *Chem. Soc. Rev.*, **41**, 7723–7760.
159 Lu, X., Bao, L., Akasaka, T., and Nagase, S. (2014) Recent progress in the chemistry of endohedral metallofullerenes. *Chem. Commun.*, **50**, 14701–14715.
160 Lu, X., Echegoyen, L., Balch, A.L. *et al.* (eds) (2014) *Endohedral Metallofullerenes: Basics and Applications*, CRC Press.
161 Shinohara, H. and Tagmatarchis, N. (2015) *Endohedral Metallofullerenes: Fullerenes with Metal inside*, John Wiley & Sons.
162 Dunk, P.W., Kaiser, N.K., Mulet-Gas, M. *et al.* (2012) The smallest stable fullerene, $M@C_{28}$ (M = Ti, Zr, U): stabilization and growth from carbon vapor. *J. Am. Chem. Soc.*, **134**, 9380–9389.
163 Che, Y., Yang, H., Wang, Z. *et al.* (2009) Isolation and structural characterization of two very large, and largely empty, endohedral fullerenes: $Tm@C_{3v}$-C_{94} and $Ca@C_{3v}$-C_{94}. *Inorg. Chem.*, **48**, 6004–6010.
164 Lu, X., Lian, Y., Beavers, C.M. *et al.* (2011) Crystallographic X-ray analyses of $Yb@C_{2v}(3)$-C_{80} reveal a feasible rule that governs the location of a rare earth metal inside a medium-sized fullerene. *J. Am. Chem. Soc.*, **133**, 10772–10775.
165 Sato, S., Nikawa, H., Seki, S. *et al.* (2012) A co-crystal composed of the paramagnetic endohedral metallofullerene $La@C_{82}$ and a nickel porphyrin with high electron mobility. *Angew. Chem. Int. Ed.*, **51**, 1589–1591.
166 Yang, H., Yu, M., Jin, H. *et al.* (2012) Isolation of three isomers of $Sm@C_{84}$ and X-ray crystallographic characterization of $Sm@D_{3d}(19)$-C_{84} and $Sm@C_2(13)$-C_{84}. *J. Am. Chem. Soc.*, **134**, 5331–5338.

167 Yamada, M., Nakahodo, T., Wakahara, T. *et al.* (2005) Positional control of encapsulated atoms inside a fullerene cage by exohedral addition. *J. Am. Chem. Soc.*, **127**, 14570–14571.

168 Hachiya, M., Nikawa, H., Mizorogi, N. *et al.* (2012) Exceptional chemical properties of Sc@C_{2v}(9)–C_{82} probed with adamantylidene carbene. *J. Am. Chem. Soc.*, **134**, 15550–15555.

169 Yamada, M., Kurihara, H., Suzuki, M. *et al.* (2014) Sc_2@C_{66} revisited: an endohedral fullerene with scandium ions nestled within two unsaturated linear triquinanes. *J. Am. Chem. Soc.*, **136**, 7611–7614.

170 Mercado, B.Q., Jiang, A., Yang, H. *et al.* (2009) Isolation and structural characterization of the molecular nanocapsule Sm_2@D_{3d}(822)-C_{104}. *Angew. Chem. Int. Ed.*, **48**, 9114–9116.

171 Zhang, J., Stevenson, S., and Dorn, H.C. (2013) Trimetallic nitride template endohedral metallofullerenes: discovery, structural characterization, reactivity, and applications. *Acc. Chem. Res.*, **46**, 1548–1557.

172 Dunsch, L. and Yang, S. (2007) Metal nitride cluster fullerenes: their current state and future prospects. *Small*, **3**, 1298–1320.

173 Stevenson, S., Rice, G., Glass, T. *et al.* (1999) Small-bandgap endohedral metallofullerenes in high yield and purity. *Nature*, **401**, 55–57.

174 Dunsch, L., Krause, M., Noack, J., and Georgi, P. (2004) Endohedral nitride cluster fullerenes: formation and spectroscopic analysis of $L_{3-x}M_xN$@C_{2n} ($0 \leq x \leq 3$; $N = 39,40$). *J. Phys. Chem. Solids*, **65**, 309–315.

175 Olmstead, M.M., de Bettencourt-Dias, A., Duchamp, J.C. *et al.* (2001) Isolation and structural characterization of the endohedral fullerene Sc_3N@C_{78}. *Angew. Chem. Int. Ed.*, **40**, 1223–1225.

176 Yang, S., Troyanov, S.I., Popov, A.A. *et al.* (2006) Deviation from the planarity – a large Dy_3N cluster encapsulated in an I_h-C_{80} cage: an X-ray crystallographic and vibrational spectroscopic study. *J. Am. Chem. Soc.*, **128**, 16733–16739.

177 Beavers, C.M., Zuo, T., Duchamp, J.C. *et al.* (2006) Tb_3N@C_{84}: an improbable, egg-shaped endohedral fullerene that violates the isolated pentagon rule. *J. Am. Chem. Soc.*, **128**, 11352–11353.

178 Yang, S., Liu, F., Chen, C. *et al.* (2011) Fullerenes encaging metal clusters—clusterfullerenes. *Chem. Commun.*, **47**, 11822–11839.

179 Popov, A.A. (2016) Endohedral clusterfullerenes, in *Graphene, Fullerenes, Nanotubes, and Nanodiamonds*, Carbon Nanomaterials Sourcebook, vol. **1** (ed. K.D. Sattler), CRC Press, pp. 303–333.

180 Kurihara, H., Lu, X., Iiduka, Y. *et al.* (2012) X-ray structures of Sc_2C_2@C_{2n} ($n = 40$–42): in-depth understanding of the core–shell interplay in carbide cluster metallofullerenes. *Inorg. Chem.*, **51**, 746–750.

181 Chen, C.-H., Ghiassi, K.B., Cerón, M.R. *et al.* (2015) Beyond the butterfly: Sc_2C_2@C_{2v}(9)-C_{86}, an endohedral fullerene containing a planar, twisted

Sc_2C_2 unit with remarkable crystalline order in an unprecedented carbon cage. *J. Am. Chem. Soc.*, **137**, 10116–10119.

182 Yang, T., Hao, Y., Abella, L. *et al.* (2015) $Sc_2O@T_d(19151)$-C_{76}: hindered cluster motion inside a tetrahedral carbon cage probed by crystallographic and computational studies. *Chem. Eur. J.*, **21**, 11110–11117.

183 Tang, Q., Abella, L., Hao, Y. *et al.* (2015) $Sc_2O@C_{2v}(5)$-C_{80}: dimetallic oxide cluster inside a C_{80} fullerene cage. *Inorg. Chem.*, **54**, 9845–9852.

184 Mercado, B.Q., Chen, N., Rodríguez-Fortea, A. *et al.* (2011) The shape of the $Sc_2(\mu_2$-S) unit trapped in C_{82}: crystallographic, computational, and electrochemical studies of the isomers, $Sc_2(\mu_2$-S)@$C_s(6)$-C_{82} and $Sc_2(\mu_2$-S)@$C_{3v}(8)$-C_{82}. *J. Am. Chem. Soc.*, **133**, 6752–6760.

185 Chen, N., Beavers, C.M., Mulet-Gas, M. *et al.* (2012) $Sc_2S@C_s(10528)$-C_{72}: a dimetallic sulfide endohedral fullerene with a non isolated pentagon rule cage. *J. Am. Chem. Soc.*, **134**, 7851–7860.

186 Yang, S., Chen, C., Liu, F. *et al.* (2013) An improbable monometallic cluster entrapped in a popular fullerene cage: YCN@$C_s(6)$-C_{82}. *Sci. Rep.*, **3**, Article number: 1487.

187 Junghans, K., Rosenkranz, M., and Popov, A.A. (2016) $Sc_3CH@C_{80}$: selective ^{13}C enrichment of the central carbon atom. *Chem. Commun.*, **52**, 6561–6564.

188 Junghans, K., Schlesier, C., Kostanyan, A. *et al.* (2015) Methane as a selectivity booster in the arc-discharge synthesis of endohedral fullerenes: selective synthesis of the single-molecule magnet $Dy_2TiC@C_{80}$ and its congener $Dy_2TiC_2@C_{80}$. *Angew. Chem. Int. Ed.*, **54**, 13411–13415.

189 Rudkevich, D.M. (2005) Molecular containers in action, in *Functional Synthetic Receptors* (eds T. Schrader and A.D. Hamilton), Wiley-VCH Verlag GmbH & Co. KGaA, pp. 257–298.

190 Shapiro, M.G., Ramirez, R.M., Sperling, L.J. *et al.* (2014) Genetically encoded reporters for hyperpolarized xenon magnetic resonance imaging. *Nat. Chem.*, **6**, 629–634.

191 Fogarty, H.A., Berthault, P., Brotin, T. *et al.* (2007) A cryptophane core optimized for xenon encapsulation. *J. Am. Chem. Soc.*, **129**, 10332–10333.

192 Joseph, A.I., Lapidus, S.H., Kane, C.M., and Holman, K.T. (2015) Extreme confinement of xenon by cryptophane-111 in the solid state. *Angew. Chem. Int. Ed.*, **54**, 1471–1475.

193 Riggle, B.A., Wang, Y., and Dmochowski, I.J. (2015) A "smart" ^{129}Xe NMR biosensor for pH-dependent cell labeling. *J. Am. Chem. Soc.*, **137**, 5542–5548.

194 Spence, M.M., Rubin, S.M., Dimitrov, I.E. *et al.* (2001) Functionalized xenon as a biosensor. *Proc. Natl. Acad. Sci. U.S.A.*, **98**, 10654–10657.

195 Chambers, J.M., Hill, P.A., Aaron, J.A. *et al.* (2009) Cryptophane xenon-129 nuclear magnetic resonance biosensors targeting human carbonic anhydrase. *J. Am. Chem. Soc.*, **131**, 563–569.

196 Aaron, J.A., Chambers, J.M., Jude, K.M. *et al.* (2008) Structure of a ^{129}Xe-cryptophane biosensor complexed with human carbonic anhydrase II. *J. Am. Chem. Soc.*, **130**, 6942–6943.

197 Ghiassi, K.B., Olmstead, M.M., and Balch, A.L. (2014) Gadolinium-containing endohedral fullerenes: structures and function as magnetic resonance imaging (MRI) agents. *Dalton Trans.*, **43**, 7346–7358.

198 Mikawa, M., Kato, H., Okumura, M. *et al.* (2001) Paramagnetic water-soluble metallofullerenes having the highest relaxivity for MRI contrast agents. *Bioconjugate Chem.*, **12**, 510–514.

199 Stevenson, S., Phillips, J.P., Reid, J.E. *et al.* (2004) Pyramidalization of Gd_3N inside a C_{80} cage. The synthesis and structure of $Gd_3N@C_{80}$. *Chem. Commun.*, 2814–2815.

200 Suzuki, M., Lu, X., Sato, S. *et al.* (2012) Where does the metal cation stay in Gd@C_{2v}(9)-C_{82}? A single-crystal X-ray diffraction study. *Inorg. Chem.*, **51**, 5270–5273.

201 Zuo, T., Walker, K., Olmstead, M.M. *et al.* (2008) New egg-shaped fullerenes: non-isolated pentagon structures of $Tm_3N@C_s$(51365)-C_{84} and $Gd_3N@C_s$(51365)-C_{84}. *Chem. Commun.*, 1067–1069.

202 Tóth, É., Bolskar, R.D., Borel, A. *et al.* (2005) Water-soluble gadofullerenes: toward high-relaxivity, pH-responsive MRI contrast agents. *J. Am. Chem. Soc.*, **127**, 799–805.

203 Laus, S., Sitharaman, B., Tóth, É. *et al.* (2007) Understanding paramagnetic relaxation phenomena for water-soluble gadofullerenes. *J. Phys. Chem. C*, **111**, 5633–5639.

204 Zhang, J., Ye, Y., Chen, Y. *et al.* (2014) $Gd_3N@C_{84}(OH)_x$: a new egg-shaped metallofullerene magnetic resonance imaging contrast agent. *J. Am. Chem. Soc.*, **136**, 2630–2636.

205 Zhang, J., Fatouros, P.P., Shu, C. *et al.* (2010) High relaxivity trimetallic nitride (Gd_3N) metallofullerene MRI contrast agents with optimized functionality. *Bioconjugate Chem.*, **21**, 610–615.

206 Westerström, R., Dreiser, J., Piamonteze, C. *et al.* (2012) An endohedral single-molecule magnet with long relaxation times: $DySc_2N@C_{80}$. *J. Am. Chem. Soc.*, **134**, 9840–9843.

207 Chen, D., Dougherty, C.A., Zhu, K., and Hong, H. (2015) Theranostic applications of carbon nanomaterials in cancer: focus on imaging and cargo delivery. *J. Control. Release*, **210**, 230–245.

208 Kewley, A., Stephenson, A., Chen, L. *et al.* (2015) Porous organic cages for gas chromatography separations. *Chem. Mater.*, **27**, 3207–3210.

209 You, J., Dou, L., Yoshimura, K. *et al.* (2013) A polymer tandem solar cell with 10.6% power conversion efficiency. *Nat. Commun.*, **4**, 1446.

210 Briggs, G.A.D., Ardavan, A. (2012) Atomic clock. US Patent 8,217,724 B2, July 10.

3

Molecular Gearing Systems

CHAPTER MENU

3.1 Introduction

Gears and related mechanical devices featuring controlled rotary motion, such as ratchets and brakes, are widely used in modern technology. Their primary function is to achieve correlated motion and torque transfer between various parts of mechanical devices (Figure 3.1). Due to the relatively complex design, their technomimetic counterparts, molecular gearing systems, were developed only a few decades ago and so far have displayed only limited functionality [1]. None of the existing molecular gearing systems are capable of torque transfer from one part of a molecule to another, but it is possible to attain correlated motion of the gearing units and to control it by chemical means, such as metal ion complexation, or by external stimuli, such as light.

Molecular gearing systems are also textbook examples of technomimetics in which the differences between the molecular devices and their conventional mechanical counterparts can be easily observed. For example, early research demonstrated that it would be impossible to completely avoid gear slippage in molecular gears [2]. Later, more complex molecular ratchets were designed to test Maxwell's demon hypothesis. These molecular ratchets are not functional since unidirectional movement cannot be achieved using random thermal motion [3, 4]. However, this limitation can be overcome in molecular

Molecular Devices: An Introduction to Technomimetics and Its Biological Applications,
First Edition. Andrei A. Gakh.

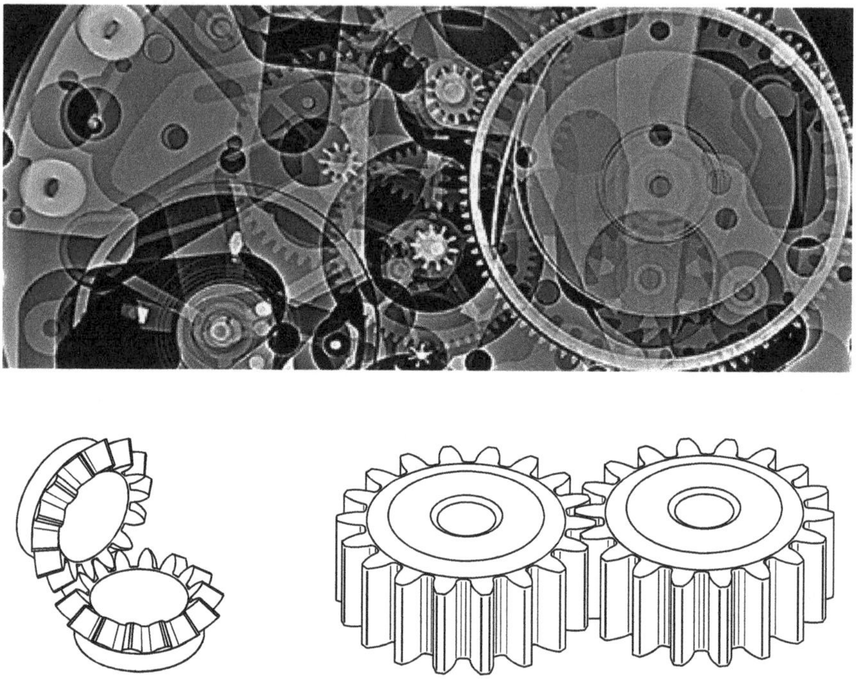

Figure 3.1 Radiograph of a pocket watch mechanism with microfocus X-ray source (top) and schematics of a bevel gear (bottom left) and a spur gear (bottom right). *Credits*: (top) Vidisco Ltd; (bottom left) Christophe Dang Ngoc Chan, Wikimedia; (bottom right) Public Domain, Wikimedia.

gearing systems powered by external sources of energy, either chemical or electromagnetic. These molecular gearing systems, upgraded with light- or chemically powered drivers, can be used for the design of rotary molecular motors [5, 6].

Construction of molecular gearing systems can be accomplished by linking appropriate gearing elements together via a single covalent bond or a metal–ligand coordination bond permitting free rotation. The best gear meshing can be achieved using gears with substantial number of teeth, but this is a difficult task for systems employing carbon-based rotary axis due to four-valence limitation of the carbon atom. The preponderance of known carbon-based molecular gearing systems are simple three-tooth gearing devices with significant performance restrains (Figure 3.2) [7].

Substantial improvement in the number of gear teeth can be accomplished using a metal atom as a "ball bearing" element. In this case, an aromatic ring, such as a planar cyclopentadienyl ring in ferrocene, can be used as a gearing unit for the design of multi-tooth gearing systems (Figure 3.2). These

Figure 3.2 Examples of molecular gearing elements: methyl fragment (left), 9-triptycyl fragment (middle), and ferrocenyl fragment (right).

organometallic systems typically have several planar gearing units assembled together via coordination with metal atoms. Unfortunately, proper angle alignment between these gearing units allowing the best meshing as well as unhindered correlated rotation around multiple metal centers is difficult to accomplish. The angle alignment task is somewhat simpler for mixed gearing systems constructed from a planar ring and a simple multiligand metal moiety as a gearing element [7].

In general, there are two major parameters affecting the performance of molecular gears. One of them is the energy barrier for rotation, and the other one is the barrier for gear slippage. Ideally, the energy barrier for rotation should be lower than 10 kcal/mol, and the barrier for gear slippage should be higher than 30 kcal/mol for a gearing system designed to operate effectively at ambient temperature. In practice, however, a compromise would have to be found since tightening of a gearing system usually results in a more constrained rotation, but at the same time leads to better meshing and improves correlation rotation parameters due to the higher energy barrier for gear slippage.

3.2 Bevel Molecular Gears

The initial interest in correlated rotation in molecular systems, which eventually led to the development of molecular gears, arose more than 50 years ago during the investigation of hexamethylbenzene (Figure 3.3) [8–10]. In this

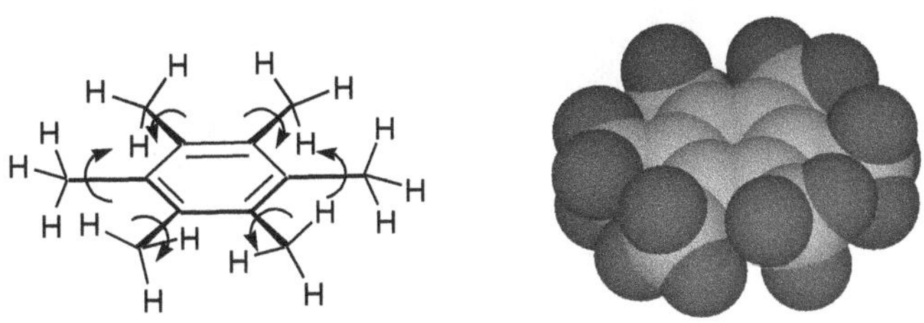

Figure 3.3 Hexamethylbenzene – a simple molecular system with a weakly correlated rotation of methyl groups.

molecule, properly aligned methyl groups can be considered as very primitive gearing elements [11]. Unfortunately, methyl groups have very poor performance due to inadequate meshing properties even under the extremely tight sterically controlled conditions [12, 13]. In addition, hexa-substituted benzenes represent practically unexploitable systems of six cyclically interconnected gears with no obvious functionality.

An attempt was made to achieve better gear meshing with branched alkyl groups (such as *tert*-butyl groups in (*tert*-butyl)silanes and (*tert*-butyl) phosphines) as gearing elements, but the results were not impressive. For example, the barrier for gear slippage of *tert*-butyl groups in tris(*tert*-butyl) silane is less than 10 kcal/mol, and the expected gear-meshed conformation is observed only in cryogenic conditions at −140 °C (Figure 3.4; note that the dynamic gearing is not possible in a system with three mutually meshed gears). The inherently shallow notches between the methyl subunits in *tert*-butyl groups make it difficult to use these groups as viable gearing elements [14].

Substantially better results were finally achieved with triptycene units as gearing elements [15, 16]. Although the triptycene moiety has only three teeth, the rigidity of its bicyclo[2.2.2]octane "spindle" combined with proper axial alignment of benzene "blades" in almost perfect D_{3h} symmetry is more than adequate to compensate for the low teeth number. In addition, triptycenes can be prepared in one step by benzyne cycloaddition to affordable 9,10-substituted anthracenes. This synthetic approach, coupled with versatile reactivity of the CH bridging groups in already formed triptycene molecules, provides for a convenient, with high yield and selectivity, pathway for the preparation of a large variety of polytriptycenes containing several interlinked triptycene fragments (Scheme 3.1).

The simplest triptycene molecular bevel gearing systems contain only two triptycene gearing units interconnected through an appropriate bridge (Figure 3.5). The most popular bridging units are usually simple divalent moieties, such as O [17], CO [18], CH_2 [17, 19], NH [20], CH=CH [21], S [22], and PH [23]. Molecular dynamics studies indicate that the best-performing bridges (such as O) have very low dynamic gearing energy barrier, typically less

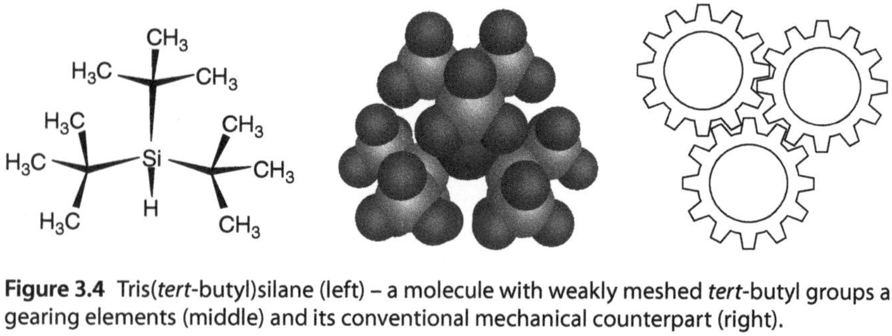

Figure 3.4 Tris(*tert*-butyl)silane (left) – a molecule with weakly meshed *tert*-butyl groups as gearing elements (middle) and its conventional mechanical counterpart (right).

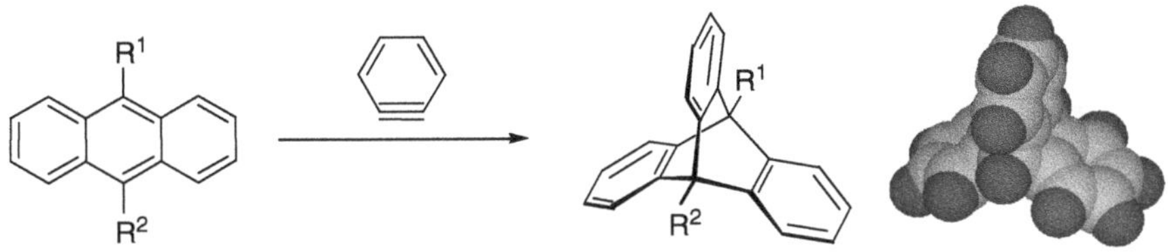

Scheme 3.1 Facile synthesis of triptycene derivatives via benzyne addition (R^1, R^2 = H, Alk, OAlk, OAr, CH_2Ar, CH_2OAlk, etc.).

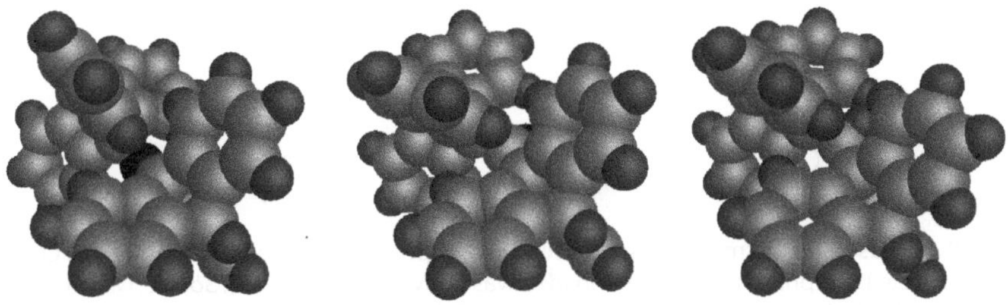

Figure 3.5 X-ray structures of bis(9-triptycyl) molecular bevel gears: bis(9-triptycyl) ether (left) [24], bis(9-triptycyl) ketone (middle) [18], and bis(9-triptycyl)methane (right) [18].

than 10 kcal/mol. The triptycene gears in these best-performing systems are mobile, practically frictionless, and exhibit rapid correlated rotation at room temperature on the NMR time scale.

Correlated rotation in unsymmetrical bevel gearing systems leads to a particular form of stereoisomerism not known in other molecules, "phase isomerism" [25]. It can be observed in the bis(9-triptycene) systems where two benzene blades, one on each triptycene unit, are labeled. Due to correlated rotation, these compounds exist in two distinct isomeric forms – as a racemic *DL* form and a *MESO* form (Figure 3.6), similar to known isomeric forms of tartaric acid. In the case of tartaric acid, these stereoisomers exist due to the high energy barrier to inversion at sp^3-hybridized carbon atom. In the cases of bis(9-triptycene) compounds, high energy barriers for gear slippage serve the same role.

The energy barrier for gear slippage depends on the nature of the bridging elements. Smaller bridges usually provide better meshing of triptycene units. Among the experimentally tested bridging moieties, the lowest barriers for gear slippage (about 24 kcal/mol) are observed for the phosphine (PH) bridge, whereas the oxygen (O) bridge displays the highest energy barriers, above 40 kcal/mol. These best molecular gearing systems with oxygen bridge are also highly efficient. According to calculations, they can perform more than 10^{21} correlated rotations between gear slippage events at room temperature [26]. Some examples of simple triptycene bevel gearing systems and their gear slippage energy barriers are presented in Figure 3.7.

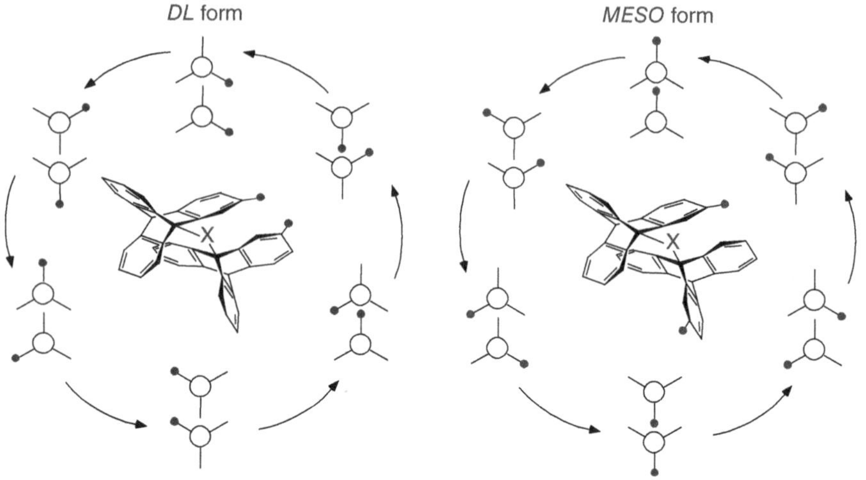

Figure 3.6 Phase isomerism in unsymmetrical bis(9-triptycene) gearing systems (X = O, CO, CH_2, NH, S, etc.). Graphical representation idea was derived from Ref. [15]. *Source:* Iwamura and Mislow 1988 [15]. Reproduced with permission of American Chemical Society.

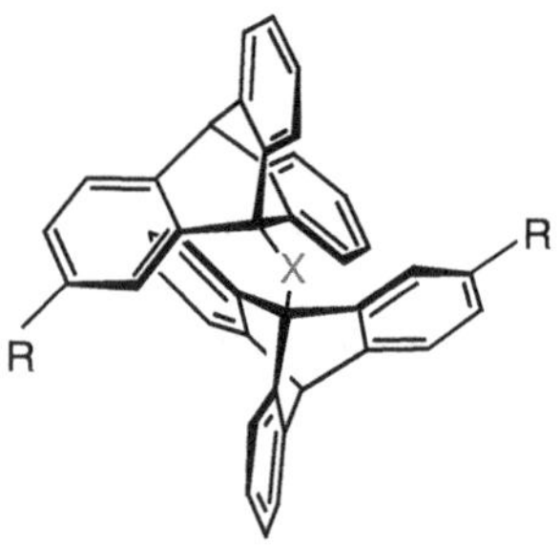

R	X	E^*, (kcal/mol)	Ref.
Cl	O	42	[17]
Cl	NH	39	[20]
Cl	CH_2	32	[17]
CH_3	*cis*-HC=CH	30	[21]
Cl	S	29	[22]
Cl	PH	24	[23]

Figure 3.7 Examples of simple triptycene bevel gearing systems and their gear slippage energy barriers (E^*, kcal/mol).

The presented data regarding the energy barriers for gear slippage clearly demonstrate that the design of these bevel gears, particularly the choice of the central bridge, makes a profound effect on the overall effectiveness of the molecular device. The best results were achieved with the second row elements, such as oxygen, nitrogen, or carbon. The third row elements, such as phosphorus and sulfur, as well as *cis*-CH=CH fragment, provide smaller slippage barriers, presumably due to a more flexible (in terms of energy) geometry of the central bridge. Existing X-ray data tend to confirm these empirical observations. For example, tight meshing of triptycene gearing units in bis(9-triptycyl) ether with a short oxygen bridge manifests itself in the increase of the C—O—C angle to almost 136°, one of the largest angles ever documented for an ether-type junction [24].

Although disadvantageous with regard to *gearing fidelity* [26], the use of the third row elements as the bridging units X (Figure 3.7) offers a unique opportunity to engage/disengage gearing in a controllable manner by changing coordination (and thus geometry) of the bridging unit via coordination with external ligands. An example of a controllable molecular gearing system with a molecular "clutch" mechanism is presented in Figure 3.8. This gearing system, bis(4-methyl-9-triptycyl)difluorosilane, was synthesized as a mixture of phase isomers in one step by the reaction of SiF_4 with 4-methyltriptycyllithium. In this molecule, two triptycene gearing elements are connected via the tetrahedral SiF_2 bridge in an "engaged" position. Similar to bis(9-triptycyl) ether discussed earlier, the C—Si—C angle (129°) is larger than what is expected for a tetrahedral silane (109.5°) due to steric interactions between the two intermeshed triptycyl gearing units. Although the energy barrier for gear slippage in the "engaged" gearing system is relatively low (about 17 kcal/mol), thus precluding chromatographic separation of *DL* and *MESO* forms, this barrier is still adequate to observe both forms by NMR means at room temperature [27].

Addition of the fluoride anion to the SiF_2 bridge changes its geometry, causing disengagement of the two triptycene gearing units. The X-ray structure analysis of the adduct (an organofluorosilicate salt) reveals that the two triptycyl groups occupy the apical positions, and the three fluorine atoms are in the equatorial plane – as expected for the almost ideal trigonal bipyramid structure. The NMR data further confirm that this trigonal bipyramid structure also exists in solution, and the two triptycyl groups rapidly and freely rotate around Si—C bonds without correlation at room temperature. This organofluorosilicate salt can be converted back to the original bis(4-methyl-9-triptycyl)difluorosilane *DL* and *MESO* isomeric mixture by treating the salt with excess of water in a reversible "molecular clutch engagement–disengagement" action sequence (Figure 3.9) [27].

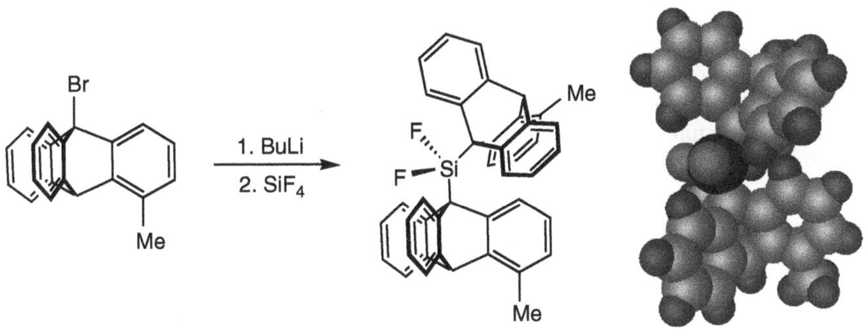

Figure 3.8 Synthesis of the bevel molecular gearing system with SiF_2 bridge (left) and its X-ray structure (right). Only *DL* form is shown for clarity.

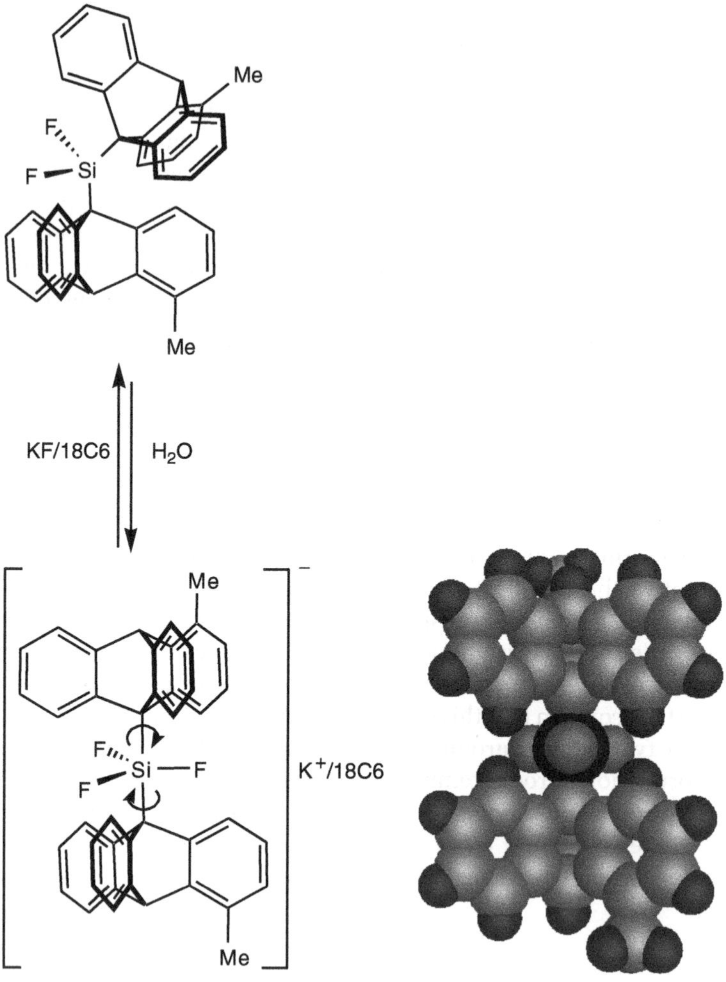

Figure 3.9 Engagement and disengagement of the triptycene gearing units in bis(4-methyl-9-triptycyl)difluorosilane (left, only the *DL* form is shown for clarity) using potassium fluoride 18-crown-6 (KF/18C6) complex as an external source of fluoride anion and the X-ray structure of bis(4-methyl-9-triptycyl)trifluorosilicate salt (right, without the K^+/18C6 countercation).

Technically speaking, a triptycene unit allows for the construction of bevel molecular gearing systems with an unlimited number of triptycene gearing units, since each triptycene unit contains two attachment sites at the bicyclo[2.2.2]octane "spindle" and hence can be linked together in a very complex system where all triptycene units will be involved in correlated rotation. A first example of these complex gearing systems containing three

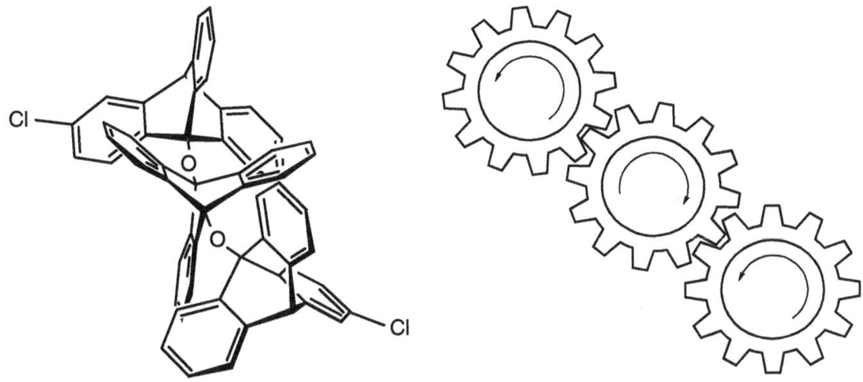

Figure 3.10 The complex molecular gearing system with three triptycene gearing elements (left) and its conventional mechanical counterpart (right).

interconnected triptycene gearing units is presented in Figure 3.10. The most interesting feature of this system is that the gear slippage energy barrier (about 43 kcal/mol) is practically indistinguishable from the energy barrier for bis(2-chloro-9-triptycyl) ether discussed earlier. This is a clear indication that proper addition of extra triptycene gearing elements would not affect the effectiveness of the extended multi-triptycene gearing systems [28].

Bevel gears are perhaps the simplest gearing devices that can be easily mimicked on the molecular level because they do not require precise gear axis alignment. A more complicated (and at the same time rewarding) task is the construction of spur molecular gears.

3.3 Spur Molecular Gears

The design of spur molecular gears requires enforced parallel alignment and precise spacing of the gearing elements, which is not a trivial task. Not surprisingly, spur molecular gearing systems began to appear in the literature in substantial numbers only recently. Even with the optimal design, achieving proper gear meshing and good gear fidelity in molecular spur gears is a formidable challenge. The main complicating factor appears to be the requirement that an entire benzene edge of one triptycene gearing unit should fit into a notch of another triptycene unit for a proper spur gear meshing – compared to only one C—H fragment in bevel triptycene gearing systems [26] (Figure 3.11). The large contact area between the benzene blades in spur molecular gears also creates conditions for undesirable π–π stacking. So far, none of the known molecular spur gear systems unequivocally demonstrate correlated rotation of the gearing elements at normal temperatures.

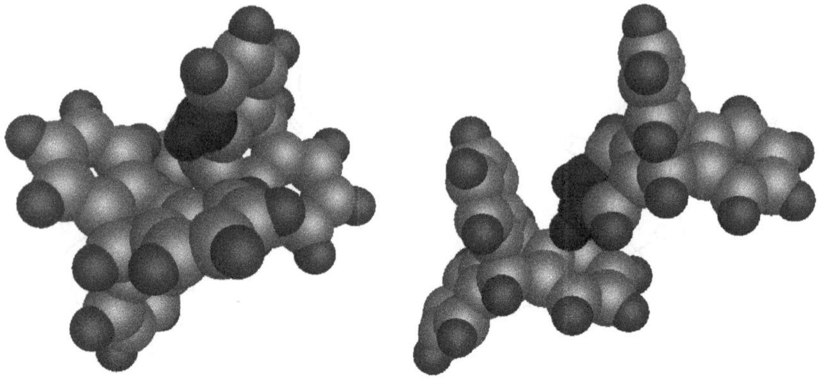

Figure 3.11 An illustration of gear meshing in bevel (left) and in spur (right) triptycene molecular gearing systems. Experimental atom coordinates were taken from X-ray structures of bis(9-triptycyl) ether [24] as an example of a bevel gear and bis(9,10-triptyceno)-32-crown-8 [29] as an example of a spur gear.

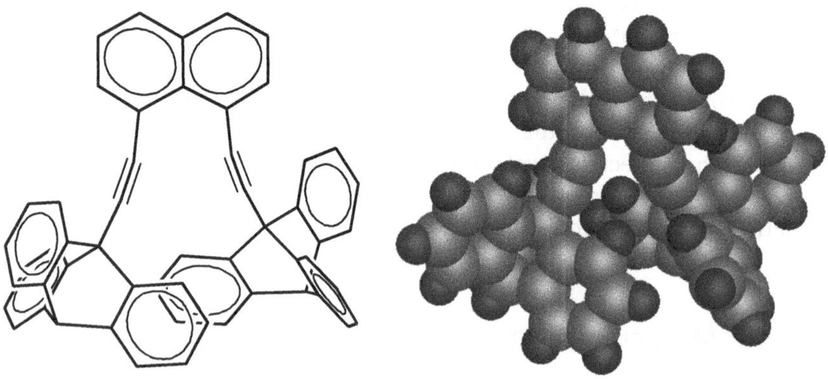

Figure 3.12 The spur molecular gearing system constructed from two triptycene units attached to the naphthalene stator via acetylene axles (left) and its X-ray structure (right).

One of the design ideas entails the use of a proper rigid "stator" (usually a fused polycyclic ring system), which can provide adequate molecular dynamic constraints while permitting free rotation of the gearing elements [26]. A straightforward practical implementation of this design idea is presented in Figure 3.12. This spur molecular gearing system utilizes two triptycene units as gearing elements attached to the naphthalene stator via extended acetylene "axles". These extended axles are important elements of this particular design. They are required to achieve unrestricted rotation of triptycene gearing units without steric interference with the naphthalene stator [30].

Variable-temperature NMR investigation of a labeled derivative of this naphthalene spur molecular gearing system shows apparent lack of correlated rotation in solution even at low temperatures. Subsequent density functional theory (DFT) calculations confirm that the energy barrier for gear slippage is indeed rather small, about 5.5 kcal/mol [30]. One of the possible explanations of this malfunction is poor alignment of the triptycene gearing elements. Previous theoretical calculations indicate that the best performance of triptycene-based spur molecular gears can be expected at axle-to-axle distances of about 8 Å [26], whereas in this particular case the axle-to-axle distance measured between the triptycene unit joint points is only about 5.6 Å. Excessive crowding of the two triptycene units is evident from the X-ray analysis, which shows a relatively wide angle (about 74°) between the triptycene gear axes (Figure 3.12) [30].

An attempt was made to relieve unfavorable overcrowding using larger anthracene stator, but with mixed results (Figure 3.13). This anthracene-based spur molecular gearing system is more sterically relaxed. The corresponding axle-to-axle angle and distance between the triptycene unit joint points show some notable improvements (23° and 6.7 Å, respectively) compared to the naphthalene stator, but the triptycene units are found to be meshed in an unfavorable C_2-type conformation [26] with significant π-stacking between the two benzene blades. No individual signals of rotational conformers were observed in NMR spectra even at −90 °C precluding determination of energy barriers for gear rotation and slippage [30].

Substantially better results could be achieved via rational design of heteroaromatic stators with precalculated optimal distances between the acetylene axles attached to the triptycene gearing units. One of the examples of such computer-optimized heteroaromatic systems is presented in Figure 3.14 [31]. Specific advantages of the rigidified bibenzimidazole as the stator unit include proper axle-to-axle distance (about 8 Å) [26] as well as almost parallel alignment of the gearing elements connected to the 4- and 4′-positions. The X-ray structure analysis confirms the optimal C_s-type conformation of these

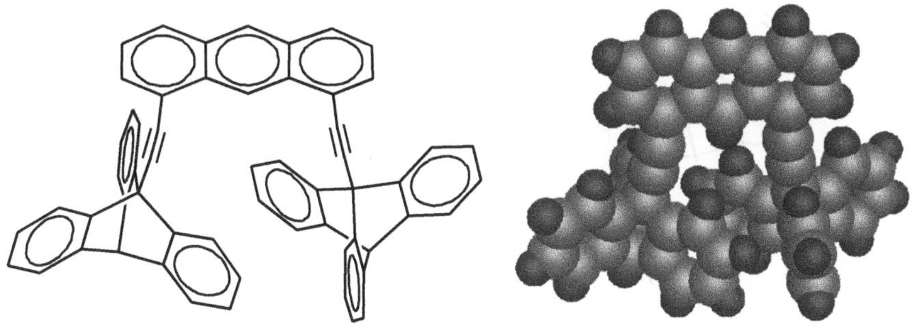

Figure 3.13 The spur molecular gearing system constructed from two triptycene units attached to the anthracene stator via acetylene axles (left) and its X-ray structure (right).

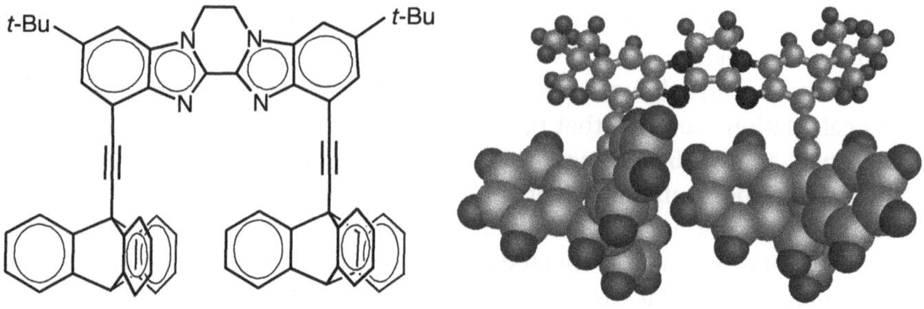

Figure 3.14 The computer-optimized spur molecular gearing system constructed on the basis of the bibenzimidazole stator (left) and its X-ray structure (right).

triptycene spur gears having almost perfect gear meshing (Figure 3.14). Despite computer-aided design optimization, variable-temperature NMR studies yield little evidence of correlated rotation of labeled triptycene gearing units in the bibenzimidazole-based spur molecular gearing systems. DFT calculations indicate that even with optimized design the energy barriers for gear slippage in these molecular gearing systems are less than 8 kcal/mol [31].

Molecular gearing systems sharing characteristics of both spur and bevel gears are also known. One of the examples of such hybrid systems is presented in Figure 3.15. The 1,1′-biindenyl stator in this system is sterically rigidified, but obviously more flexible than the previously reported fused polycyclic stators. Two triptycene gearing elements are attached directly to the stator with a moderate angle between their axes, which is typical for spur gear design. At the same time, the side edge rather than the front edge of the benzene blade is involved in gear meshing – a feature more frequently found in bevel gears. Despite high steric strain, triptycene gearing units rotate freely in solution at room temperature. Moreover, even at −80 °C NMR peaks are only slightly broadened. No

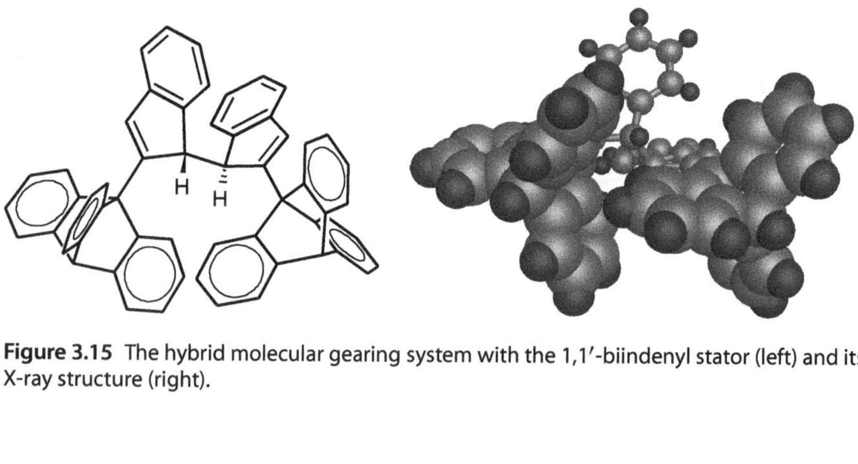

Figure 3.15 The hybrid molecular gearing system with the 1,1′-biindenyl stator (left) and its X-ray structure (right).

attempts were made to evaluate the thermodynamic parameters of a plausible correlated gear rotation in this molecule [32].

A rare example of a hybrid molecular gearing system employing pentiptycene gearing elements attached to *ortho*-positions of a benzene stator via extended acetylene axles is presented in Figure 3.16. The main advantage of pentiptycene over triptycene gearing units is the presence of an additional tooth, which theoretically can provide smoother gearing motion. However, the pentiptycene gearing units are less symmetrical compared to the triptycene gearing units and might require additional gear design adjustments to realize this advantage. This particular hybrid system has a relatively wide axle-to-axle angle. The distance between the centers of the pentiptycene gearing elements is more than 8.2 Å. Variable-temperature NMR shows no evidence of correlated rotation due to a relatively small gear slippage energy barrier [33].

A recent example of a molecular gearing system with semiflexible stator is presented in Figure 3.17. This gearing system was constructed on the basis of 2,6-pyridodicarboxamide stator. The structure is sterically relaxed. The angle

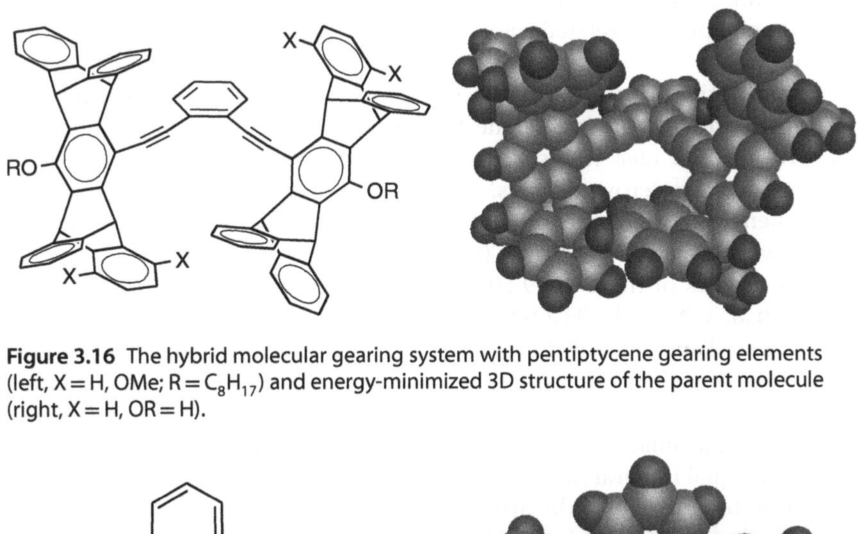

Figure 3.16 The hybrid molecular gearing system with pentiptycene gearing elements (left, X = H, OMe; R = C_8H_{17}) and energy-minimized 3D structure of the parent molecule (right, X = H, OR = H).

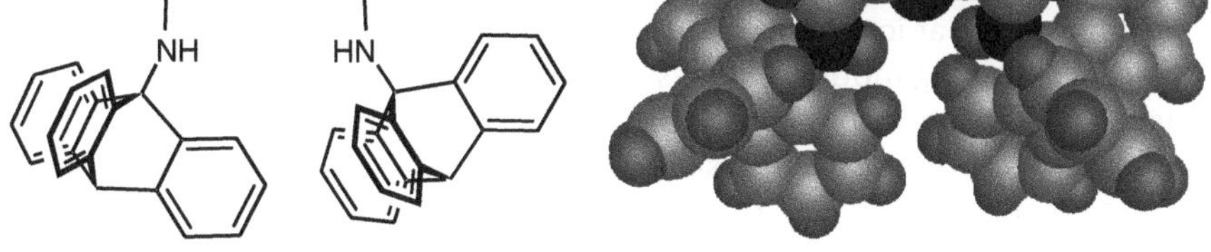

Figure 3.17 The molecular gearing system with the semiflexible 2,6-pyridodicarboxamide stator (left) and its X-ray structure (right).

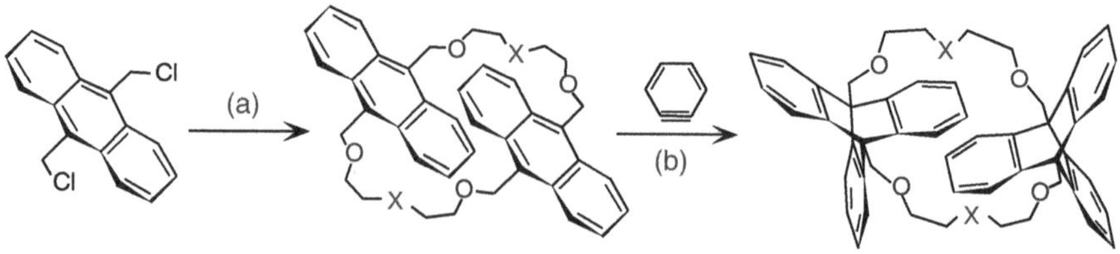

Scheme 3.2 Synthesis of bis(9,10-triptyceno)crown ethers, **2T26C6** (X = O), and **2T32C8** (X = OCH_2CH_2O); (a): (1) 5.5 equiv. $(HOCH_2CH_2)_2X$, 2.1 equiv. KOH, 10–12 h, 80 °C; (2) 1.2 equiv. $ClCH_2C_{14}H_8CH_2Cl$, 2.3 equiv. KOH in *t*-BuOH, reflux 20 h; (b): *o*-$H_2NC_6H_4COOH$, *i*-$C_5H_{11}ONO$ in CH_2Cl_2, reflux 20 h.

between the triptycene unit axes is wide, more than 120°. The distance between the centers of the triptycene gearing elements is approximately 9.0 Å in the CH_3CN solvate crystals. Despite the substantial spatial separation between the two triptycene gears, phase isomers were successfully detected by NMR at low temperatures. There is, however, computational evidence that the actual gear dynamics in this system is closer to rotation and rocking vibration mode rather than to pure correlated rotation found in some other molecular gearing systems [34].

Another design idea for the construction of spur molecular gearing systems entails double linking of the two triptycene gearing units from both ends. This design idea was successfully implemented with the synthesis of bis(9,10-triptyceno)crown ethers. These compounds were synthesized by double benzyne addition to bis(9,10-anthraceno)crown ethers according to the following synthetic procedure (Scheme 3.2) [1].

The structures of these bis(9,10-triptyceno)crown ethers (**2T26C6**, X = O, and **2T32C8**, X = OCH_2CH_2O) feature two triptycene gears interconnected by two polyoxyalkane links at their axle joints. A detailed variable-temperature NMR study shows that triptycene gearing units in **2T32C8** undergo fast internal rotation in solution at the temperatures above 60 °C (Figure 3.18). At these temperatures, aromatic hydrogens of the triptycene units appear as a pair of peaks with equal integral intensities. Lowering the temperature results in dramatic peak broadening of the aromatic hydrogens. At subzero temperatures, these broad peaks resolve into individual signals of several static conformers that have adequate stability on the NMR time scale. The presence of several stable conformers at low temperatures precludes a straightforward evaluation of molecular gearing motion in this system [1].

Further elucidation of the structural characteristics of these two bis(9,10-triptyceno)crown ethers (**2T26C6** and **2T32C8**) was performed using single-crystal X-ray crystallography [1, 29]. Intramolecular gear meshing was observed for both compounds in the solid state, but in different conformations, thus underscoring the importance of proper molecular design in achieving the best functionality of molecular gearing systems.

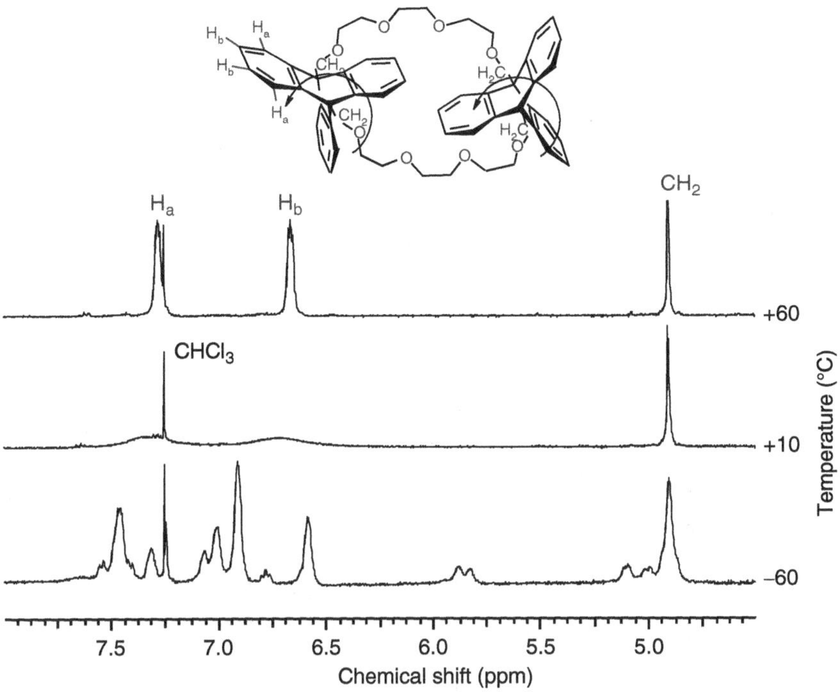

Figure 3.18 Variable-temperature ^{1}H NMR spectra of **2T32C8**.

The larger bis(9,10-triptyceno)crown ether **2T32C8** is sterically relaxed due to its longer 12-membered links. The axle-to-axle distance between the two triptycene gearing units is approximately 8.1 Å, and the angle between the axles is very small (less than 15°). Both parameters are close to optimal for spur molecular gears favoring the desirable C_s-type conformation [26]. The benzene blade of one triptycene unit resides deeply in the cavity created by two benzene rings of the opposite triptycene unit. Two polyoxyalkane links are pushed outside and do not interfere with the gearing motion (Figure 3.19) [1, 29].

The structure of **2T26C6** with shorter nine-membered links is more sterically congested (Figure 3.20). The distance between the axles of the two triptycene gearing units is about 6.9 Å, and the axles are exactly parallel. Shorter axle-to-axle distances usually favor less desirable C_2-type conformations [26], which proved to be the case for **2T26C6**. Similar to **2T32C8**, the polyoxyalkane links in **2T26C6** are pushed away, and the internal cavity is occupied by the two benzene rings precisely parallel to each other. Strong π–π interactions [35] are evident from the very short distance (about 3.5 Å) between these benzene rings.

Tight, symmetrical structure of **2T26C6** facilitates not only intramolecular but also intermolecular π-stacking between the triptycene units in crystal. The

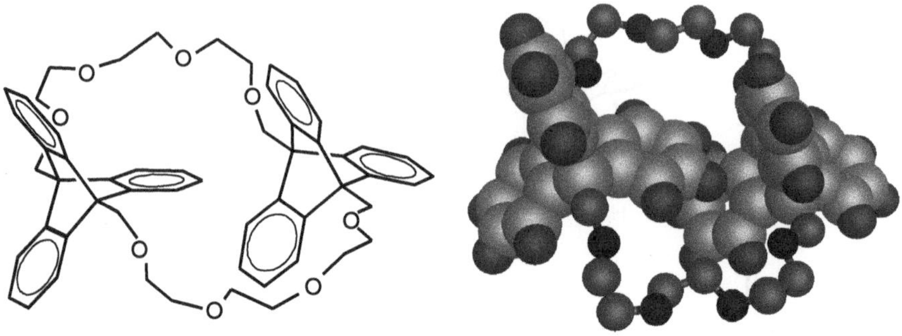

Figure 3.19 Bis(9,10-triptyceno)crown ether **2T32C8** (left) and its X-ray structure (right, hydrogen atoms of the polyoxyalkane links were removed for clarity).

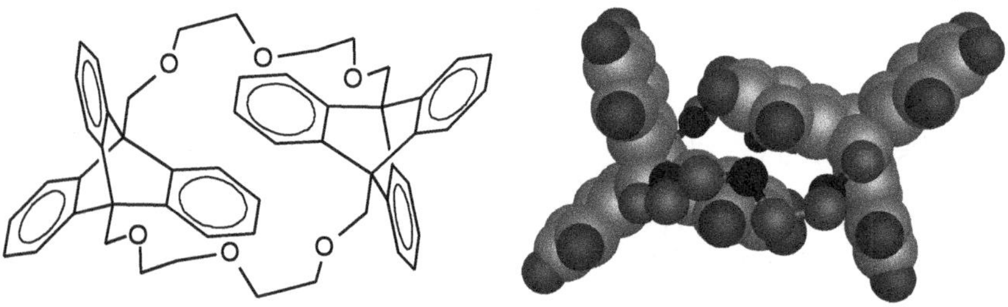

Figure 3.20 Bis(9,10-triptyceno)crown ether **2T26C6** (left) and its X-ray structure (right, hydrogen atoms of the polyoxyalkane links were removed for clarity).

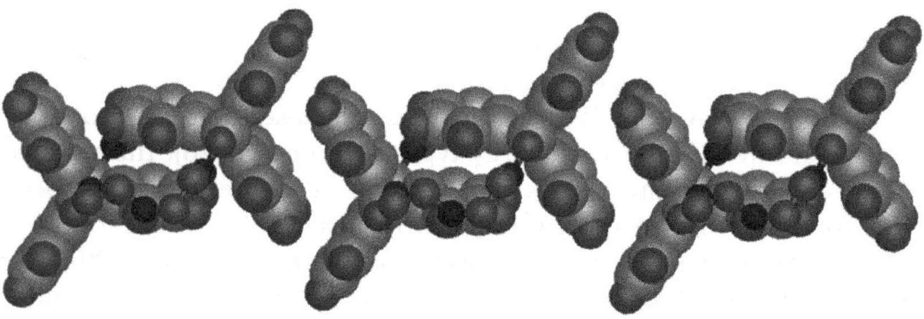

Figure 3.21 Self-organization and gear meshing of **2T26C6** via intermolecular π–π stacking in the crystal. Hydrogen atoms of the 2,5,8-trioxanonane links were removed for clarity.

intermolecular distance between the benzene rings (about 3.7 Å) is only slightly larger than the intramolecular distance, thus enabling tight gear meshing through self-assembly. The individual **2T26C6** molecules in crystal organize themselves in parallel with slight shift along the triptycene axis to maximize the π–π interactions [36] between the triptycene units (Figure 3.21) [1, 29].

So far, no attempts were made to use all-carbon gearing elements to construct molecular gears with different transmission ratios due to design complexity limitations. These limitations can be easily overcome in organometallic gearing systems.

3.4 Organometallic Molecular Gears

Perhaps the simplest organometallic gearing system can be constructed from a multiligand metal unit attached directly to a planar cyclic moiety, which could serve as a secondary gearing element. Depending on the number of ligands involved in steric interactions with peripheral substituents attached to the cyclic gear, the organometallic bevel gearing systems can have different transmission ratios. In the simplest case, these peripheral substituents could be just hydrogen atoms – providing proper alignment of the multiligand metal gearing unit.

One of the notable examples of this simple organometallic gearing system design is presented in Figure 3.22 [36]. This is a very special type of molecular gearing system, which employs pericyclic reactions to achieve perpetual motion over the hydrocarbon frame of cycloocta-2,4,6-triene. This motion at slightly elevated temperatures is facilitated by relatively weak carbon–phosphorus bonds. Functionally, this system acts as a molecular bevel gear, where the motion of the phosphorous atom over the eight-toothed hydrocarbon ring is correlated with the rotary motion of the attached $W(CO)_5$ unit as a pinion (smaller) gear. The presence of hexacoordinated wolfram atom with four carbonyl groups acting as four gear teeth by meshing with eight hydrogen atoms of cycloocta-2,4,6-triene gear ring allows for the design of the molecular gear with the exact 2:1 transmission ratio.

The compound was prepared using the reaction of *in situ* generated phosphinidene $MePW(CO)_5$ with cyclooctatetraene. The reaction produces two isomers, of which only the *anti*-isomer, with the $W(CO)_5$ group oriented away from the ring, can be considered as a molecular gear. Variable-temperature

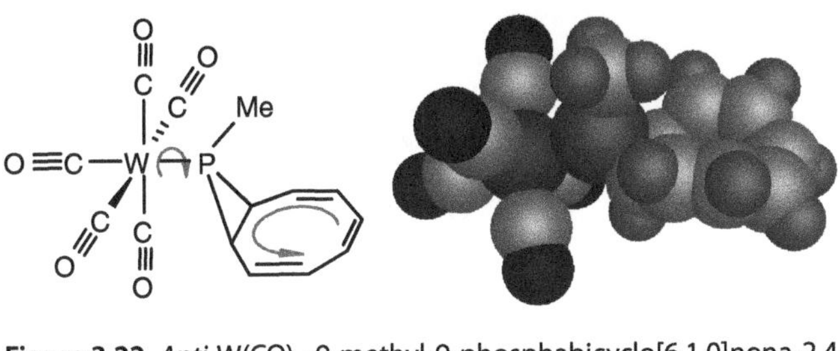

Figure 3.22 *Anti*-$W(CO)_5$-9-methyl-9-phosphabicyclo[6.1.0]nona-2,4,6-triene as organometallic 2:1 molecular gearing system (left) and its X-ray structure (right).

NMR of this *anti*-isomer shows that the rotation of the $W(CO)_5$ fragment is fast on the NMR time scale in solution even below −80 °C with the energy barrier less than 10 kcal/mol. Gear rotation dynamics in this system is further evaluated using computational methods, since the symmetry of the rotating fragments makes it impossible to determine this dynamics in full detail experimentally. The calculated [1.7]-shift energy barrier is about 13 kcal/mol, which is within the range of the experimentally determined value of about 17 kcal/mol. Correlated rotation is confirmed by molecular dynamics simulations. It is important to note that this gearing system can be operated effectively only within a limited temperature range due to competitive conversion to another isomer at the temperatures above 100 °C [36].

It would also be prudent to note that the design of the abovementioned molecular gearing system is not ideal and can be improved further. The energy barrier for gear slippage is too low. This complicates the full synchronization of the phosphorus atom movement and the rotation of the $W(CO)_5$ gearing unit. For a better transfer of the circumambulatory motion, the transition metal gearing element must be meshed more tightly with the hydrocarbon gear.

The gearing motion in the abovementioned system is quite sensitive with regard to other substituents attached to the cycloocta-2,4,6-triene ring. For example, the presence of two methyl groups in the ring forces the $MePW(CO)_5$ fragment to migrate between the only two available positions at ambient temperature (Scheme 3.3). Circumambulatory rotation past the methyl groups requires temperatures above 50 °C. In essence, the presence of these two methyl groups forces this gear system to perform a rocking motion rather than a cyclic motion at the temperatures below 50 °C [36].

The use of metallocenes significantly extends the capabilities of molecular gearing systems. In these metallocene-based gearing systems, a coordinated metal atom could serve as a ball bearing element, and the metallocene ring could serve as a gearing element. The use of the ball bearing junction allows for the construction of the molecular gearing system with gear ratios other

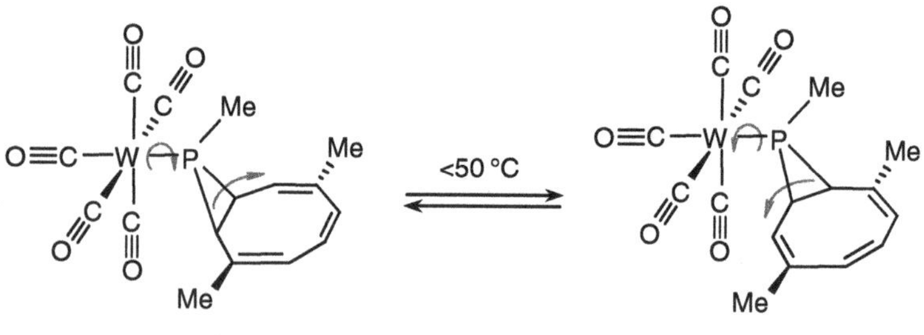

Scheme 3.3 The rocking motion of the $MePW(CO)_5$ gearing unit in *anti*-$W(CO)_5$-3,7,9-trimethyl-9-phosphabicyclo[6.1.0]nona-2,4,6-triene.

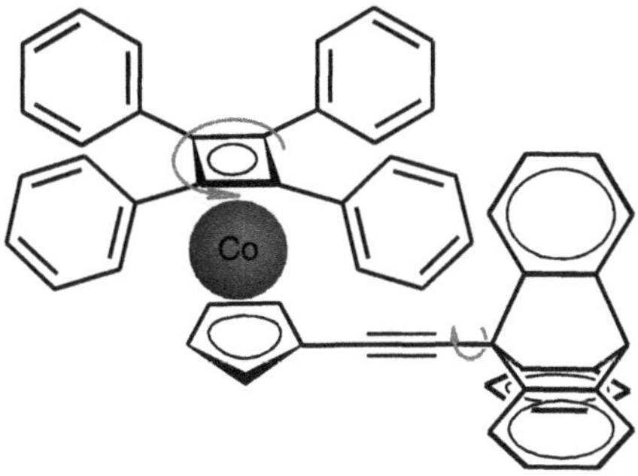

Figure 3.23 The metallocene-based molecular gearing system with the 3:4 gear ratio.

than 1:1, which is typical for bis-triptycene systems. In one of the examples presented in Figure 3.23, the gear ratio is 3:4, even though correlated rotation in this particular molecular gearing system cannot be unequivocally confirmed experimentally due to the symmetry of the gearing elements. The compound was prepared in five steps with 19% overall yield. According to the reported procedure, the reaction of sodium carbomethoxycyclopentadienylide and diphenylacetylene in the presence of Co^+ source yields the carbomethoxy derivative of (cyclopentadienyl)(tetraphenylcyclobutadiene)cobalt. This carbomethoxy derivative is converted to aldehyde and then to alkyne by a Wittig reaction with chloromethyltriphenylphosphonium followed by BuLi-promoted elimination of HCl. The last two steps entail Sonogashira cross-coupling with 9-iodoanthracene and addition of benzyne. Variable-temperature NMR experiments demonstrate fast rotation of the triptycene gear on the NMR time scale, thus establishing a probable case of correlated rotation. This conclusion is further supported by molecular mechanics modeling [37].

More complex metallocene molecular gearing systems with several metallocene units are also known. One of the examples of such systems having two metallocene fragments attached to one triptycene unit is presented in Figure 3.24. The gearing rotation of the ferrocene fragments around the central triptycene moiety proceeds in a step-by-step fashion described in the original publication in terms of "molecular dials". The energy barrier for interconversion between the two rotamers was found to be about 17 kcal/mol [38].

The use of metal atoms as ball bearings provides an attractive venue for the construction of sophisticated multilevel molecular gearing systems. One of such systems is presented in Figure 3.25. This gearing system contains two functionalized porphyrin rings as lower and upper decks interconnected by La atom as a ball bearing element. An additional porphyrin gearing element with Rh atom as another ball bearing is attached to the lower deck through complexation with pyridine junction unit. Computer simulation data indicate

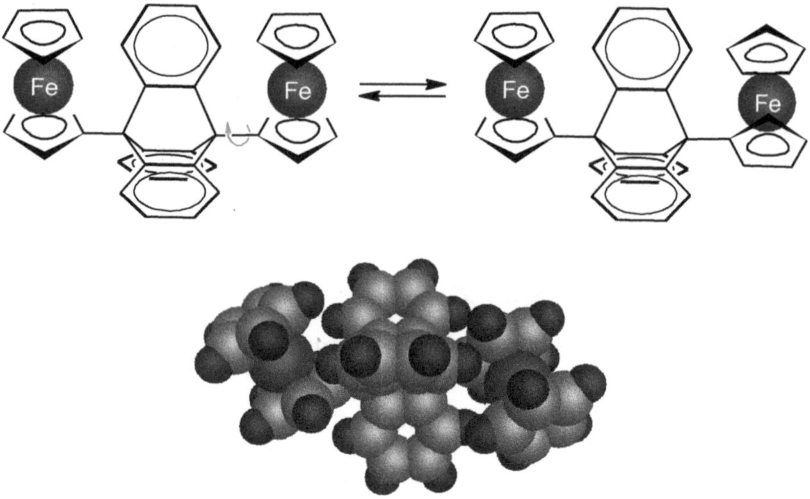

Figure 3.24 The triptycene-based molecular system with two metallocene "dials" (top) and its X-ray structure (bottom).

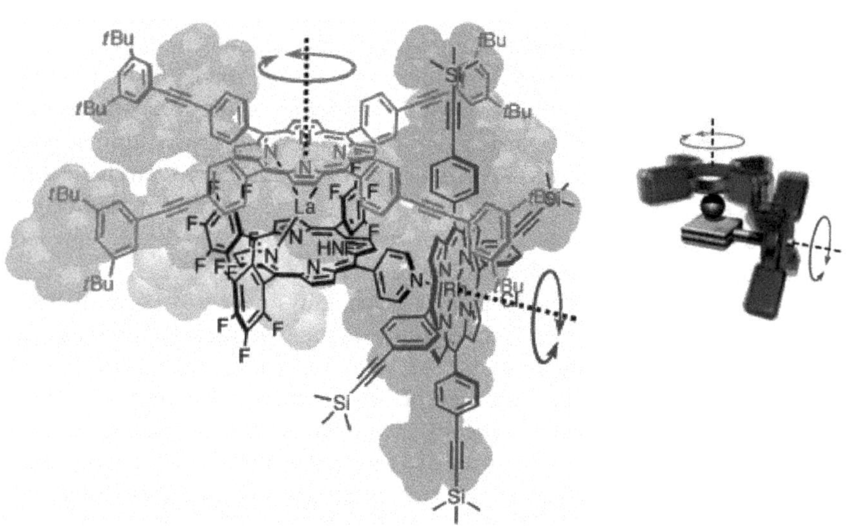

Figure 3.25 The complex double-decker porphyrin-based bevel molecular gearing system. *Source*: Adapted from Ogi *et al.* 2010 [39]. Reproduced with permission of John Wiley & Sons.

a possibility of a correlated rotation between the upper deck and the side gearing unit. Experimental NMR data lend some credence to this possibility due to almost identical coalescence temperatures of the signals attributed to different porphyrin rings involved in the gearing motion [39].

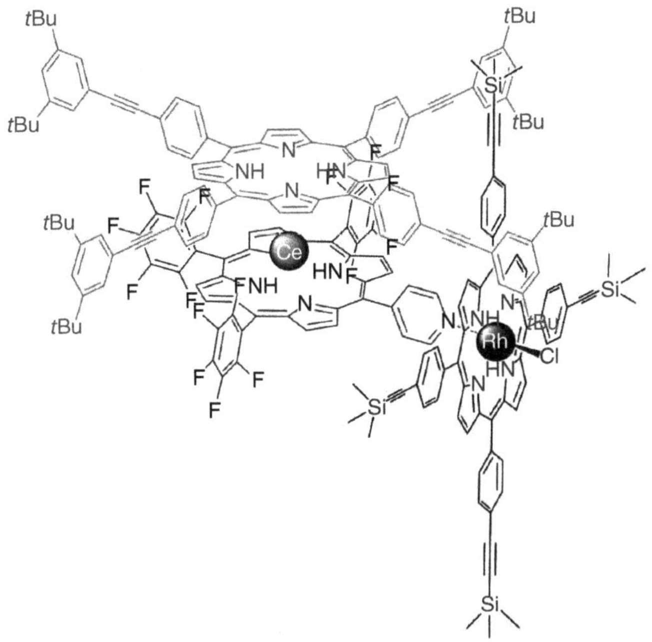

Figure 3.26 The double-decker Ce porphyrin-based rotary system without the gearing effect. *Source*: Erbas-Cakmak *et al.* 2015 [4]. http://pubs.acs.org/doi/full/10.1021/acs.chemrev.5b00146?src=recsys. Licensed under CC BY 4.0.

Replacement of the larger La ball bearing atom with a smaller Ce atom (ionic radii of La^{3+} is 1.18 Å, Ce^{4+} – 0.97 Å) results in a complete loss of the correlation between the rotation of the upper deck unit and the side gear unit (Figure 3.26) [40]. One of the factors unfavorably influencing the rotation dynamics of this system could be the inverted dome-like shape of the porphyrin rings in the Ce double-decker unit [41].

The versatile porphyrin-based ball bearing double-decker framework employed in the abovementioned molecular gearing systems permits further modifications, leading to the design of molecular systems with several side gearing units. An example of one of the systems with two side gearing elements is presented in Figure 3.27 [42]. Despite visual proximity, the two side gear units are not enmeshed, thus allowing proper functionality of this complex multigear system.

Similar design ideas were used for the manufacturing of organometallic molecular gearing systems with a cavitand stator (Figure 3.28) [43]. The rotation dynamics of this quadruple molecular gearing system and its analogs was investigated by means of variable-temperature ^1H NMR in solution and by single-crystal X-ray diffraction analysis. X-ray studies show that four

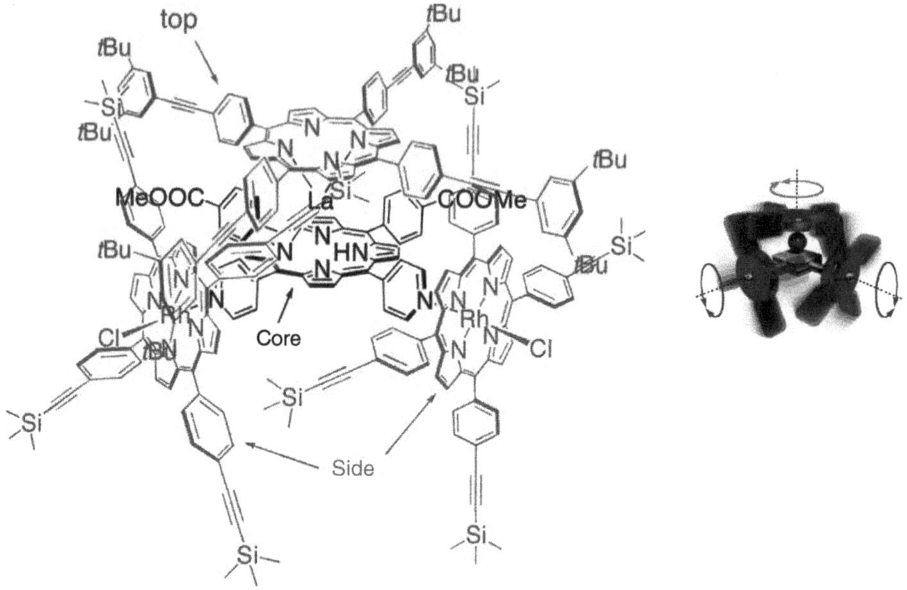

Figure 3.27 The molecular gearing system with two auxiliary gearing elements. *Source*: Adapted from Ogi *et al*. 2011 [42]. Reproduced with permission of John Wiley & Sons.

porphyrin gears enmeshed with one another in a cyclical manner. NMR studies demonstrate that the porphyrin gearing subunits rotate quickly at 298 °K on the NMR time scale. The energy barrier for rotation was found to be about 7 kcal/mol. Unfortunately, the energy barrier for gear slippage cannot be determined experimentally due to the symmetry of the gearing subunits. Organometallic molecular gearing systems with four triptycene gears attached to a dirhodium metal oxide cluster are also known [44].

Even more complex transmission-like organometallic systems can be constructed using supramolecular chemistry methodology. One of the examples of such systems containing four gearing rings in a sandwich-type molecular superstructure with three layers of metal ions between the gearing rings is presented in Figure 3.29. Two of these gearing rings are rotors, and another two are transmitters. There are also nine Ag^+ atoms serving as transmission elements and also holding the system together. What is special about this design is that some sort of molecular torque (in this case – helicity) is being transmitted through the several layers of transmitting elements with no covalent bonds between these layers. The middle layer of Ag^+ atoms can be replaced with Hg^{2+} atoms without the loss of functionality [45].

A set of truly unique molecular caterpillar systems was constructed recently. The major distinctive feature of these systems compared to all other gearing systems presented in this chapter is that correlated rotation of the two

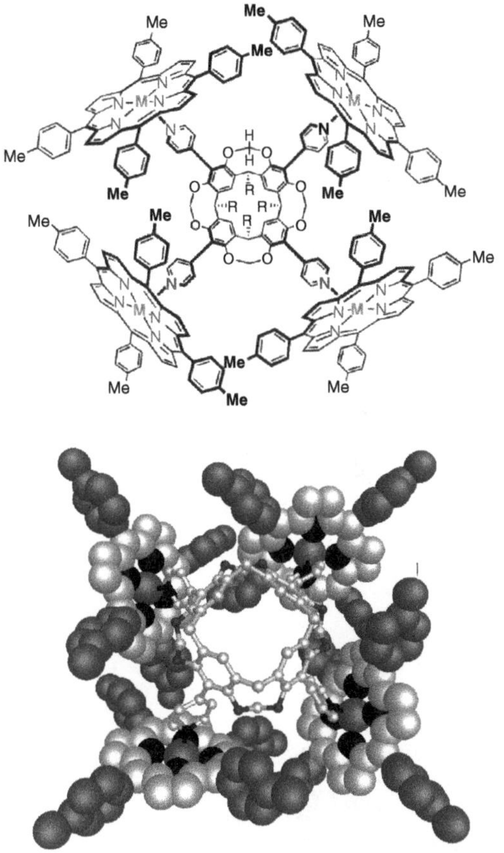

Figure 3.28 The molecular gearing systems with the cavitand stator and four porphyrin gearing elements (left, R = $CH_2CH(CH_3)_2$ or $(CH_2)_6CH_3$; M = Rh(III)Cl) and the X-ray structure (right, hydrogen atoms, chloride counteranions, and $CH_2CH(CH_3)_2$ groups were removed for clarity). *Source*: Adapted from Nakamura *et al.* 2016 [43]. Reproduced with permission of American Chemical Society.

hexapyridine gearing elements is achieved using a belt-like macrocyclic track assembled from interconnected porphyrin rings. The design and mode of action of one of these molecular caterpillar systems with eight porphyrin rings is presented in Figure 3.30. Detailed motion dynamic studies of this system were performed using ^1H EXSY NMR. The data show a clear preference of correlated template – macrocyclic caterpillar track motion over the competing complexation–decomplexation process [46].

A larger molecular caterpillar system with ten porphyrin rings has adequate spacing between the two hexapyridine gearing units to accommodate a metal cation. The complexation halts the caterpillar-type motion with the formation

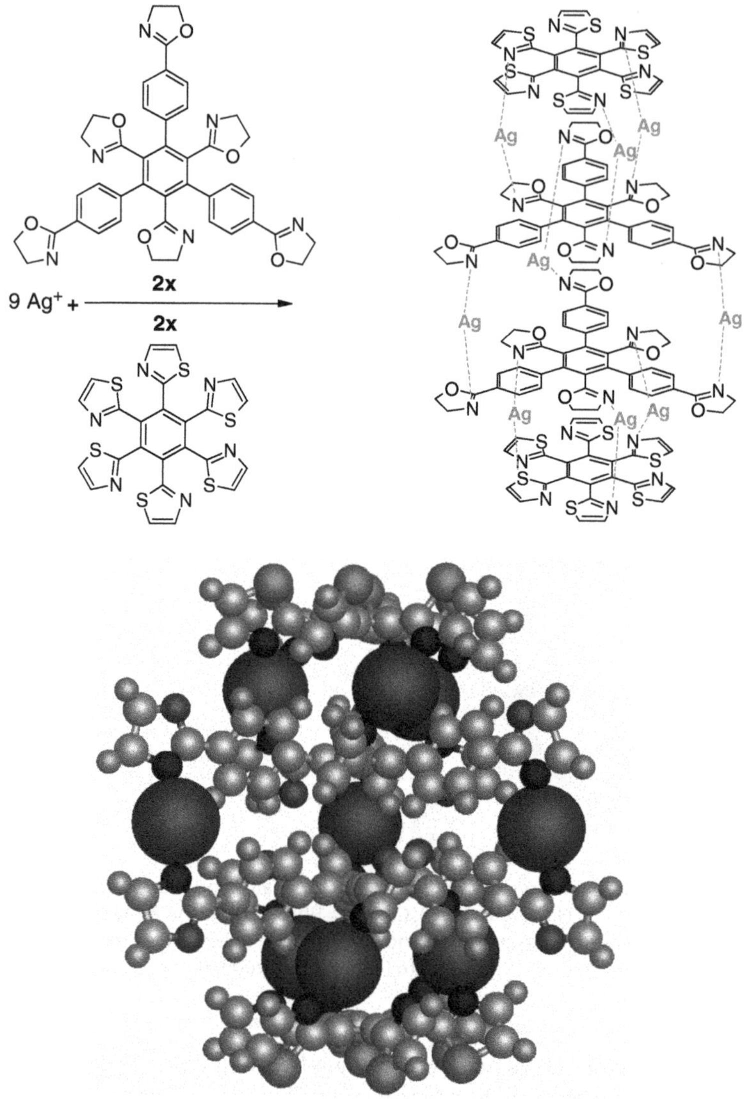

Figure 3.29 Synthesis of the supramolecular system with four gearing rings and three layers of metal ions in a sandwich-type arrangement (top) and its X-ray structure (bottom).

of a rigid, frozen structure. The structure of the molecular caterpillar complex with $PdCl_2$ is depicted in Figure 3.31. A metal-free molecular caterpillar system with ten porphyrin tracks actually moves faster than the smaller eight-track system described earlier due to less steric congestion between the two hexapyridine gearing elements [46].

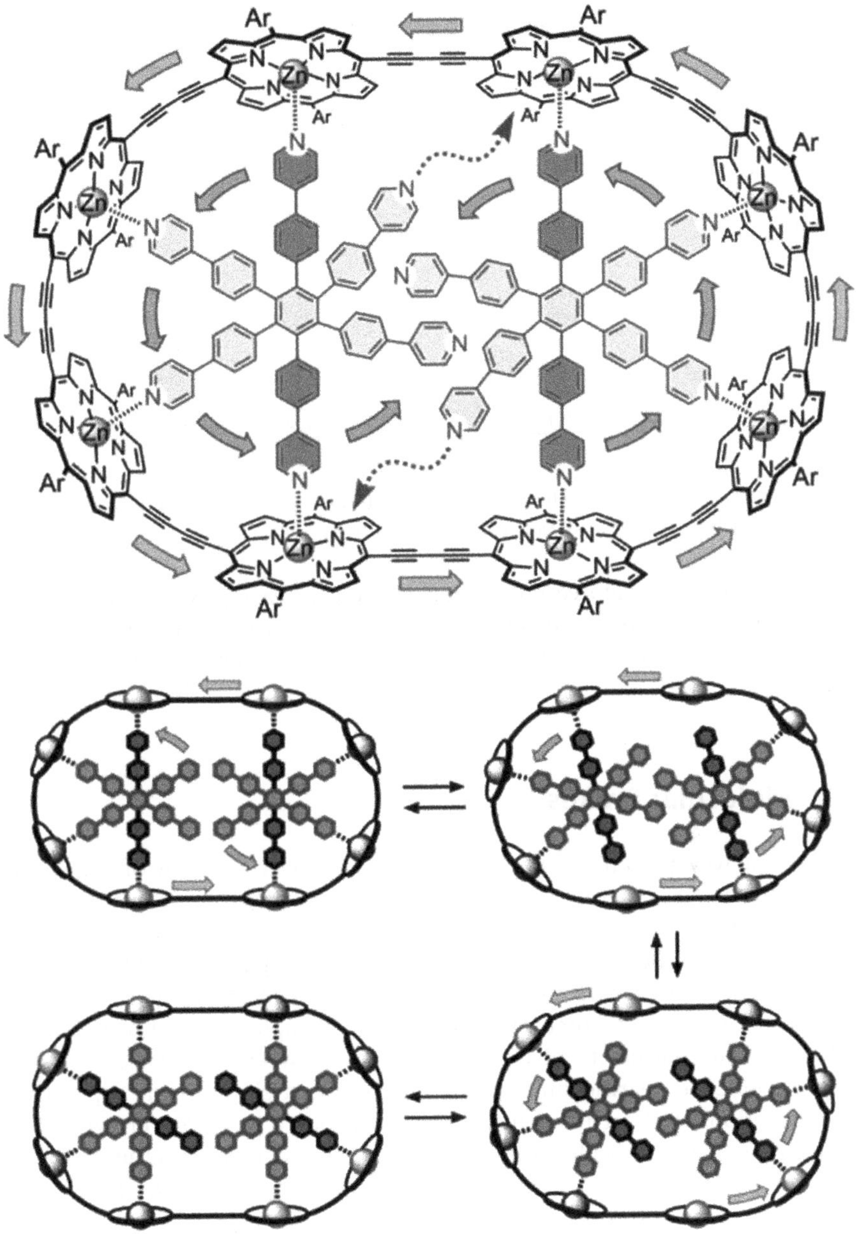

Figure 3.30 Schematics of the molecular caterpillar system (top) and its mode of action (bottom). *Source*: Adapted from Liu *et al*. 2015 [46]. Reproduced with permission of John Wiley & Sons.

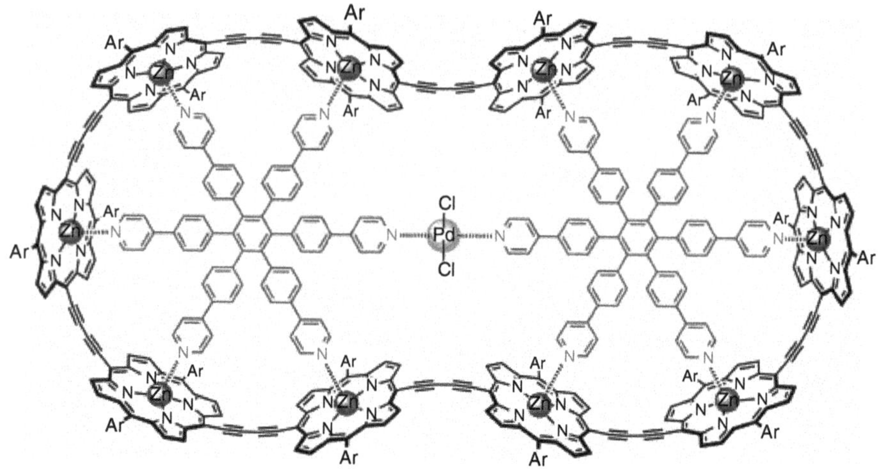

Figure 3.31 The structure of the frozen molecular caterpillar complex with $PdCl_2$. *Source:* Liu *et al.* 2015 [46]. Reproduced with permission of John Wiley & Sons.

The majority of the above-mentioned molecular gearing systems lack dedicated control elements that can be used to regulate the rotary dynamics. Construction of more complex devices requires precise external motion control functionality achievable only in specially designed molecular brake systems.

3.5 Molecular Brakes

The easiest way to achieve the external control over rotation of a gearing element is the use of reversible complexation, such as complexation with a heavy metal cation. This type of control can be accomplished in a "paddlewheel on a rope" family of compounds called paddlanes [47]. In the simplest case, the gearing element can be a triptycene "cogwheel" incorporated in a large crown ether ring. Subsequent complexation of a metal ion within the internal cavity of the crown ether between the benzene blades of the triptycene gearing unit would interfere and slow down "rope skipping" rotation of the triptycene cogwheel through the large crown ether macrocyclic "rope". This idea was successfully implemented with the synthesis of mono(9,10-triptyceno)crown ethers presented in Scheme 3.4. These compounds are related to bis(9,10-triptyceno)crown ethers discussed earlier (see Section 3.3) [1, 48, 49].

The size of the macrocyclic ring plays a critical role in the rotation dynamics of these systems. Too tight ring systems (such as **T22C6**) do not display rope skipping, even at 140 °C. On the other hand, both **T25C7** and **TB28C8**

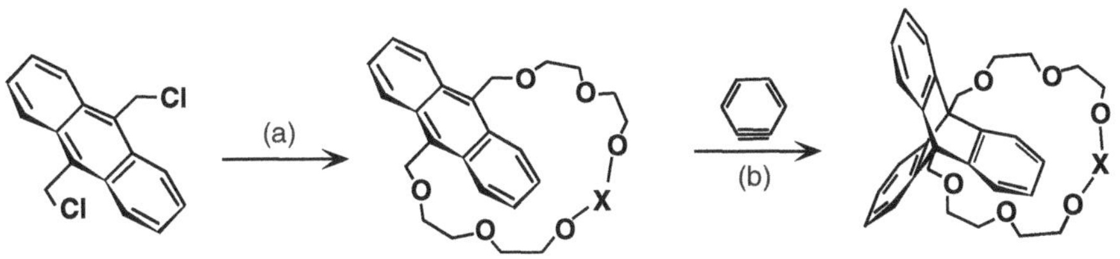

Scheme 3.4 Synthesis of mono(9,10-triptyceno)crown ethers: **T22C6**, X = CH_2CH_2; **T25C7**, X = $(CH_2CH_2)_2O$; **TB28C8**, X = *o*-$(CH_2CH_2O)_2C_6H_4$. Reagents and conditions: (a) 1.3 equiv. $(HOCH_2CH_2OCH_2CH_2O)_2X$, 2.1 equiv. KOH in *t*-BuOH, reflux 6–8 h; (b) *o*-$H_2NC_6H_4COOH$, *i*-$C_5H_{11}ONO$ in CH_2Cl_2, reflux 6 h.

with larger macrocyclic rings clearly show rope skipping dynamics. The results of variable-temperature NMR studies of **T25C7** are presented in Figure 3.32. Aromatic hydrogens H_a and H_b (as well as H_c and H_d) become equivalent at elevated temperatures, indicating fast motion of the triptycene gear through the crown ether macrocycle on the NMR time scale [1].

Subsequent NMR studies demonstrate that in more relaxed mono(9,10-tripyceno)crown ethers, such as **T25C7** and **TB28C8**, the rope skipping rotation can be controlled by complexation with appropriate metal cations, such as silver and thallium. The best results were achieved with **T25C7** and Tl^+ salts in CH_3OH-d_4, although even in this case the control effect was moderate due to poor binding properties of (9,10-triptyceno)crown ethers [1].

Single-crystal X-ray analyses of **T25C7** and its Tl^+ complex confirm the results of the variable-temperature NMR experiments. The crown ether moiety in **T25C7** has distinctively elliptical shape (Figure 3.33). As a result, only four oxygen atoms in Tl^+ complex of **T26C7** are involved in the complexation, and these four oxygen atoms are not at the optimal positions. The metal cation also displays weak cation–π interactions [50] with the triptycene gearing unit (Figure 3.33) [1, 48].

Better external control over gear rotation has been achieved in triptycene–2,2′-bipyridine systems. In noncoordinated state, two pyridine subunits provide enough flexibility for the triptycene gearing unit to rotate freely at 30 °C on the NMR time scale. Addition of a bonding element (in this case – Hg^{2+}) causes significant rotation rate decrease, since in the Hg^{2+} complex the two pyridine subunits are locked. The lock can be released by subsequent addition of ethylenediaminetetraacetic acid (EDTA) (Scheme 3.5) [51].

A similar design idea entails incorporation of a "sliding latch" as a control element in a molecular gearing system (Scheme 3.6). The system employs a sliding shift of a molecular latch to block the rotation of the triptycene cogwheel in the engaged position [52].

One practical implementation of this design idea takes advantage of the known controlled migration of organometallic fragments over a fused polyaromatic scaffold (Scheme 3.7). This particular "cogwheel-latch" system utilizes

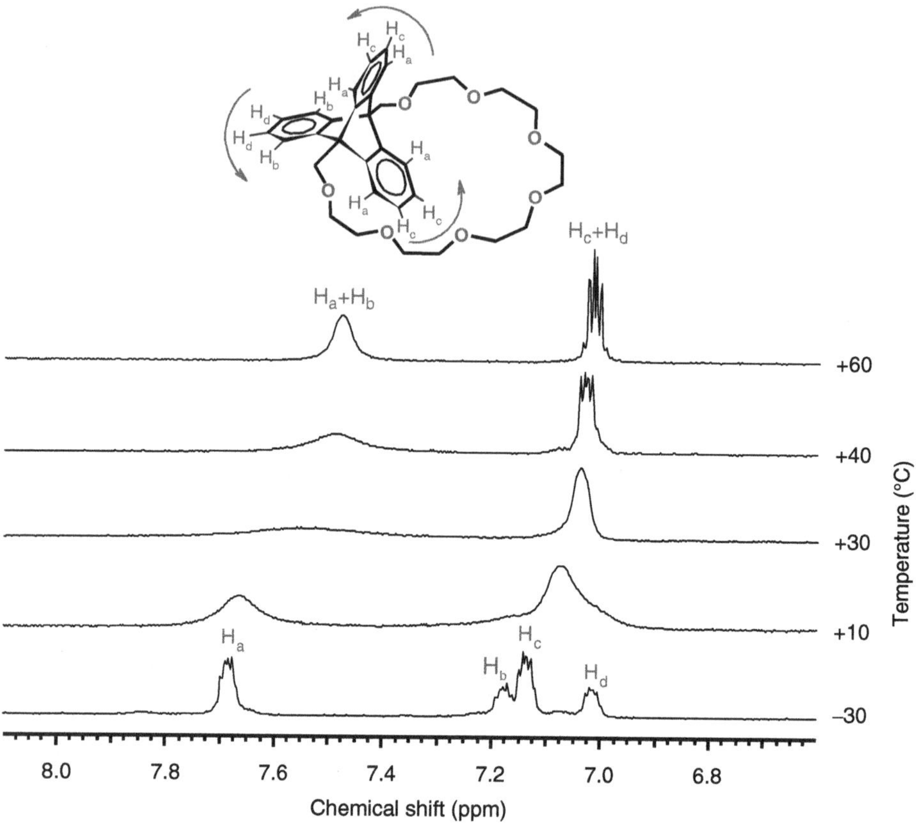

Figure 3.32 Variable-temperature ^1H NMR spectra of **T25C7**.

a metal-carbonyl tripod as a sliding element and a triptycene moiety as a cogwheel to achieve the desired molecular brake functionality. The migration of organometallic fragments ML_n over the surfaces of fused polyaromatic molecules (haptotropic shifts) is typically directed by the relative affinities of the ML_n fragment toward the docking sites of the individual aromatic rings. In the case of indene, an ML_3 tripod, located at the six-membered ring (η^6-complex), upon deprotonation shifts toward the more negatively charged five-membered ring (η^5-complex) of the indenide anion [52].

The X-ray structure analysis of the η^6-complex with $Cr(CO)_3$ shows a sterically relaxed molecule with seemingly unrestricted rotation around the indene–triptycene bond (Figure 3.34). This rotation in η^5-complex with $Re(CO)_3$ is blocked by steric interactions between the organometallic latch and the triptycene gearing unit due to their close proximity to each other (Figure 3.34) [52].

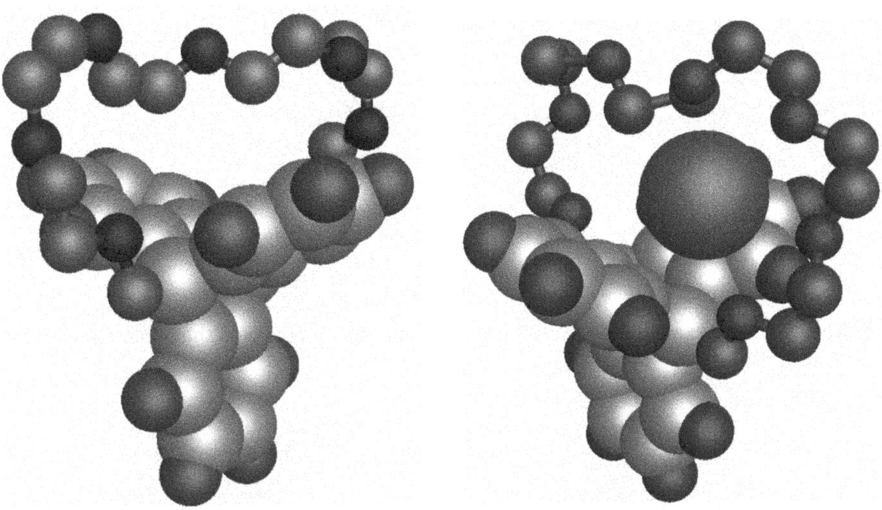

Figure 3.33 The X-ray structures of **T25C7** (left) and its Tl^+ complex (right). Hydrogen atoms of the crown ether chains were removed for clarity.

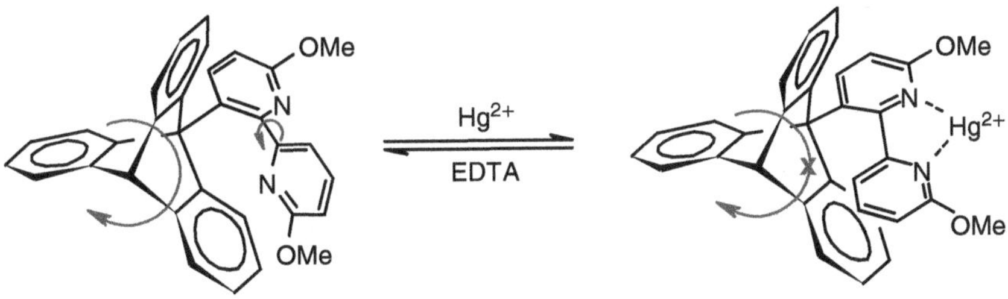

Scheme 3.5 The molecular brake system with the 2,2′-bipyridyl control unit (EDTA – ethylenediaminetetraacetic acid).

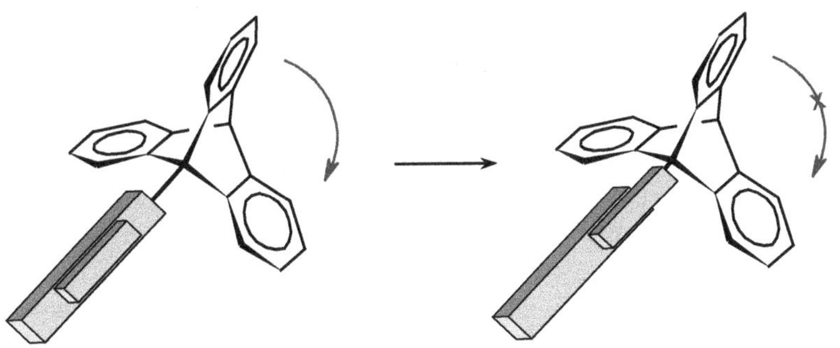

Scheme 3.6 The molecular brake system with the latch-type control unit.

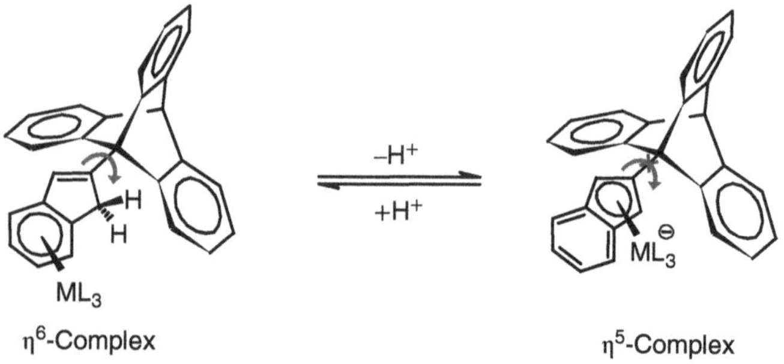

Scheme 3.7 The example of the molecular latch control unit utilizing the haptotropic shift in the indene system, $ML_3 = Cr(CO)_3$.

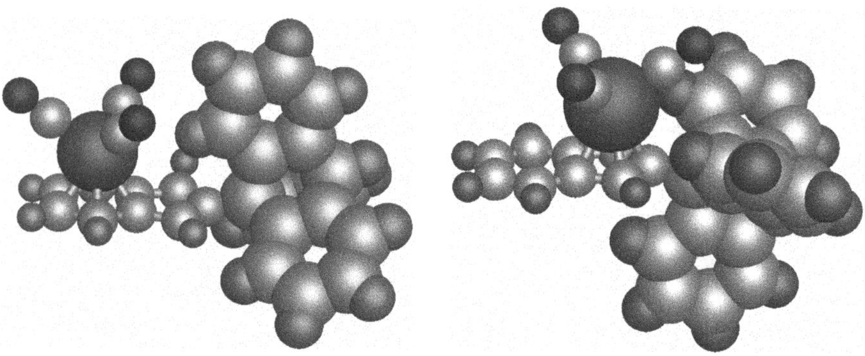

Figure 3.34 X-ray structures of [η^6-2-(9-triptycyl)-indene]tricarbonylchromium (left) and [η^5-2-(9-triptycyl)-indene]tricarbonylrhenium (right).

The performance of the cogwheel-latch $Cr(CO)_3$ molecular gearing system was further investigated in solution using ^{13}C NMR. The free rotation of the triptycene cogwheel in the η^6-complex was confirmed by the presence of only two peaks of the benzene ring junction carbon atoms. When the latch is engaged by the addition of a strong base leading to the formation of η^5-complex, the signals of these carbon atoms are split in four peaks in a 2:1 ratio. The computer simulation of the $Cr(CO)_3$ system reveals that the organometallic latch engagement (rearrangement of the η^6-complex into the η^5-complex) increases the energy barrier for the triptycene gearing unit rotation by about 12 kcal/mol, which corresponds to an approximately 10^8-fold rotation rate decrease [52].

Light-controlled molecular brakes have certain advantages over chemically controlled systems due to the fast response and the absence of waste by-products, but complete on–off switching is usually problematic. The simplest light-controlled systems usually utilize *cis–trans* isomerization of a

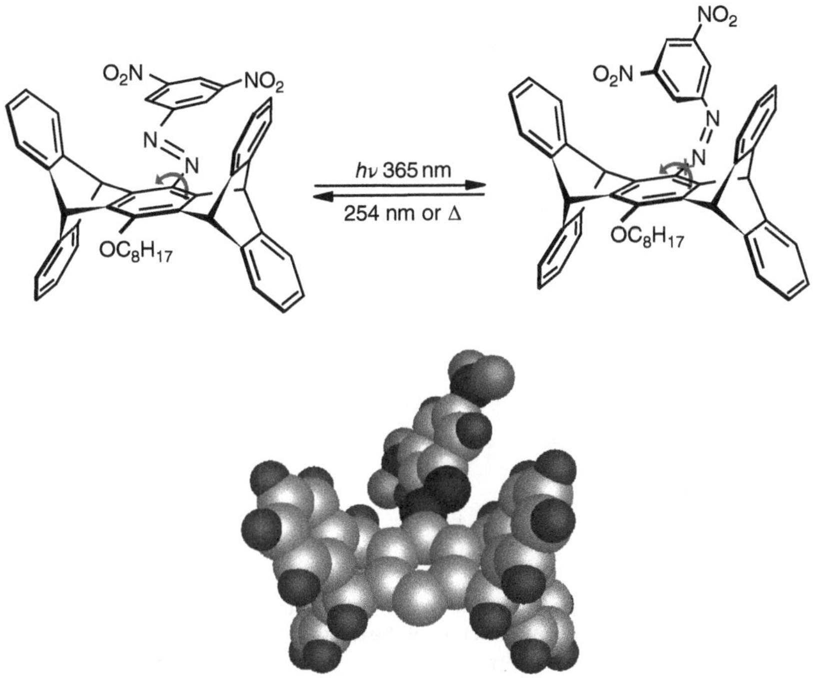

Figure 3.35 An example of the pentiptycene-based light-controlled molecular brake with the azobenzene controller (top) and X-ray structure of the brake in "off" position (bottom, OC_8H_{17} fragment was removed for clarity).

double bond within a dedicated control unit. Several recent examples include pentiptycene derivatives with azobenzene controllers (Figure 3.35). On–off switching in these molecular brakes can be accomplished by UV irradiation at different wavelengths, but the best switching efficiency approaching 78% of net isomeric excess can be attained by the combination of thermal (*cis*-to-*trans*) and light (*trans*-to-*cis*) stimuli. The achievable rotation rate difference between the "on" and "off" positions (braking power) is about 10^8-fold at ambient temperature [53].

A very similar idea was used to construct a stilbene-based molecular brake system (Figure 3.36), but with inferior results. Stilbene control units do not respond to a thermal stimulus due to a relatively high energy barrier for *cis*-to-*trans* interconversion, and light-induced reversible isomerization produces only about 20% net difference in isomeric composition even in optimal conditions. On a positive side, the braking power of the stilbene controller is comparable to the azobenzene control unit described earlier – up to 10^9-fold at ambient temperature [54]. Further improvements of the stilbene control element can be achieved by replacing the nitro groups with other moieties.

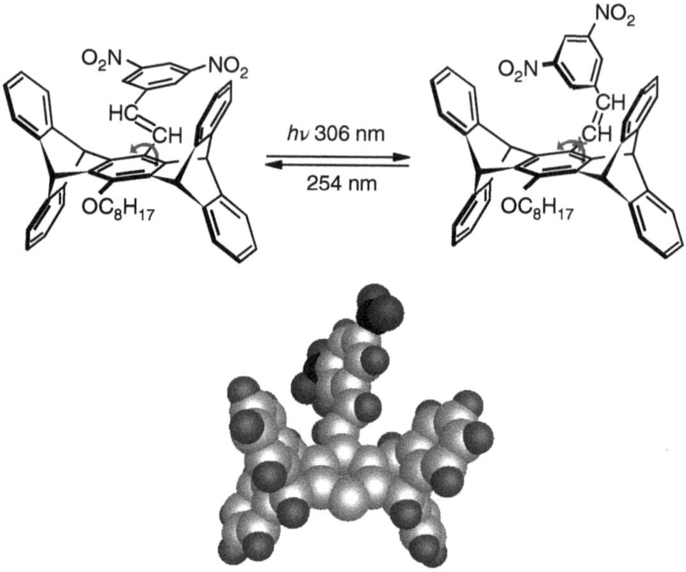

Figure 3.36 An example of the pentiptycene-based light-controlled molecular brake with the stilbene controller (top) and X-ray structure of the brake in "off" position (bottom, OC_8H_{17} fragment was removed for clarity).

The best light switching efficiency (up to 46% net difference in isomeric composition) was attained with bulky alkyl groups, such as *i*-Pr and *t*-Bu [55].

Electrochemical controllers were also successfully used in various molecular brake systems [56, 57]. One of them is presented in Scheme 3.8 [56]. In "on" position (Scheme 3.8), the rotation of triptycene gears proceeds with a significant energy barrier, in excess of 20 kcal/mol. Electrochemical oxidation of the tetrathiafulvalene vinylogue control unit (Scheme 3.8) reduces the energy barrier and allows fast rotation of triptycene gearing elements at room temperature on the NMR time scale [56].

Pentiptycene-based molecular brakes with dual mode controllers (redox and light) are also known [58, 59]. One of them is presented in Scheme 3.9 [58]. The photochemical *E* to *Z* (brake "off" to "on") switching in this system is essentially irreversible, so the brake release would have to be performed by different means – in this case electrochemically. The braking power of the controller is relatively small (only about 500) due to a limited steric change during the switching, but the dual switching mode (photochemical–electrochemical) provides an excellent switching efficiency of about 85% [58]. The incorporation of a chiral center into the control unit raises a possibility of unidirectional rotation, thus potentially converting this molecular brake system into a molecular motor (see also Section 3.6) [60].

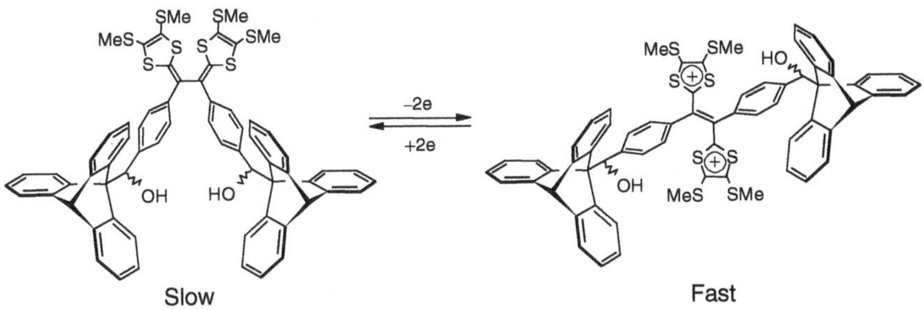

Scheme 3.8 The example of the molecular brake with the tetrathiafulvalene vinylogue redox controller.

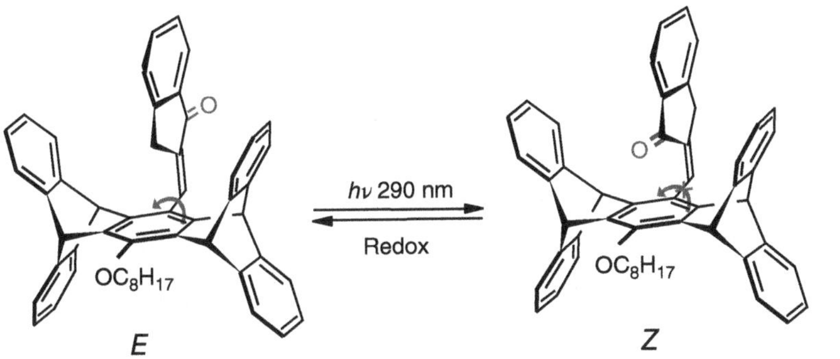

Scheme 3.9 The example of the pentiptycene-based molecular brake with the dual mode (photochemical–electrochemical) controller.

More complex "ABS" antilock molecular braking systems with two independent control units were reported recently [61, 62]. One of the examples of the system with ABS "turn-off" fluorescent alarm functionality is presented in Scheme 3.10 [62]. It employs a stilbene-type controller as a main light-activated switch and a pH-dependent secondary switch, which in an engaged (nonprotonated) position interacts with the fluorescent stilbene moiety via photoinduced electron transfer (PET) and slows down the braking process. Protonation of the secondary switch disrupts the PET and speeds up the braking process. The free (nonengaged) stilbene control unit also produces enhanced fluorescence as an ABS turn-off signal.

Very complex molecular gearing systems with rotation control are also known. Some examples of such systems with four circular intermeshed triptycene gears attached to a dirhodium metal oxide cluster are presented in Figure 3.37. In these systems, the energy barrier for rotation can be controlled by a proper ligand choice. The highest energy barriers are observed for bulky ligands, such as 2-ethylpyridine (18.0 kcal/mol), 2-methylpyridine (14.5 kcal/mol), and ethyl ether (11.4 kcal/mol). Compact ligands, such as

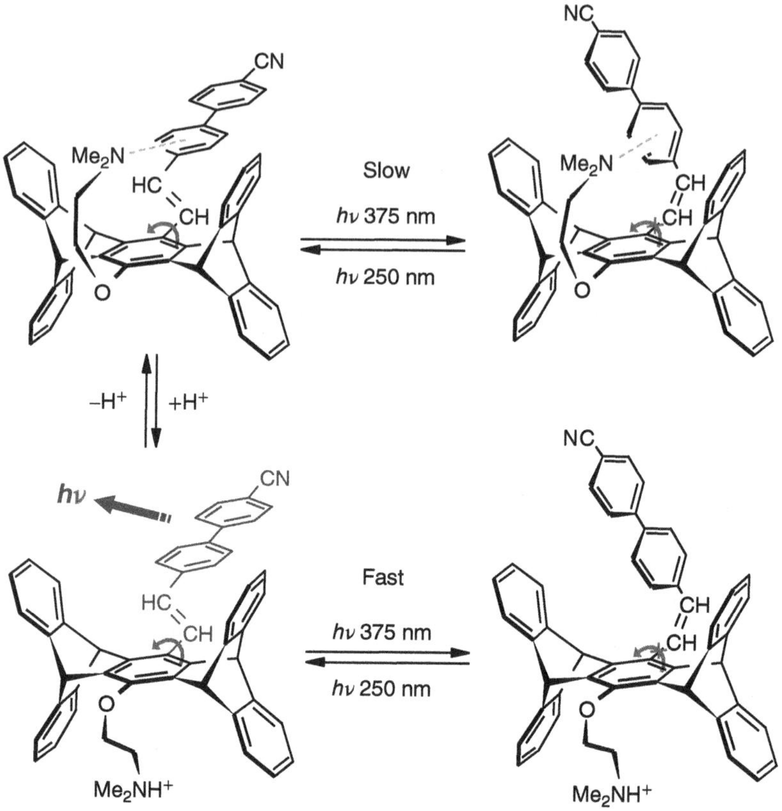

Scheme 3.10 The example of the ABS antilock molecular braking system with the ABS turn-off fluorescent alarm functionality.

pyridine, 4-dimethylaminopyridine, and methyl isonicotinate, allow for an almost barrier-free rotation on the NMR time scale. No attempts were made to determine experimentally the level of presumable correlated rotation of the triptycene gearing elements in these systems [44].

The preponderance of the molecular gearing systems with brake control units described earlier lacks directional rotation control, which can be implemented with the use of chiral controllers. This will essentially convert them into molecular ratchets and motors presented in Section 3.6.

3.6 Molecular Ratchets and Motors

Molecular ratchets represent a very special case of molecular gearing systems, and undoubtedly deserve a special attention. As a conventional mechanical

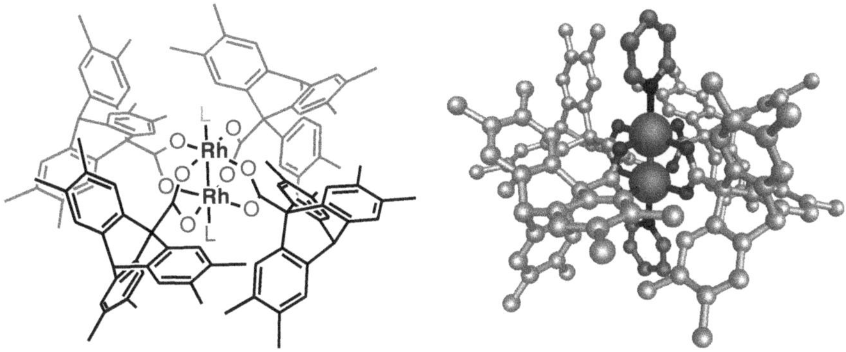

Figure 3.37 The examples of the complex molecular gearing systems with the ligand-based rotation control (left, L = Py, 2-MePy, 2-EtPy, 4-Me_2NPy, 4-MeO_2CPy, Et_2O) and X-ray structure of one of these systems (right, L = Py, hydrogen atoms were removed for clarity).

device, a simple ratchet has two fundamental elements: a ratchet gear and a pawl (Figure 3.38). The asymmetry of the ratchet gear teeth allows for only a unidirectional movement of the gear. So far, only a few unsuccessful attempts were made to create a technomimetic version of a conventional ratchet [60, 63]. These failures highlight fundamental differences between the conventional and molecular engineering: unlike mechanical ratchets, simple molecular ratchets are impossible to construct because this would violate the Second Law of Thermodynamics [3, 5], even though there are several ways to get around some of the limitations in more complex ratchet-type systems [4].

An early design of a model molecular ratchet employs triptycene as a ratchet gearing element and [4]helicene as an asymmetrical pawl element. There is no dedicated molecular spring mechanism forcing the pawl against the ratchet gear, even though elastic triptycene–helicene junction bond could be considered as a rudimental spring element. The synthesis of the compound has been

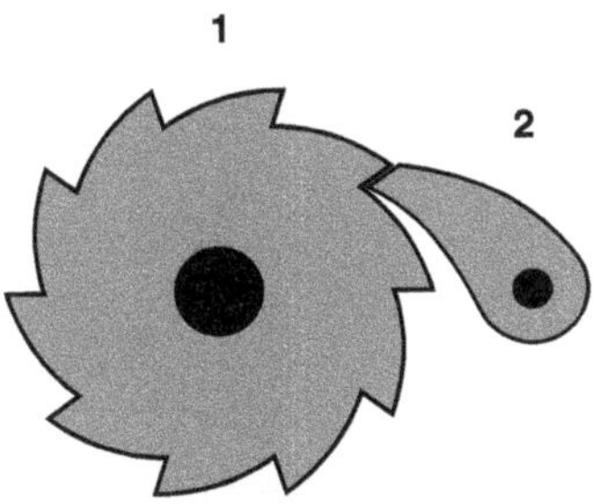

Figure 3.38 A schematic representation of a simple mechanical ratchet: 1 – a ratchet gear, 2 – a pawl. *Source*: Adapted from: "Ratchet Drawing" by Dr. Schorsch, Wikimedia.

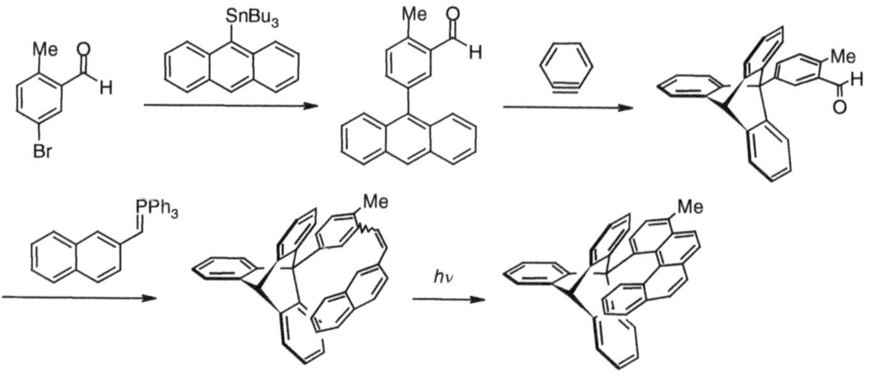

Scheme 3.11 Synthesis of triptycyl[4]helicene as a model molecular ratchet system.

accomplished in four steps starting from 2-methyl-5-bromobenzaldehyde (Scheme 3.11) [63, 64].

Molecular mechanics calculations clearly indicate that the energy diagram for rotation of the triptycene gear in this model molecular ratchet system is asymmetric. Clockwise rotation of the triptycene "cogwheel" forces the helicene "pawl" further and further out, until the maximum is reached and subsequently abruptly resolved when one of the triptycene "teeth" gets out of contact with the helicene pawl (Figure 3.39) [64].

It would be logical to expect that the clockwise rotation of the triptycene gear could be achieved much more easily than the counterclockwise rotation due to the much steeper energy barrier slope for the counterclockwise motion. Experimental evaluation of this system is performed by spin polarization NMR, which is capable of distinguishing between unidirectional and bidirectional rotations. These experiments unequivocally indicate that, contrary to

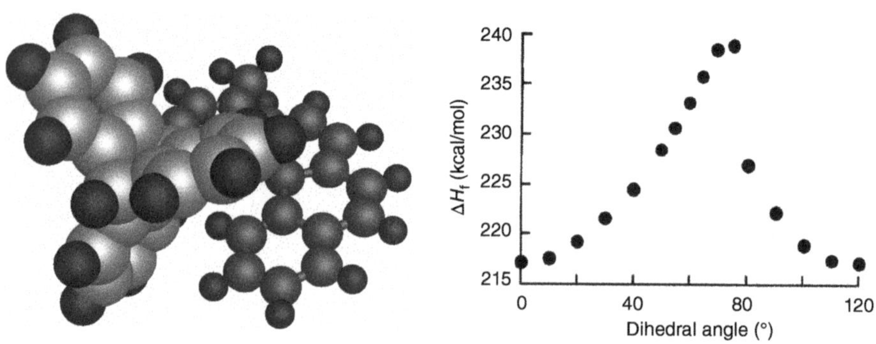

Figure 3.39 Energy-minimized structure of triptycyl[4]helicene (left) and calculated energy of rotation around the triptycene–helicene bond (right). *Source*: Kelly *et al.* 1998 [64]. Reproduced with permission of American Chemical Society.

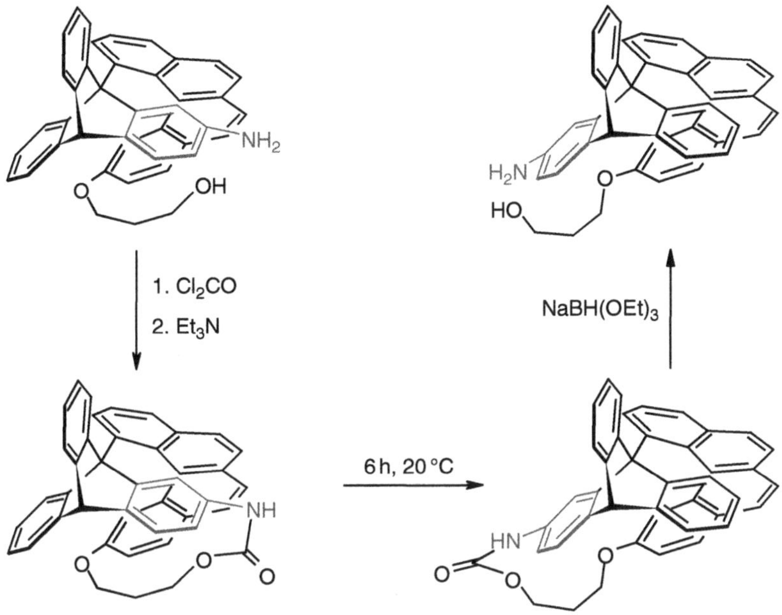

Scheme 3.12 Unidirectional rotation in the molecular rotary system utilizing the energy of a chemical reaction.

conventional mechanical systems, the triptycene gear rotates in both directions equally. Unidirectional rotation of the triptycene gear requires an external source of energy. This was eventually accomplished with the construction of a modified system presented in Scheme 3.12 [65].

This molecular gearing system with chemical energy-driven unidirectional rotation represents a rudimental model of a rotary chemical motor, even though it can only perform a 120° rotation [65, 66]. Nevertheless, it serves as a prototype of more advanced molecular rotary devices [7, 67], utilizing different sources of energy, such as light [68] or electric field [69].

3.7 Applications

Although molecular gearing systems did not yet deliver any practical applications employing their "classic" mode of action (i.e., by transmission of a mechanical torque or by synchronized motion of coupled gearing elements), unconventional applications already appeared in the literature. One of the applications involves a design of a new type of chemical sensors based on cooperative binding. It utilizes a simple gear correlation effect that can be achieved using two uncoupled identical gearing elements: if two opposite "teeth" are aligned, the positions of the other teeth elements should be aligned

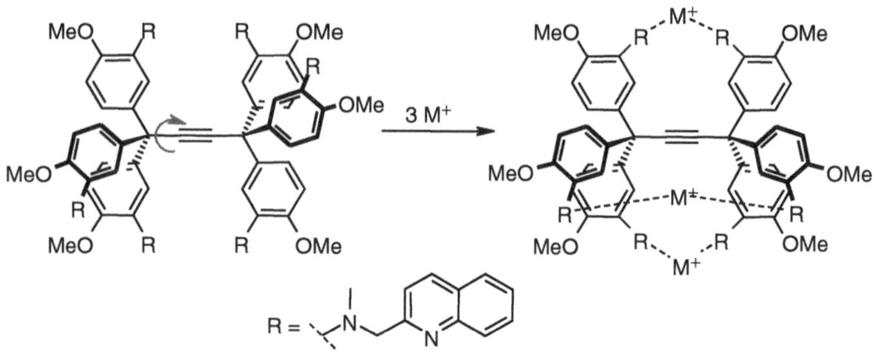

Scheme 3.13 The chemical sensing system utilizing cooperative binding between the trityl gearing groups ($M^+ = Ag^+$).

too. This design idea was successfully used to construct the so-called "pinwheel receptors" [70]. The structure of these receptors is fairly simple. They contain two trityl gearing elements attached to an acetylene "pin", which provides an appropriate face-to-face alignment of the gearing elements. Each tooth of the gearing elements has a binding site for a metal cation, which also serves as a fluorescent detection element. Complexation with a metal cation proceeds in such a way that two binding sites from the opposite gears are involved in the binding (Scheme 3.13). In the absence of metal cations, trityl groups exhibit free, uncoordinated rotation around the pin. First binding with a metal cation brings this uncoordinated motion to a halt, since two trityl gearing elements are now linked together via a metal cation. The loss of entropy, as well as unfavorable steric interactions, makes the first binding event less efficient. However, the second and the third binding events occur in the already preorganized system, and these binding events are much more efficient. The final result is cooperative binding, when the binding constants for each subsequent event are progressively increasing. Typically, this cooperative binding mechanism leads to a very sharp transition from the nonbonded to the fully bonded receptor. As a result, these fluorescent sensors offer a better sensitivity compared to traditional chemosensors [71].

An advanced version of this pinwheel receptor utilizes only two teeth of each trityl gear for the binding, while the third tooth is used to support a fluorescent probe. Although more complicated in design, such systems can provide better read-out properties due to separation of binding and fluorescent elements (Scheme 3.14) [72].

Pinwheel receptors were also adapted to accommodate more complex organic ions. An example of one of these chemosensors designed to detect dicarboxylates is presented in Scheme 3.15. In this system, guanidinium groups serve as carboxylate-specific binding elements. The system also features a

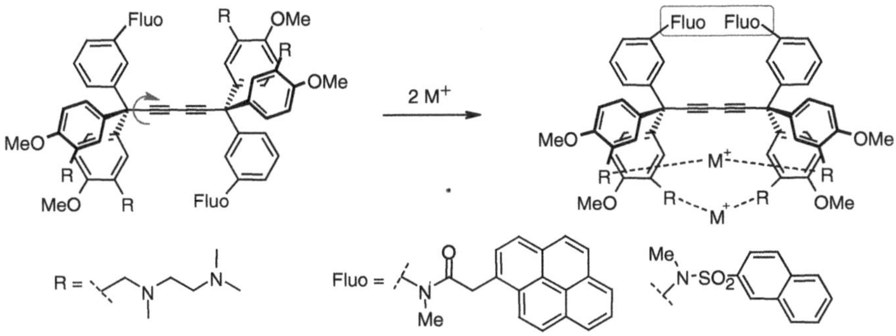

Scheme 3.14 The advanced pinwheel receptors with separate binding and fluorescent elements ($M^+ = Ag^+$).

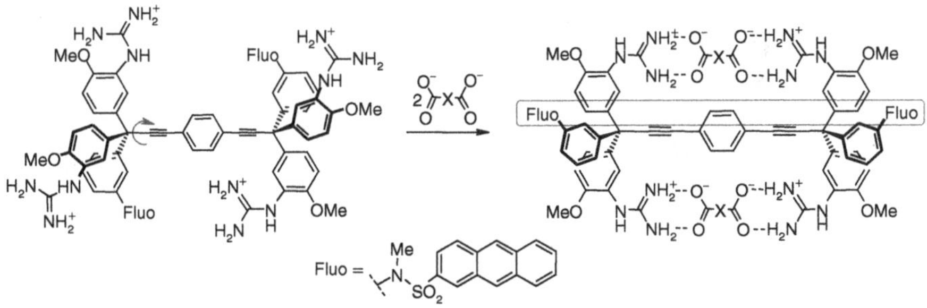

Scheme 3.15 Pinwheel chemosensor for the detection of dicarboxylates in water (X = none, CH_2, CH_2CH_2, *cis*-CH=CH, *trans*-CH=CH, *o*-C_6H_4, etc.).

longer pin, which is required due to a relatively large size of the dicarboxylate dianions. This chemosensor has good selectivity and can differentiate mono- and dicarboxylates. The best binding affinity was achieved with phthalate dianion. In this particular case, the binding is facilitated by a relatively high hydrophobicity of the phthalate dianion [73].

References

1 Gakh, A.A., Sachleben, R.A., and Bryan, J.C. (1997) Molecular gearing systems. *Chemtech*, **27**, 26–33.
2 Mislow, K. (1989) Molecular machinery in organic chemistry. *Chemtracts Org. Chem.*, **2**, 151–174.
3 Musser, G. (1999) Taming Maxwell's demon. *Sci. Am.*, **280**, 24.
4 Erbas-Cakmak, S., Leigh, D.A., McTernan, C.T., and Nussbaumer, A.L. (2015) Artificial molecular machines. *Chem. Rev.*, **115**, 10081–10206.

5 Kelly, T.R. and Sestelo, J.P. (2001) Rotary motion in single-molecule machines, in *Molecular Machines and Motors*, Structure and Bonding (ed. J.-P. Sauvage), Springer, Berlin, Heidelberg, pp. 19–53.
6 Kelly, T.R. (2001) Progress toward a rationally designed molecular motor. *Acc. Chem. Res.*, **34**, 514–522.
7 Kottas, G.S., Clarke, L.I., Horinek, D., and Michl, J. (2005) Artificial molecular rotors. *Chem. Rev.*, **105**, 1281–1376.
8 Schnepp, O. and McClure, D.S. (1957) Near ultraviolet spectrum of crystalline hexamethyl benzene and the structure of the hexamethyl benzene molecule. *J. Chem. Phys.*, **26**, 83–92.
9 Schnepp, O. (1958) Changes in the near ultraviolet spectrum of crystalline hexamethylbenzene accompanying a low temperature phase transition. *J. Chem. Phys.*, **29**, 56–61.
10 Schnepp, O. (2004) Low-temperature phase transition of crystalline hexamethylbenzene. *II. J. Chem. Phys.*, **30**, 48–52.
11 Iroff, L.D. (1980) On the gearing of methyl groups in hexamethylbenzene. *J. Comput. Chem.*, **1**, 76–80.
12 Metzger, J.V., Chanon, M.C., and Roussel, C.M. (1996) The gear effect: self organization of structural fragments induced by release of nonbonded interactions – consequences and reactivity. Part I. History, definition of gear effect, thiazoline-2-thione model, triptycene derivatives, in *Reviews on Heteroatom Chemistry*, vol. **15**, MYU, pp. 161–202.
13 Metzger, J.V., Chanon, M.C., and Roussel, C.M. (1997) The gear effect: self organization of structural fragments induced by release of nonbonded interactions – consequences and reactivity. Part II. Other models, alkyl substituted ethylenes, arenes and heteroarenes, incidence on chemical reactivity, creation of gear effect, in *Reviews on Heteroatom Chemistry*, vol. **16**, MYU, pp. 1–45.
14 Hounshell, W.D., Iroff, L.D., Wroczynski, R.J., and Mislow, K. (1978) Correlated rotation of *tert*-butyl groups in tri-*tert*-butylsilane. *J. Am. Chem. Soc.*, **100**, 5212–5213.
15 Iwamura, H. and Mislow, K. (1988) Stereochemical consequences of dynamic gearing. *Acc. Chem. Res.*, **21**, 175–182.
16 Chen, C.-F. and Ma, Y.-X. (2013) Iptycenes and their derivatives in molecular machines, in *Iptycenes Chemistry: From Synthesis to Applications*, Springer, Berlin, Heidelberg, pp. 211–230.
17 Kawada, Y. and Iwamura, H. (1983) Correlated rotation in bis(9-triptycyl)methanes and bis(9-triptycyl) ethers. Separation and interconversion of the phase isomers of labeled bevel gears. *J. Am. Chem. Soc.*, **105**, 1449–1459.
18 Johnson, C.A., Guenzi, A., Nachbar, R.B. *et al.* (1982) Crystal and molecular structure of bis(9-triptycyl) ketone and bis(9-triptycyl)methane. *J. Am. Chem. Soc.*, **104**, 5163–5168.

19 Guenzi, A., Johnson, C.A., Cozzi, F., and Mislow, K. (1983) Dynamic gearing and residual stereoisomerism in labeled bis(9-triptycyl)methane and related molecules. Synthesis and stereochemistry of bis(2,3-dimethyl-9-triptycyl)methane, bis(2,3-dimethyl-9-triptycyl)carbinol, and bis(1,4-dimethyl-9-triptycyl)methane. *J. Am. Chem. Soc.*, **105**, 1438–1448.
20 Kawada, Y., Yamazaki, H., Koga, G. *et al.* (1986) Bis(9-triptycyl)amines, a missing link between the corresponding methanes and ethers. An unconventional synthesis and influence of nitrogen configurational inversion on the coupled disrotatory trajectory. *J. Org. Chem.*, **51**, 1472–1477.
21 Kawada, Y., Sakai, H., Oguri, M., and Koga, G. (1994) Preparation of and dynamic gearing in *cis*-1,2-bis(9-triptycyl)ethylene. *Tetrahedron Lett.*, **35**, 139–142.
22 Kawada, Y., Ishikawa, J., Yamazaki, H. *et al.* (1987) Correlation of the two torsional degrees of freedom about the bonds connecting the bridgehead carbons to the sulfur atom in bis(9-triptycyl) sulfide. *Tetrahedron Lett.*, **28**, 445–448.
23 Kawada, Y., Kimura, Y., Yamazaki, H. *et al.* (1994) Further studies on the dependence of the barriers for gear slippage on the joint group in bis(9-triptycyl)X type molecules. Bis(9-triptycyl)phosphine, the missing link in the series along the third row of the periodic table. *Chem. Lett*, **23**, 1311–1314.
24 Iwamura, H., Ito, T., Ito, H. *et al.* (1984) Crystal and molecular structure of bis(9-triptycyl) ether. *J. Am. Chem. Soc.*, **106**, 4712–4717.
25 Kawada, Y. and Iwamura, H. (1981) Phase isomerism in gear-shaped molecules. *Tetrahedron Lett.*, **22**, 1533–1536.
26 Frantz, D.K., Baldridge, K.K., and Siegel, J.S. (2009) Application of structural principles to the design of triptycene-based molecular gears with parallel axes. *Chim. Int. J. Chem.*, **63**, 201–204.
27 Setaka, W., Nirengi, T., Kabuto, C., and Kira, M. (2008) Introduction of clutch function into a molecular gear system by silane-silicate interconversion. *J. Am. Chem. Soc.*, **130**, 15762–15763.
28 Koga, N., Kawada, Y., and Iwamura, H. (1983) Recognition of the phase relationship between remote substituents in 9,10-bis(3-chloro-9-triptycyloxy)triptycene molecules undergoing rapid internal rotation cooperatively. *J. Am. Chem. Soc.*, **105**, 5498–5499.
29 Bryan, J.C., Sachleben, R.A., Gakh, A.A., and Bunick, G.J. (1999) Molecular gears: structures of (9,10-triptyceno) crown ethers. *J. Chem. Crystallogr.*, **29**, 513–521.
30 Toyota, S., Shimizu, T., Iwanaga, T., and Wakamatsu, K. (2011) Structures and conformational analysis of 1,8-bis(9-triptycylethynyl)anthracene and its derivatives as prototypes of molecular spur gears. *Chem. Lett.*, **40**, 312–314.

31 Frantz, D.K., Linden, A., Baldridge, K.K., and Siegel, J.S. (2012) Molecular spur gears comprising triptycene rotators and bibenzimidazole-based stators. *J. Am. Chem. Soc.*, **134**, 1528–1535.
32 Nikitin, K., Müller-Bunz, H., Ortin, Y. *et al.* (2008) Twin triptycyl spinning tops: a simple case of molecular gearing with dynamic C_2 symmetry. *Eur. J. Org. Chem.*, **2008**, 3079–3084.
33 Kao, C.-Y., Hsu, Y.-T., Lu, H.-F. *et al.* (2011) Toward a four-toothed molecular bevel gear with C_2-symmetrical rotors. *J. Org. Chem.*, **76**, 5782–5792.
34 Wang, G., Ma, L., Xiang, J. *et al.* (2015) 2,6-Pyridodicarboxamide-bridged triptycene molecular transmission devices: converting rotation to rocking vibration. *J. Org. Chem.*, **80**, 11302–11312.
35 Hunter, C.A. and Sanders, J.K.M. (1990) The nature of π–π interactions. *J. Am. Chem. Soc.*, **112**, 5525–5534.
36 Bulo, R.E., Allaart, F., Ehlers, A.W. *et al.* (2006) Circumambulatory rearrangement with characteristics of a 2:1 covalent molecular bevel gear. *J. Am. Chem. Soc.*, **128**, 12169–12173.
37 Stevens, A.M. and Richards, C.J. (1997) A metallocene molecular gear. *Tetrahedron Lett.*, **38**, 7805–7808.
38 Nikitin, K., Müller-Bunz, H., Ortin, Y. *et al.* (2010) Molecular dials: hindered rotations in mono- and diferrocenyl anthracenes and triptycenes. *J. Am. Chem. Soc.*, **132**, 17617–17622.
39 Ogi, S., Ikeda, T., Wakabayashi, R. *et al.* (2010) A bevel-gear-shaped rotor bearing a double-decker porphyrin complex. *Chem. Eur. J.*, **16**, 8285–8290.
40 Ogi, S., Ikeda, T., and Takeuchi, M. (2012) Synthetic molecular gear based on double-decker porphyrin complexes. *J. Inorg. Organomet. Polym. Mater.*, **23**, 193–199.
41 Ikeda, M., Takeuchi, M., Sugasaki, A. *et al.* (2000) Strong positive allosterism which appears in molecular recognition with cerium(IV) double decker porphyrins: correlation between the number of binding sites and hill coefficients. *Supramol. Chem.*, **12**, 321–345.
42 Ogi, S., Ikeda, T., Wakabayashi, R. *et al.* (2011) Mechanically interlocked porphyrin gears propagating two different rotational frequencies. *Eur. J. Org. Chem.*, **2011**, 1831–1836.
43 Nakamura, M., Kishimoto, K., Kobori, Y. *et al.* (2016) Self-assembled molecular gear: a 4:1 complex of Rh(III)Cl tetraarylporphyrin and tetra(*p*-pyridyl)cavitand. *J. Am. Chem. Soc.*, **138**, 12564–12577.
44 Sanada, K., Ube, H., and Shionoya, M. (2016) Rotational control of a dirhodium-centered supramolecular four-gear system by ligand exchange. *J. Am. Chem. Soc.*, **138**, 2945–2948.
45 Hiraoka, S., Okuno, E., Tanaka, T. *et al.* (2008) Ranging correlated motion (1.5 nm) of two coaxially arranged rotors mediated by helix inversion of a supramolecular transmitter. *J. Am. Chem. Soc.*, **130**, 9089–9098.

46 Liu, S., Kondratuk, D.V., Rousseaux, S.A.L. *et al.* (2015) Caterpillar track complexes in template-directed synthesis and correlated molecular motion. *Angew. Chem. Int. Ed.*, **54**, 5355–5359.
47 Vögtle, F. and Mew, P.K.T. (1978) Triptycenophanes—synthesis via "double nonbenzylic sulfone pyrolysis". *Angew. Chem. Int. Ed. Engl.*, **17**, 60–62.
48 Gakh, A.A., Sachleben, R.A., Bryan, J.C., and Moyer, B.A. (1995) A facile synthesis and X-ray structure determination of the first triptycenocrown ethers. *Tetrahedron Lett.*, **36**, 8163–8166.
49 Bryan, J.C., Gakh, A.A., and Sachleben, R.A. (2002) (9,10-Triptyceno)benzo-28-crown-8. *Acta Crystallogr. Sect. E: Struct. Rep. Online*, **58**, o129–o131.
50 Dougherty, D.A. (1996) Cation-π interactions in chemistry and biology: a new view of benzene, Phe, Tyr, and Trp. *Science*, **271**, 163–168.
51 Kelly, T.R., Bowyer, M.C., Bhaskar, K.V. *et al.* (1994) A molecular brake. *J. Am. Chem. Soc.*, **116**, 3657–3658.
52 Nikitin, K., Müller-Bunz, H., Ortin, Y., and McGlinchey, M.J. (2009) A molecular paddlewheel with a sliding organometallic latch: syntheses, X-ray crystal structures and dynamic behaviour of [$Cr(CO)_3$\{η^6-2-(9-triptycyl)indene\}], and of [$M(CO)_3$\{η^5-2-(9-triptycyl)indenyl\}] (M=Mn, Re). *Chem. Eur. J.*, **15**, 1836–1843.
53 Tan, W.S., Chuang, P.-Y., Chen, C.-H. *et al.* (2015) Light-gated molecular brakes based on pentiptycene-incorporated azobenzenes. *Chem. Asian J.*, **10**, 989–997.
54 Yang, J.-S., Huang, Y.-T., Ho, J.-H. *et al.* (2008) A pentiptycene-derived light-driven molecular brake. *Org. Lett.*, **10**, 2279–2282.
55 Sun, W.-T., Huang, Y.-T., Huang, G.-J. *et al.* (2010) Pentiptycene-derived light-driven molecular brakes: substituent effects of the brake component. *Chem. Eur. J.*, **16**, 11594–11604.
56 Chen, G. and Zhao, Y. (2014) Redox-regulated rotary motion of a bis(9-triptycyl)-TTFV system. *Org. Lett.*, **16**, 668–671.
57 Yang, C.-H., Prabhakar, C., Huang, S.-L. *et al.* (2011) A redox-gated slow-fast-stop molecular rotor. *Org. Lett.*, **13**, 5632–5635.
58 Chen, Y.-C., Sun, W.-T., Lu, H.-F. *et al.* (2011) A pentiptycene-derived molecular brake: photochemical $E \rightarrow Z$ and electrochemical $Z \rightarrow E$ switching of an enone module. *Chem. Eur. J.*, **17**, 1193–1200.
59 Kao, C.-Y., Lee, I.-T., Prabhakar, C., and Yang, J.-S. (2014) Light- and redox-gated molecular brakes consisting of a pentiptycene rotor and an indole pad. *J. Chin. Chem. Soc.*, **61**, 507–516.
60 Kao, C.-Y., Lu, H.-F., Chao, I., and Yang, J.-S. (2014) A rotary molecular motor gated by electrical energy. *Org. Lett.*, **16**, 6100–6103.
61 Sun, W.-T., Huang, S.-L., Yao, H.-H. *et al.* (2012) An antilock molecular braking system. *Org. Lett.*, **14**, 4154–4157.

62 Sun, W.-T., Huang, G.-J., Huang, S.-L. *et al.* (2014) A light-gated molecular brake with antilock and fluorescence turn-on alarm functions: application of singlet-state adiabatic cis → trans photoisomerization. *J. Org. Chem.*, **79**, 6321–6325.

63 Kelly, T.R., Tellitu, I., and Sestelo, J.P. (1997) In search of molecular ratchets. *Angew. Chem. Int. Ed. Engl.*, **36**, 1866–1868.

64 Kelly, T.R., Sestelo, J.P., and Tellitu, I. (1998) New molecular devices: in search of a molecular ratchet. *J. Org. Chem.*, **63**, 3655–3665.

65 Kelly, T.R., Silva, R.A., Silva, H.D. *et al.* (2000) A rationally designed prototype of a molecular motor. *J. Am. Chem. Soc.*, **122**, 6935–6949.

66 Kelly, T.R., Cai, X., Damkaci, F. *et al.* (2007) Progress toward a rationally designed, chemically powered rotary molecular motor. *J. Am. Chem. Soc.*, **129**, 376–386.

67 Kay, E.R., Leigh, D.A., and Zerbetto, F. (2007) Synthetic molecular motors and mechanical machines. *Angew. Chem. Int. Ed.*, **46**, 72–191.

68 Koumura, N., Zijlstra, R.W.J., van Delden, R.A. *et al.* (1999) Light-driven monodirectional molecular rotor. *Nature*, **401**, 152–155.

69 Tierney, H.L., Murphy, C.J., Jewell, A.D. *et al.* (2011) Experimental demonstration of a single-molecule electric motor. *Nat. Nanotechnol.*, **6**, 625–629.

70 Raker, J., Secor, K., Moran, R. *et al.* (2005) Design and synthesis of cooperative "pinwheel" fluorescent sensors, in *Strategies and Tactics in Organic Synthesis*, vol. **6** (ed. M. Harmata), Academic Press, pp. 395–414.

71 Glass, T.E. (2000) Cooperative chemical sensing with bis-tritylacetylenes: pinwheel receptors with metal ion recognition properties. *J. Am. Chem. Soc.*, **122**, 4522–4523.

72 Raker, J. and Glass, T.E. (2001) Cooperative ratiometric chemosensors: pinwheel receptors with an integrated fluorescence system. *J. Org. Chem.*, **66**, 6505–6512.

73 Raker, J. and Glass, T.E. (2002) Selectivity via cooperative interactions: detection of dicarboxylates in water by a pinwheel chemosensor. *J. Org. Chem.*, **67**, 6113–6116.

4

Molecular Belts and Tubes

CHAPTER MENU

4.1 Introduction

Belts and tubes are widely used in conventional engineering due to their rigidity and well-defined three-dimensional (3D) characteristics, mainly as construction and transmission elements. Long tubes are key components of modern technology for the transportation and handling of gas and liquid materials. Short tubes and belts with variable geometry are used as clamps, and extended belts are used as torque transmission elements. These are just a few application examples of belts and tubes as ubiquitous man-made devices (Figure 4.1).

The same primary uses can be envisioned for molecular belts and tubes, which are intriguing molecular objects due to their elegant 3D structures offering a unique set of physical and chemical properties. Their elaborate synthesis and rich coordination chemistry are frequently reported in the emerging scientific fields of molecular engineering, self-organization, and recognition. Besides their own application appeal, molecular belts and tubes can be further employed as building blocks for the fabrication of more complex technomimetic molecules. The unique structural features of tubular molecules offer a capability to transport internal payload while insulating it from the chemical environment. They also have a rare combination of relative across-axis flexibility combined with the relative along-axis stiffness, making them attractive components of advanced materials. Certain long aromatic

Molecular Devices: An Introduction to Technomimetics and Its Biological Applications,
First Edition. Andrei A. Gakh.

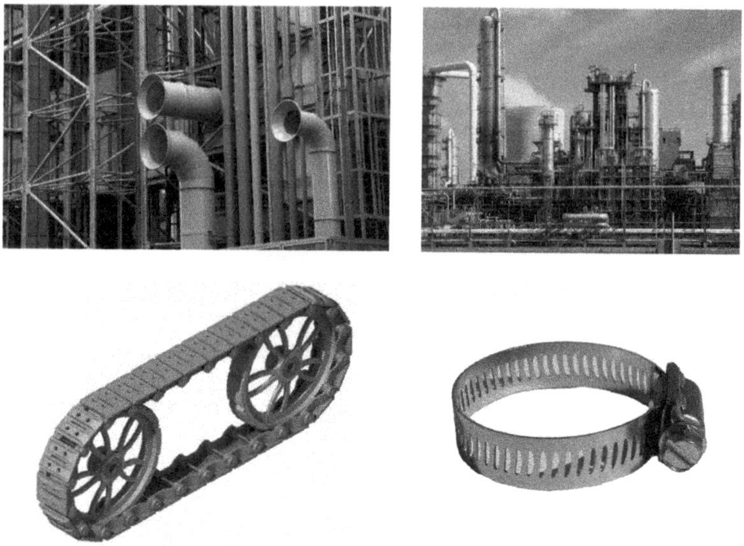

Figure 4.1 Examples of conventional man-made tubes and belts. *Source*: Courtesy of Tellepsen Corp., Robosoft Systems.

molecular tubes can also be considered as molecular wires for molecular electronic devices.

The rational synthesis of all-carbon molecular tubes (Figure 4.2) remains elusive. These technomimetic molecules can be produced by a variety of "physical" methods (such as chemical vapor deposition), but with little control over their structure and chirality. Only recently, template-directed self-assembly

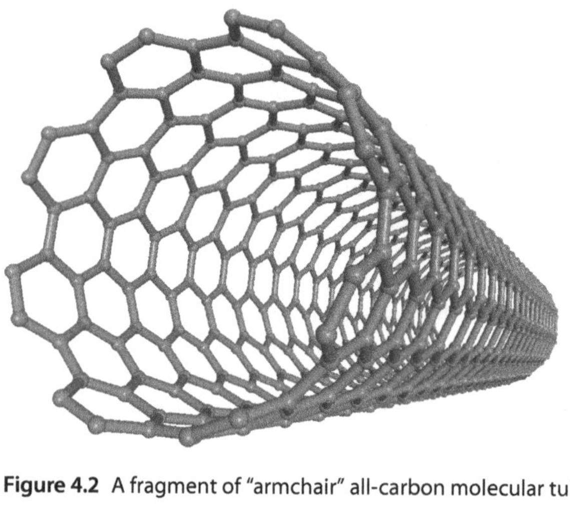

Figure 4.2 A fragment of "armchair" all-carbon molecular tube.

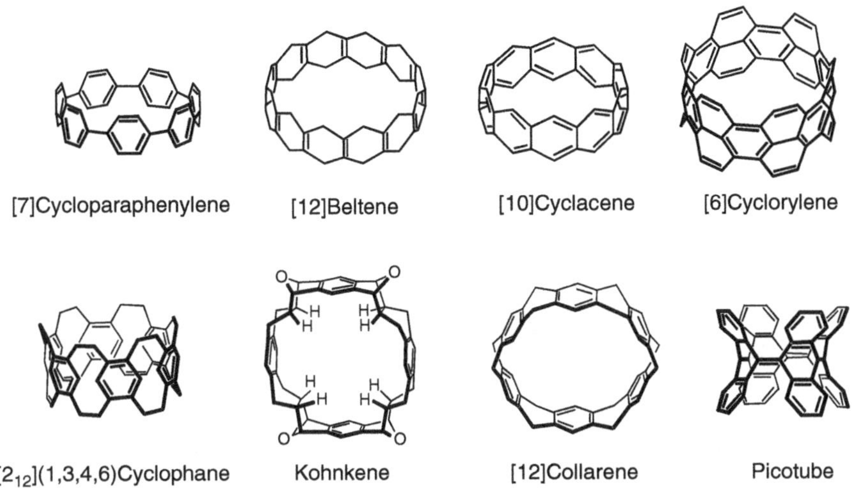

Figure 4.3 Examples of molecular belts and tubes.

of molecular tubes with precise predetermined structures became feasible (see Chapter 1).

So far, the portfolio of synthetic molecular belts and tubes consists mainly of molecular belts and short fragments of molecular tubes. They are accessible by a variety of modern synthetic methods. Several examples of belt- and tube-shaped molecules are shown in Figure 4.3 [1–11].

The "true" molecular belts can be identified by the presence of upper and lower edges. Examples of these true molecular belts include [*n*]beltenes, [*n*]cyclacenes, [*n*]cyclorylenes, $[2_n]$(1,3,4,6)cyclophanes, and [*n*]collarenes (Figure 4.3). The other molecular belts without clearly defined edges, such as [*n*]cycloparaphenylenes, as well as picotubes (Figure 4.3), can be considered as pseudobelts. In some long molecular belts, the distance between the top and the bottom edges can exceed the diameter of the belt. These extra wide belts can be considered as molecular tubes, even though this distinction is arbitrary. So far, the progress in the construction of molecular tubes has been inhibited by the complexity of synthetic procedures required for the fabrication of these sophisticated molecular objects, with a notable exception of cucurbit[*n*]urils (see Section 4.4), which can be produced quite efficiently via self-assembly methodology.

4.2 True Belts

While the construction of true molecular belts can technically be done using almost any linkable moieties, synthetic tasks can be greatly facilitated by the use of prefabricated molecular blocks with proper spatial characteristics.

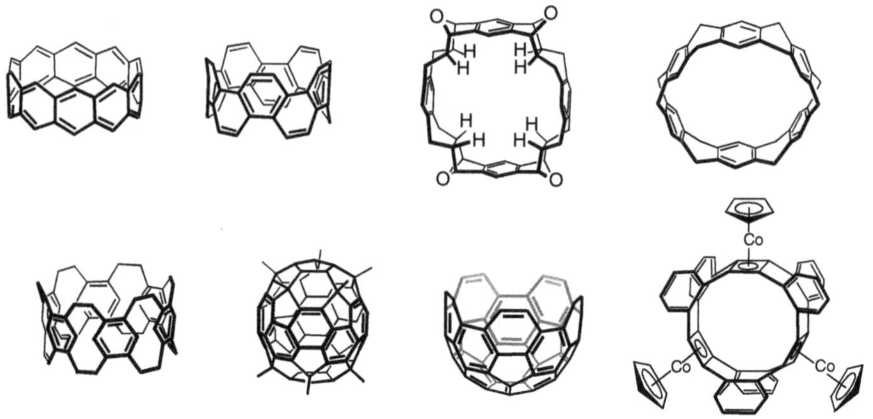

Figure 4.4 Examples of true molecular belts and tubes.

Not surprisingly, many true molecular belts were made from benzene units, fused together ([*n*]cyclacenes, [*n*]cyclophenacenes), interconnected with two aliphatic links ([3.3]*meta*cyclophanes, ([*n*]collarenes), or incorporated in a 3D fullerene-type network. Some examples of these true belts are presented in Figure 4.4.

It may appear that the simplest approach toward true molecular belts is cyclization of molecular ribbons made of properly aligned aromatic rings and fitted with appropriate functional end units. Unfortunately, this straightforward approach is technically the most challenging one since it requires tedious multistep fabrication of properly functionalized ribbon molecules. The task can be simplified by the use of "iterative" or "repetitive" synthetic methodology [12]. It is a powerful tool allowing semiautomation of reaction procedures for the preparation of complex precursor compounds (see Chapter 1).

Repetitive Diels–Alder [4 + 2] cycloaddition reactions are frequently used for the preparation of curved molecular ribbons due to high yield and good stereoselectivity, which are critical features for the success of multistep procedures involving simultaneous generation of several chiral centers. An outline of one of these synthetic strategies is presented in Scheme 4.1. This strategy is based on repetitive Diels–Alder reactions of suitable bis-dienes and bis-dienophiles until the "head bites the tail" of a growing curved ribbon strip. An attractive feature of this strategy is the ability to close the belt ring using the same set of consecutive Diels–Alder cycloaddition reactions used for the construction of the ribbon. The intermediate molecular ribbon should have an appropriate level of the curvature to allow the final ring closure [13, 14].

A feasibility of this approach was successfully demonstrated by an elaborate synthesis of poly-oxy molecular belts based on the cycloaddition reactions of 2,3,5,6-tetramethylene-7-oxabicyclo[2.2.1]heptane as bis-diene

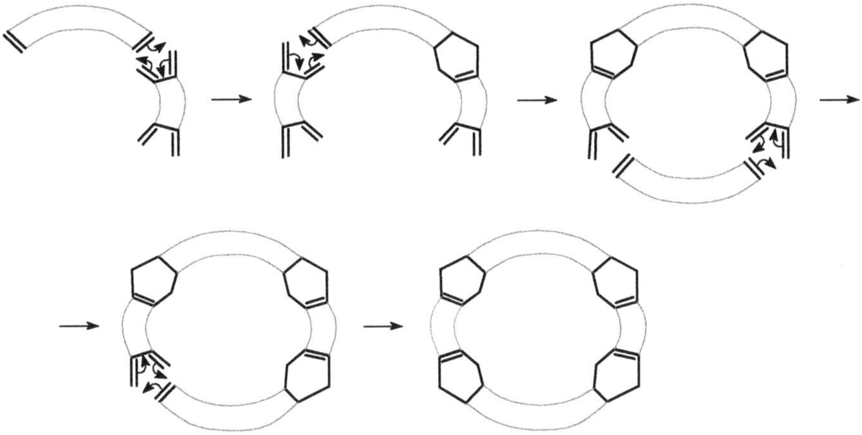

Scheme 4.1 The general strategy for the preparation of a molecular belt using repetitive Diels–Alder reactions between bis-dienes and bis-dienophiles [13].

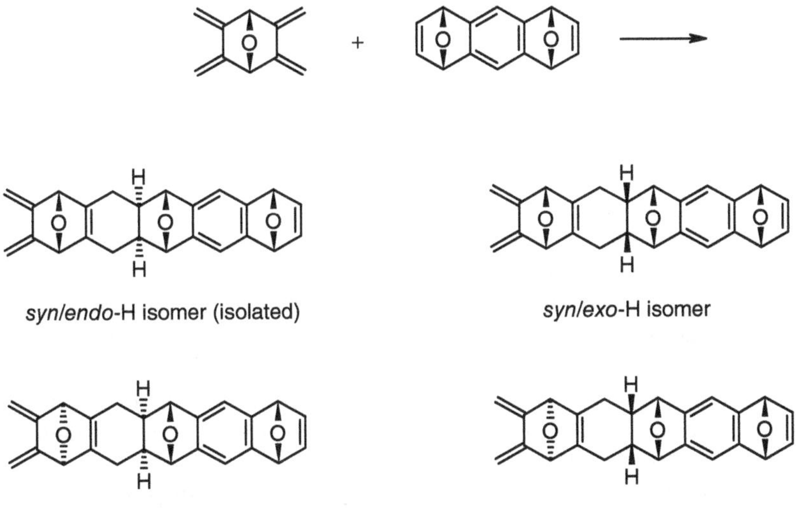

Scheme 4.2 Structures of four possible diastereoisomers that theoretically can be produced by the cycloaddition reaction of 2,3,5,6-tetramethylene-7-oxabicyclo[2.2.1]heptane with 1,4:5,8-diepoxy-l,4,5,8-tetrahydroanthracene.

with 1,4:5,8-diepoxy-l,4,5,8-tetrahydroanthracene as bis-dienophile. There are eight different ways to connect these two components together, but, due to the symmetry of the reacting molecules, only four diastereoisomeric 1:1 adducts can theoretically be obtained. In practice, however, only *syn/endo*-H isomer was isolated from this reaction (Scheme 4.2). Molecular modeling

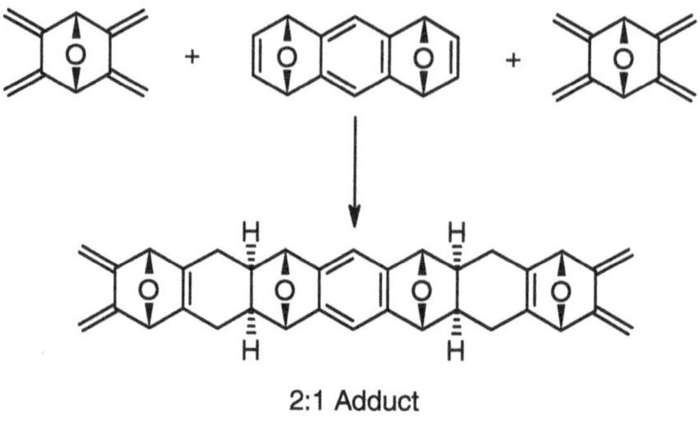

Scheme 4.3 The synthesis of the 2:1 Diels–Alder product.

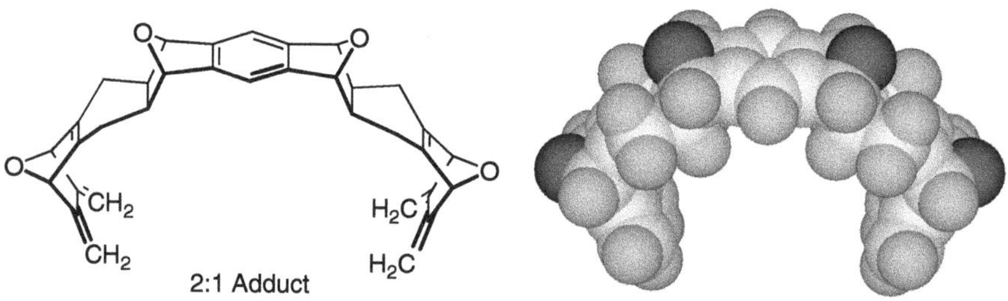

Figure 4.5 The 2:1 *syn/endo* Diels–Alder adduct as a curved ribbon molecule.

studies indicate that steric and stereoelectronic factors are responsible for the observed high diastereoselectivity of this reaction [13, 15].

The reaction sequence outlined earlier can be continued with the formation of the 2:1 product, which is actually coproduced together with the 1:1 product shown above in approximately 3:1 ratio (61% and 24%, respectively). Similar to the 1:1 product, the only one stereoisomer of the 2:1 product was isolated from the reaction mixture, highlighting the high stereoselectivity of this cycloaddition reaction (Scheme 4.3) [13, 15].

Due to the *syn/endo* configuration of the 2:1 Diels–Alder adduct, this ribbon molecule has adequate curvature. It can serve as a key precursor for subsequent synthesis of cyclic molecular belt compounds (Figure 4.5).

The ring closure of the 2:1 Diels–Alder adduct presented earlier is not very efficient and requires the use of high pressure and elevated temperatures. Even under these conditions, the yield of a [12]cyclacene derivative is only 20–36%, thus underscoring the difficulty of the cyclization process (Scheme 4.4). This compound can also be prepared directly one-pot from 2,3,5,6-tetramethylene-7-oxabicyclo[2.2.1]heptane and 1,4:5,8-diepoxy-1,4,5,8-tetrahydroanthracene mentioned earlier, albeit in a very low yield [15, 16]. It is worth noting here that the ring closure leading to

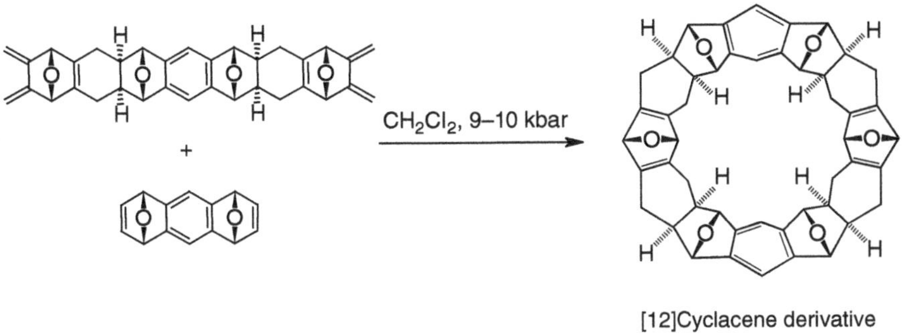

Scheme 4.4 The synthesis of the [12]cyclacene derivative.

the larger [14]cyclacene derivative proceeds in high yield and does not require high pressure conditions [16].

The single-crystal X-ray structure of this [12]cyclacene derivative, obtained from a chloroform solution, confirms the expected high symmetry of this elegant belt-like molecule with all six oxygen atoms pointing outward (Figure 4.6). Two benzene rings are found to be parallel and separated by about 7.9 Å. A clathrate-like inclusion of chloroform molecules between the layers of the molecular belts is also observed [15].

Two oxygen atoms can be removed from the [12]cyclacene derivative by reaction with $TiCl_4/LiAlH_4$ at ambient temperature (Scheme 4.5). This new compound has a rigid, almost square cross-section cavity, with a water molecule trapped inside (Figure 4.7) [17]. The removal of the remaining four oxygen atoms was achieved in a mixture of acetic anhydride and hydrochloric acid at elevated temperatures. The dehydration reaction was accompanied by acid-catalyzed isomerization leading to more thermodynamically stable unsymmetrical hydrocarbon with two naphthalene, one anthracene, and one benzene rings in 56% yield (Scheme 4.5). Birch reduction yielded symmetrical dodecahydro[12]cyclacene, which was named "[12]collarene" in the original

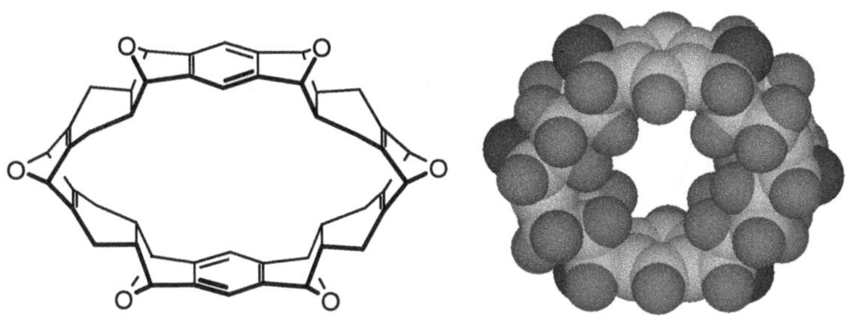

Figure 4.6 The [12]cyclacene derivative with six outward oxygen bridges (left) and its X-ray crystal structure (right).

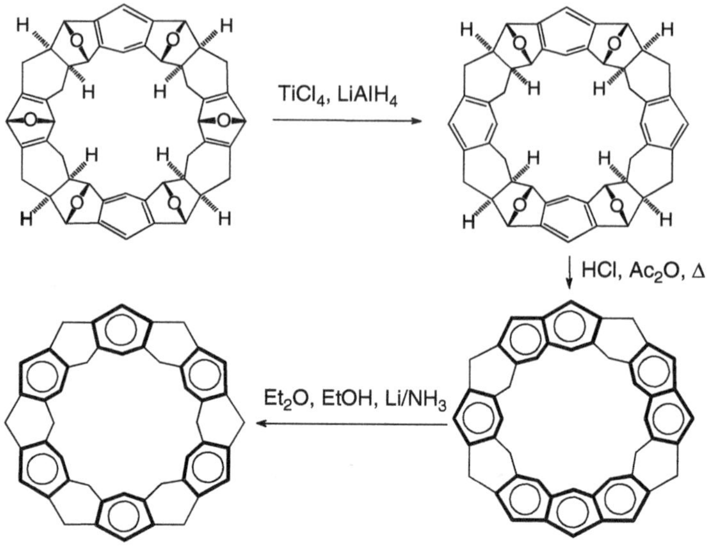

Scheme 4.5 The synthesis of [12]collarene.

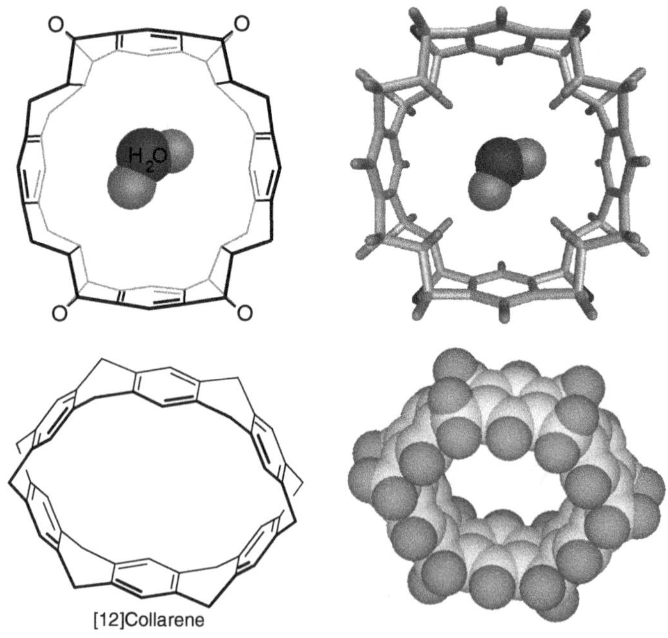

Figure 4.7 The X-ray structure of the [12]cyclacene derivative with four oxygen bridges (top) and the energy-minimized structure of [12]collarene (bottom). Water molecule is partially disordered in the crystal structure of the [12]cyclacene derivative (top).

publication (Scheme 4.5) [17]. This belt-like molecule has six isolated benzene rings and is presumably the most stable compound among other partially hydrogenated [12]cyclacenes. The parent compound, [12]cyclacene, remains elusive despite numerous attempts to make it [5].

The three-way interconnected belt molecules, which can be envisioned as junction elements between three interconnected molecular tubes, can be prepared from tris-dienophiles using the same synthetic methodology outlined earlier. The synthetic procedure entails the sequence of Diels–Alder reactions of the tris-dienophile and the bis-diene followed by cyclization reaction with one more molecule of tris-dienophile (Scheme 4.6). As in the case of the [12]cyclacene derivative presented earlier, the final cyclization reaction requires high pressure at elevated temperatures. Only one isomer of this "trinacrene" was isolated from the reaction mixture in a very low yield. The structure of trinacrene was determined by fast atom bombardment (FAB) mass spectrometry and ^{1}H NMR spectroscopy [18, 19].

The other all-carbon [12]cyclacene derivatives were successfully prepared with the use of pressure-induced Diels–Alder macrocyclization reactions.

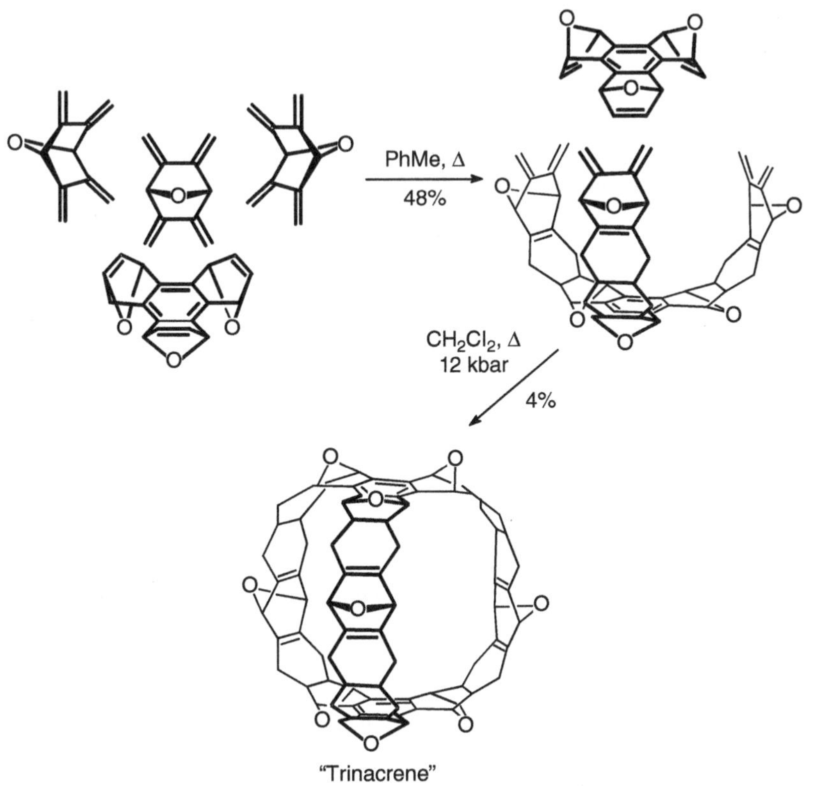

Scheme 4.6 The synthesis of trinacrene from bis-diene and tris-dienophile molecules.

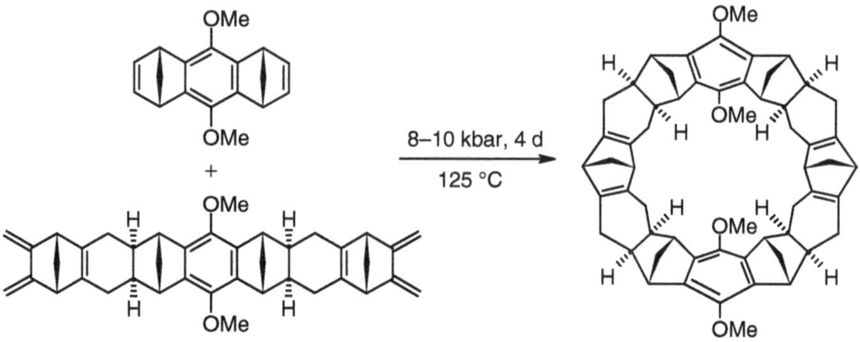

Scheme 4.7 The synthesis of the [12]cyclacene derivative with six methylene bridges.

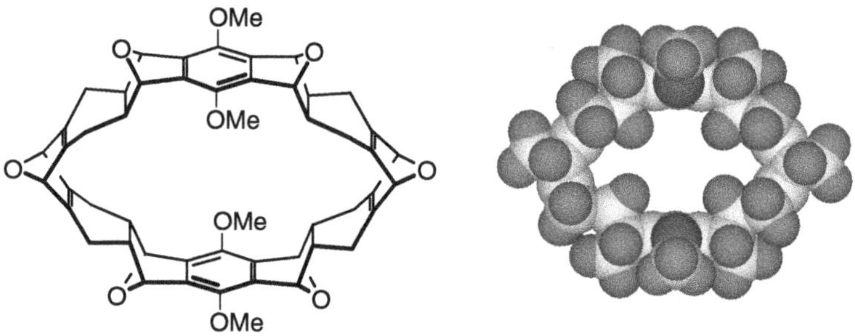

Figure 4.8 The energy-minimized structure of the [12]cyclacene derivative with six methylene bridges.

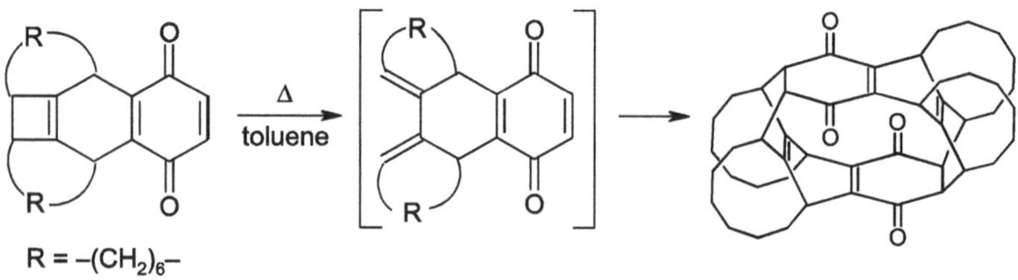

Scheme 4.8 The synthesis of the [6]beltene derivative via Diels–Alder dimerization.

One of the examples of substituted [12]cyclacenes with six methylene bridges (instead of six oxygen bridges) is presented in Scheme 4.7 and Figure 4.8 [20].

Similar Diels–Alder macrocyclization methodology was also successfully employed for the synthesis of a smaller [6]beltene derivative (Scheme 4.8). In this case, both bis-diene and bis-dienophile fragments belong to the same molecule, which typically results in the formation of large quantities of polymeric materials. The efficient synthesis of the dimeric [6]beltene derivative can be achieved in diluted solutions. In optimized condition, the yield of this dimer can be as high as 70%. The X-ray crystal structure analysis reveals that the molecule has somewhat flattened geometry with a relatively small internal cavity (Figure 4.9) [21].

Larger [8]cyclacene derivatives were similarly prepared by Diels–Alder cyclization of a tricyclic bis-dienophile and a tricyclic diene (Scheme 4.9). To suppress the formation of polymeric by-products, this "macroannulation" reaction was carried out via a simultaneous slow addition of bis-diene and bis-dienophile solutions into a reactor with boiling dioxane. In these conditions, the desired [8]cyclacene derivative was obtained in 69% yield. Attempts to convert this compound into the parent [8]cyclacene were not successful (Scheme 4.9) [22, 23].

Diels–Alder macrocyclization methodology was also successfully employed for the preparation of cyclododeciptycene derivatives by fusing two properly functionalized triptycene precursors (Figure 4.10) [24]. This

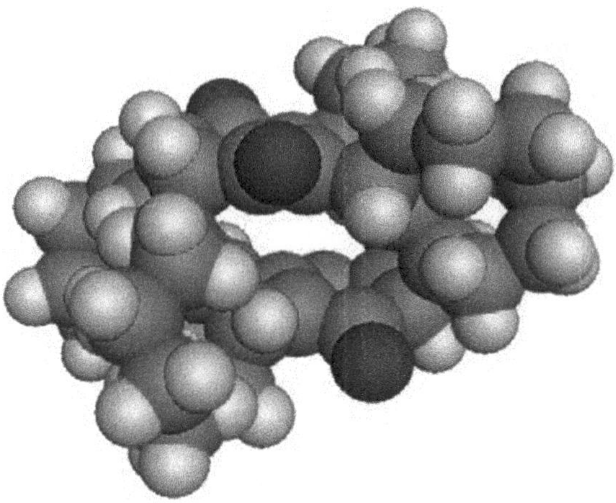

Figure 4.9 The X-ray structure of the [6]beltene derivative.

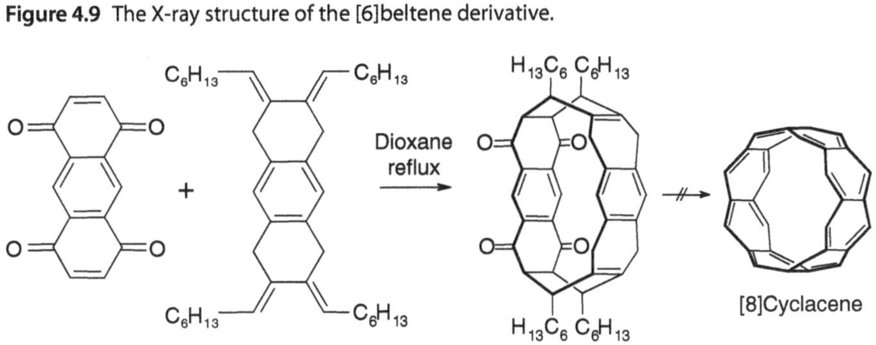

Scheme 4.9 The synthesis of the [8]cyclacene derivative.

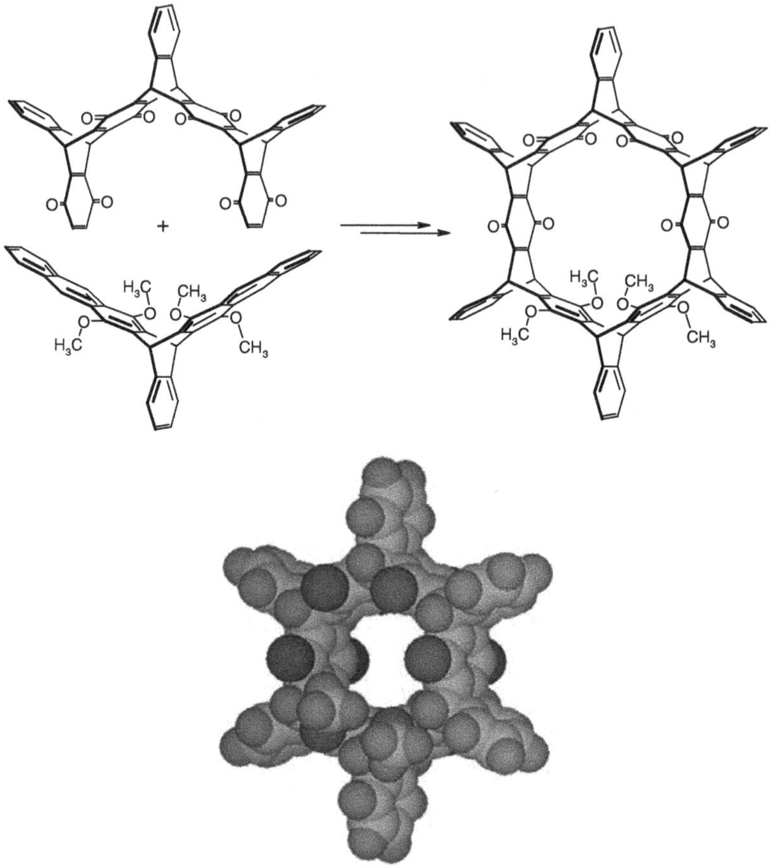

Figure 4.10 The synthesis of the cyclododeciptycene derivative (top) and its X-ray crystal structure (bottom).

cyclododeciptycene can be considered as a [12]cyclacene derivative with six exocyclic 1,2-phenylene fragments. Due to the specific gear-like symmetry and shape, this molecule can be useful for the construction of molecular gearing systems (see Chapter 3).

The wider "double-stranded" belt molecules containing fluoranthene units were similarly prepared using Diels–Alder dimerization of the appropriately functionalized monomeric intermediates (Scheme 4.10). This cyclization reaction is remarkable due to the high yields attained even when carried out in bulk, which usually favors linear polymerization [25, 26].

While structurally similar to the [12]cyclacene derivatives described previously, these cyclic fluoranthene derivatives have quite different 3D characteristics. For example, the double-stranded belt molecule with four

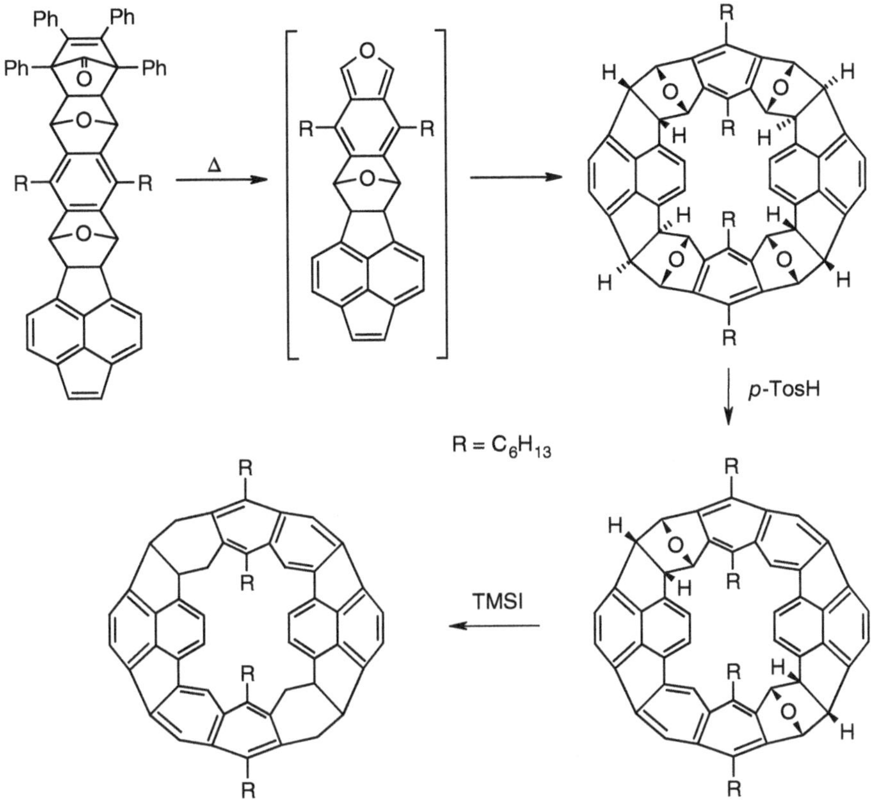

Scheme 4.10 The synthesis of the double-stranded belt molecules.

oxygen bridges has four *endo* hydrogen atoms pointing inside the cavity (Figure 4.11). The belt itself is "squashed": almost planar aromatic subunits are connected by curved cycloaliphatic links [25].

The removal of two oxygen atoms in acidic conditions leads exclusively to the double-stranded belt molecule with two oxygen bridges and all-*exo* hydrogen atoms attached to the cycloaliphatic links (Figure 4.12). This belt molecule with two oxygen bridges is even more flattened than the corresponding tetraoxy derivative. The distance between the slightly bended aromatic ring systems is only about 4.1 Å. Obviously, the internal cavity is too small to accommodate guest molecules [25].

The complete removal of all oxygen bridges was achieved with trimethylsilyl iodide in nonanhydrous conditions. The reaction yielded only poorly soluble poly-iodinated products in the absence of trace amounts of water. This all-hydrocarbon belt molecule has a distinct elliptical shape and a slightly larger internal cavity (Figure 4.13) [26].

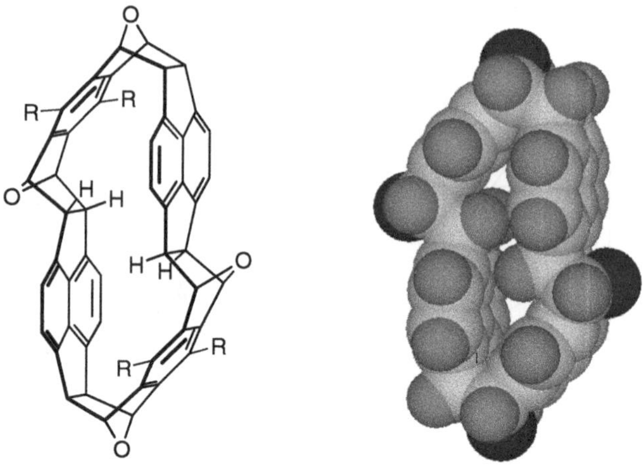

Figure 4.11 The double-stranded belt molecule with four oxygen bridges (left, $R = n\text{-}C_6H_{13}$) and its X-ray crystal structure (right, $n\text{-}C_6H_{13}$ fragments were replaced with hydrogen atoms for clarity).

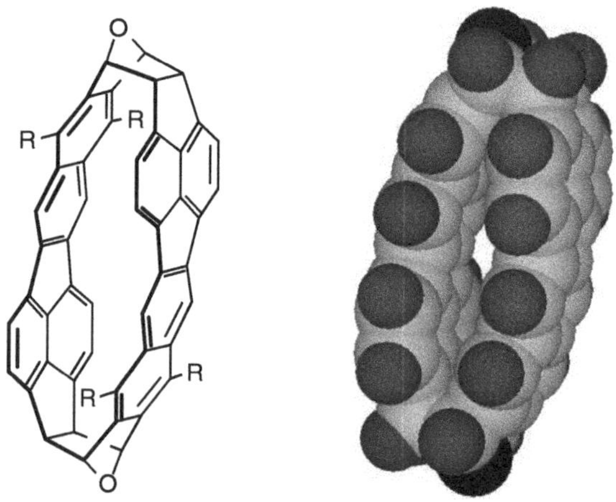

Figure 4.12 The double-stranded belt molecule with two oxygen bridges (left, $R = n\text{-}C_6H_{13}$) and its X-ray crystal structure (right, $n\text{-}C_6H_{13}$ fragments were replaced with hydrogen atoms for clarity).

The synthesis of fully conjugated double-stranded belt molecules is a formidable task. Despite several attempts, no bulk samples of these elusive compounds were ever made. Their existence was tentatively confirmed only by mass-spectroscopic means (Figure 4.14) [27, 28]. Semiempirical AM1 calculations indicate that the build-up of the so-called macrocyclic strain energy is the main factor affecting synthetic availability and chemical stability of these elegant belt molecules [25].

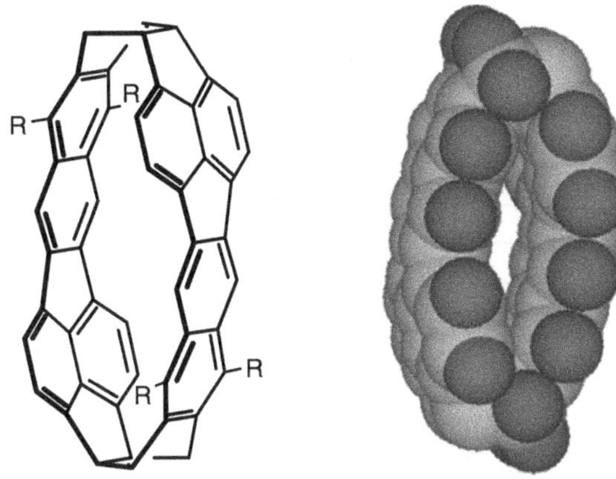

Figure 4.13 The all-carbon double-stranded belt molecule (left, R = n-C_6H_{13}) and its X-ray crystal structure (right, n-C_6H_{13} fragments were replaced with hydrogen atoms for clarity).

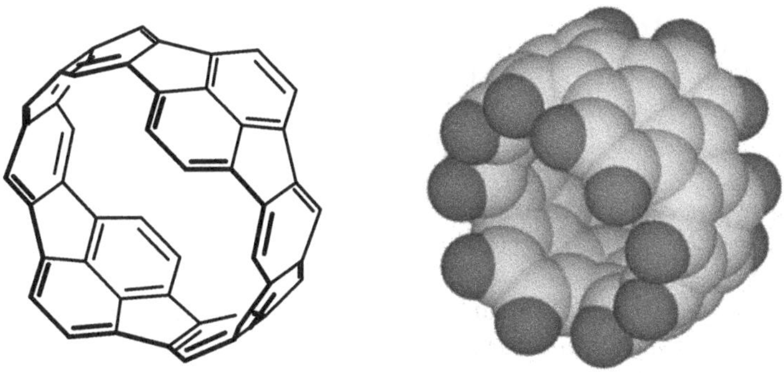

Figure 4.14 The energy-minimized structure of a fully conjugated double-stranded belt molecule.

Unsuccessful attempts to prepare these fully conjugated all-carbon molecular belts were in sharp contrast with the synthesis of fullerenes via evaporation of graphite, which yields predominantly aromatized molecular carbon clusters. One of the possible explanations of this difference is 3D stabilization of fullerenes, which, unlike molecular belts, cannot be easily flattened with the release of the abovementioned macrocyclic strain energy. A good example of such 3D stabilization is the successful synthesis of a stable, fully conjugated [10]cyclophenacene molecular belt incorporated in a functionalized fullerene C_{60} framework (Scheme 4.11). This first fully conjugated molecular belt, a derivative of [10]cyclophenacene, was initially reported in 2003 [29]. Instead of attempting to assemble a free-standing [10]cyclophenacene belt from smaller building blocks, the successful synthetic strategy invokes stepwise isolation of the fully conjugated [10]cyclophenacene

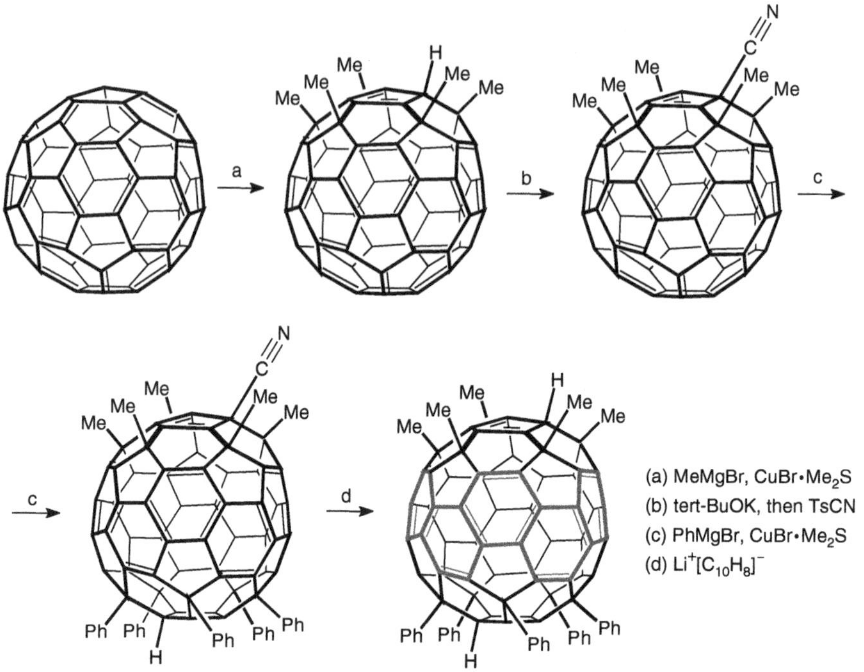

Scheme 4.11 The synthesis of the fully conjugated capped [10]cyclophenacene derivative.

pattern around the equator of the fullerene C_{60} ball by converting all of the attached carbon atoms into saturated quaternary centers. According to this synthetic strategy, fullerene C_{60} is pentamethylated (Scheme 4.11). This pentamethylated derivative is then treated with *t*-BuOK and *p*-toluenesulfonyl cyanide (TsCN) to yield the corresponding cyano derivative (Scheme 4.11). Pentaphenylation of this cyano compound with phenylcopper affords the "capped" [10]cyclophenacene derivative (Scheme 4.11). The cyanide group is finally removed with lithium-naphthalenide to give a highly luminescent all-carbon [10]cyclophenacene derivative $C_{60}Me_5Ph_5H_2$.

The crystal structure of an oxygenated derivative with [10]cyclophenacene substructure incorporated in a fullerene C_{60} framework reveals important details regarding the hitherto unknown stand-alone [*n*]cyclophenacenes (Figure 4.15). The fully conjugated [10]cyclophenacene belt has significant bond length alterations: the double bonds at the edges are short (1.36–1.37 Å), while the equator bonds are longer (1.40–1.43 Å). This X-ray data do not support the hypothesis of ideal aromatic pattern with equal bonds and could be represented by a restricted set of Kekulé structures, one of which is depicted in Figure 4.15 [29].

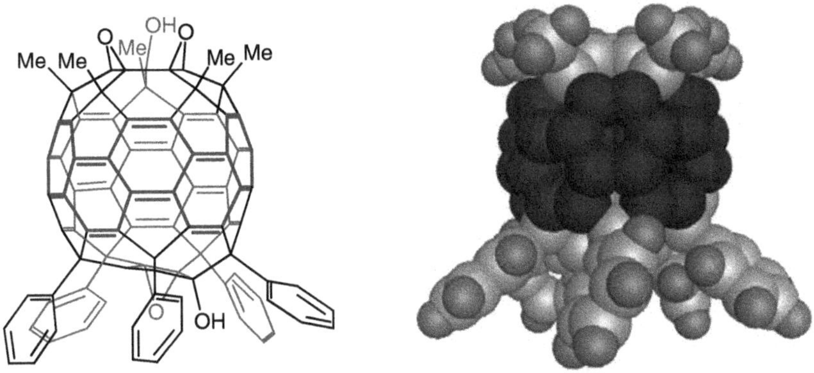

Figure 4.15 The X-ray structure of the oxygenated derivative (left) with [10]cyclophenacene belt incorporated in the fullerene C_{60} framework. Oxygen atoms (right) were removed for clarity.

Even partial, one-side 3D stabilization of [10]cyclophenacene is possible. This was recently confirmed by the synthesis of a "half-fullerene" one-side capped belt molecule. The compound was prepared in three steps starting from corannulene (Scheme 4.12). This capped belt molecule was found to be stable to air, heat, and light [30].

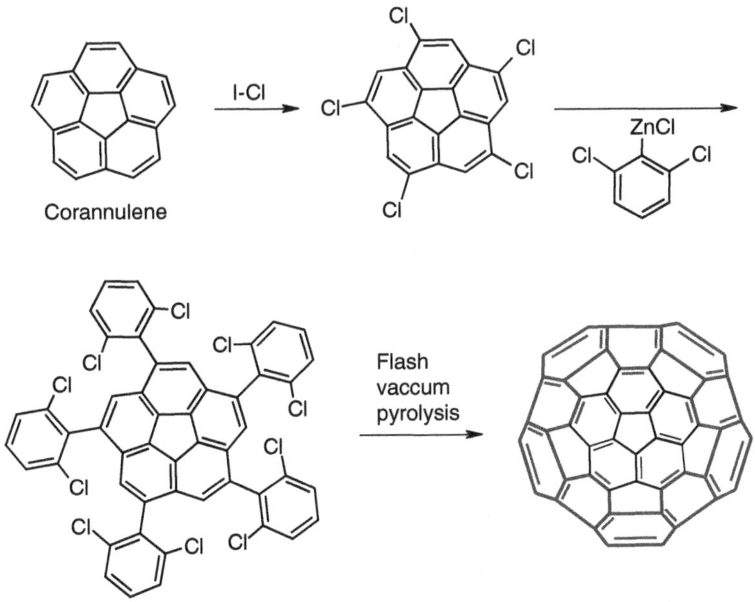

Scheme 4.12 The synthesis of the 3D-stabilized [10]cyclophenacene derivative starting from corannulene.

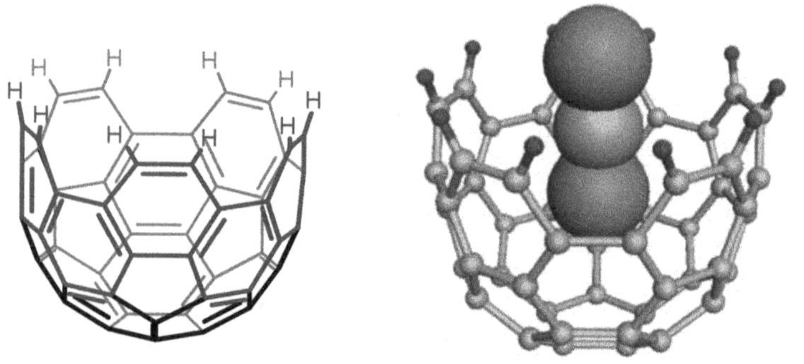

Figure 4.16 The 3D-stabilized [10]cyclophenacene derivative (left) and the X-ray crystal structure of its complex with CS_2 (right).

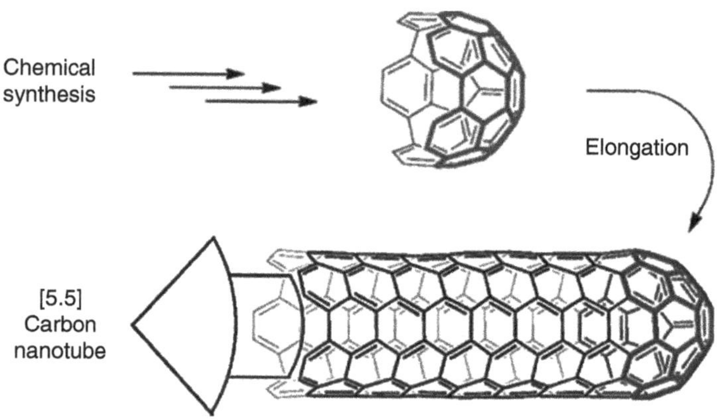

Scheme 4.13 The proposed rational synthesis of [5.5]carbon nanotube using the side-capped [10]cyclophenacene template. *Source*: Adapted from Scott *et al.* 2001 [30]. Reproduced with permission of American Chemical Society.

The X-ray crystal structure of this 3D-stabilized [10]cyclophenacene derivative resembles a capped short tube with internal cavity large enough to accommodate small guest molecules. One of such complexes (with CS_2) is presented in Figure 4.16. The carbon–carbon bonds on the top edge of the molecule are shorter (1.36–1.37 Å) than the interior bonds, which display a distinctive alternating pattern (1.39–1.47 Å) [30].

This one-side capped [10]cyclophenacene can theoretically be used for the subsequent preparation of [5.5]carbon nanotube (Scheme 4.13) [30]. This possibility is confirmed by successful synthesis and isolation of fullerene C_{90}-D_{5h}(1), which can be considered as [5.5]carbon nanotube capped from both ends with corannulene hemispheres (Figure 4.17) [31].

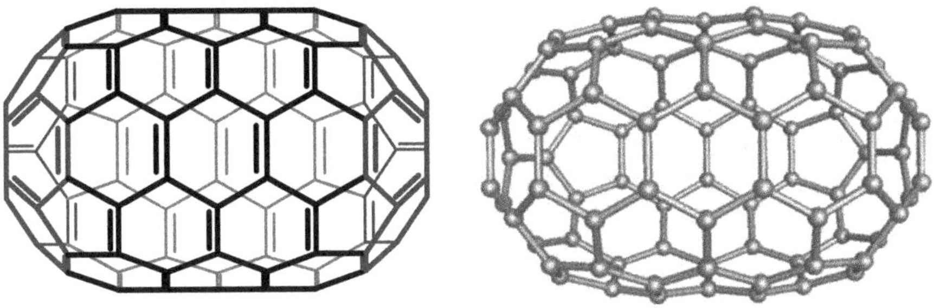

Figure 4.17 The X-ray structure of fullerene C_{90}-D_{5h}(1).

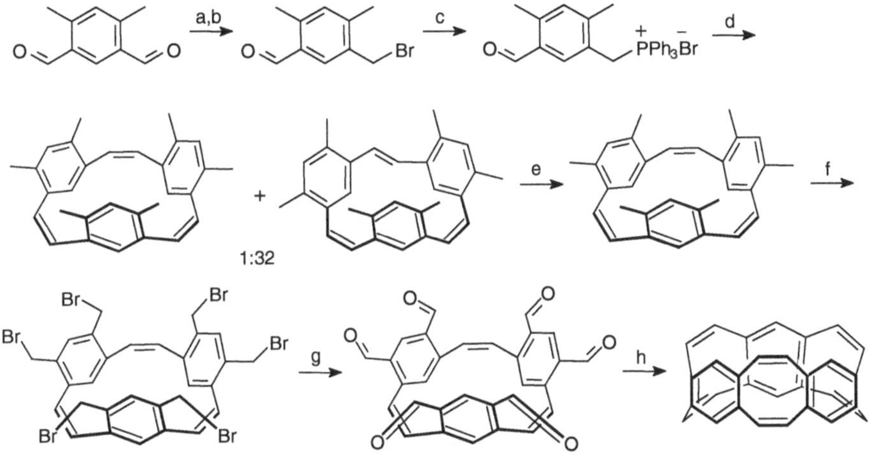

Scheme 4.14 The synthesis of $[6.8]_3$cyclacene: (a) $NaBH_4$; (b) HBr, HOAc; (c) PPh_3; (d) LiOEt; (e) *hν*; (f) NBS, dibenzoylperoxide; (g) 2-iodoxybenzoic acid; (h) $TiCl_3$, Zn–Cu.

An alternative approach, aimed at overcoming macrocyclic strain energy limitations in a fully conjugated belt molecule, was successfully demonstrated with the synthesis of chemically and thermally stable $[6.8]_3$cyclacene. In this belt molecule, the strain is reduced via the incorporation of three flexible eight-membered rings. The synthesis of $[6.8]_3$cyclacene represents an example of consecutive ring-by-ring synthetic methodology, which entails the construction of a secondary rim from a set of periphery functional groups attached to already assembled cyclic substructure (Scheme 4.14). This synthetic strategy is not very efficient and requires eight steps starting from 4,6-dimethylisophthalaldehyde. Nevertheless, the developed synthetic path offers a general route to higher members of $[6.8]_n$cyclacene family. The structural features of $[6.8]_3$cyclacene permit to consider this molecule as a model for a novel type of all-carbon molecular tubes [32, 33].

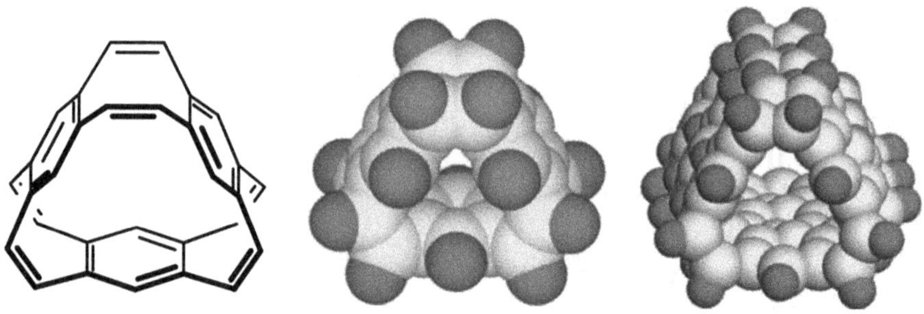

Figure 4.18 The X-ray crystal structure (middle) and packing (right) of $[6.8]_3$cyclacene (left).

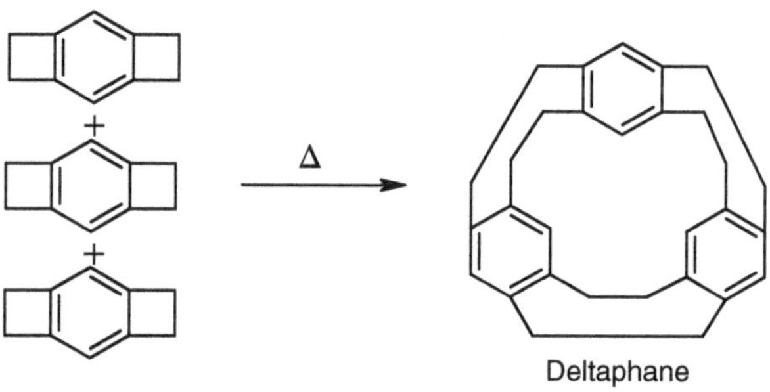

Scheme 4.15 The synthesis of deltaphane.

In crystal, individual $[6.8]_3$cyclacene molecules stack atop each other, thus producing infinite tubular structures (Figure 4.18). Unfortunately, the internal cavities of these belt molecules are too small to accommodate common elongated guests inside the tubular assemblies [32, 33].

The partially hydrogenated derivative of $[6.8]_3$cyclacene, which was named "deltaphane" in the original publication, was first prepared almost 30 years ago; long before the fully conjugated $[6.8]_3$cyclacene was described earlier. Deltaphane was isolated in 2.3% yield from the products of thermolysis of benzo[1,2;4,5]dicyclobutene at 425 °C in a stream of nitrogen (Scheme 4.15). It has triangular prism-like structure reflected in its trivial name. In solution, the compound exists in a highly symmetrical conformation with only three ^{13}C NMR signals even at low temperatures [34].

The X-ray crystal structure analysis of deltaphane confirms the high symmetry of the molecule established by ^{13}C NMR in solutions. All 30 carbon atoms lie essentially in three triangular prism planes. Some minor twisting of the methylene fragments was observed, most likely due to unfavorable nonbonding H—H interactions (Figure 4.19) [34].

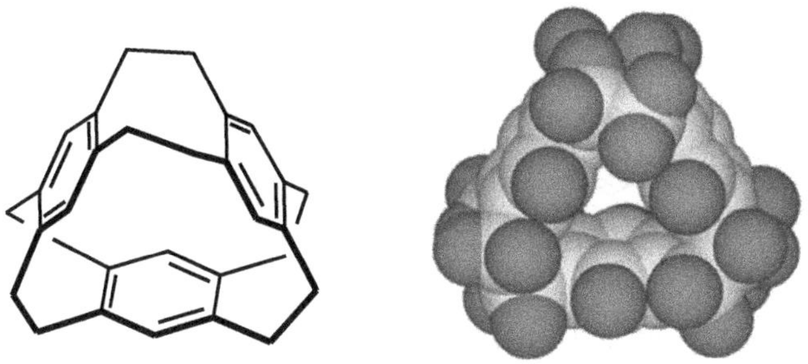

Figure 4.19 The X-ray crystal structure of deltaphane.

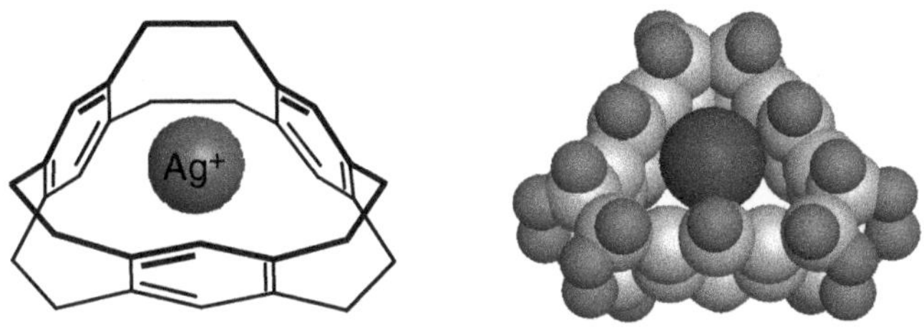

Figure 4.20 The X-ray crystal structure of the deltaphane–Ag^+ complex.

The internal cavity of deltaphane is large enough to accommodate small guest cations, such as Ag^+. The crystal structure of the deltaphane–silver triflate complex is presented in Figure 4.20. The silver atom resides asymmetrically (close to the end of the internal cavity). The average distance of Ag^+ cation from the three benzene ring carbon atoms is about 2.4 Å, with the Ag–C vectors almost perpendicular to the aromatic rings. This particular position of Ag^+ cation is a clear indication of predominantly π–Ag^+ interactions. The asymmetrical structure of the complex was also confirmed in solutions by means of low temperature ^{13}C NMR spectroscopy [34].

As noted previously, the incorporation of flexible macrocyclic subunits into molecular belts allows for substantial reduction in strain energy, thus making these belts accessible via traditional synthetic approaches. This strategy was successfully employed for the synthesis of [3.3]*meta*cyclophane-based molecular belts using macrocyclization reaction of properly designed molecular precursors equipped with four terminal functional groups [35]. The synthesis of the versatile [3.3]*meta*cyclophane precursor with four CH_2Br groups was described previously (see Chapter 1). The ring closure of this [3.3]*meta*cyclophane can be achieved by macrocyclization reaction with an appropriate tetramine (Scheme 4.16) [36].

The same [3.3]*meta*cyclophane precursor can be used for the synthesis of a wide range of belt-like molecules ranging from the smallest molecular belts containing only three benzene rings to large macrocyclic belts containing up

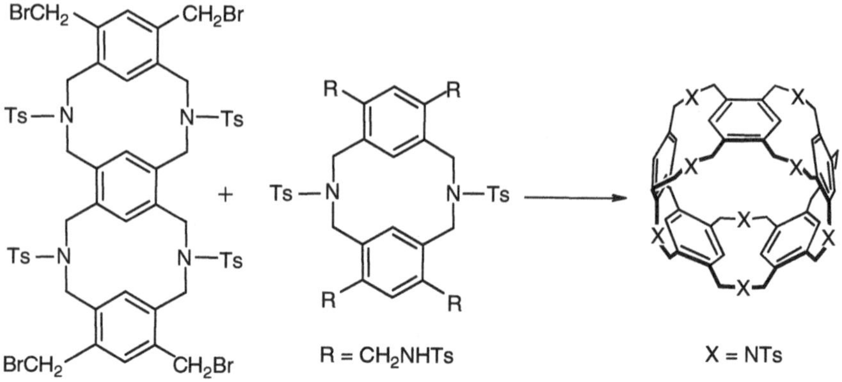

Scheme 4.16 The synthesis of the belt-shaped macrocyclic [3.3]*meta*cyclophane derivative.

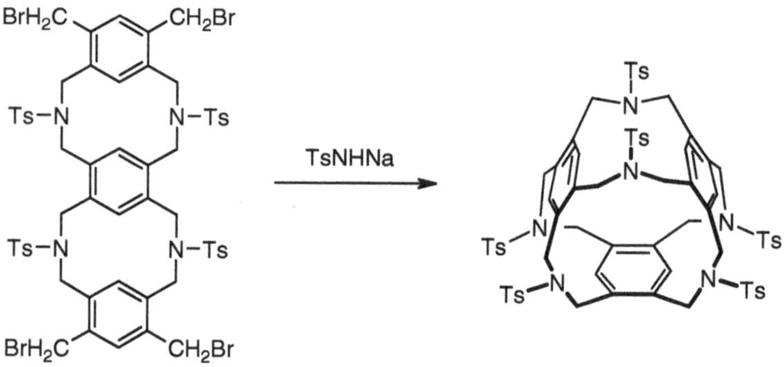

Scheme 4.17 The synthesis of *meta*-deltaphane with $CH_2N(Ts)CH_2$ links.

to 21 interconnected benzene rings. The smallest molecular belt was prepared in almost quantitative yield from the abovementioned fourfold functionalized [3.3]*meta*cyclophane and four equivalents of tosylamide monosodium salt (Scheme 4.17). The single-crystal X-ray analysis of this "*meta*-deltaphane" (the original deltaphane molecule was discussed earlier) [34] confirms its belt-like spatial features. However, longer links between the benzene rings result in more distorted structure [37].

Larger porphyrin rings were frequently employed for the construction of belt molecules [38]. One of the notable examples is a large molecular belt containing four porphyrin rings interconnected with eight pyridine links (Figure 4.21). These pyridine links are tilted outward with dihedral angles of about 45° on average. The inner diameter of this belt is about 14 Å (based on Ni—Ni distance). Not surprisingly, the inside space is filled with several solvent molecules in the crystal [39].

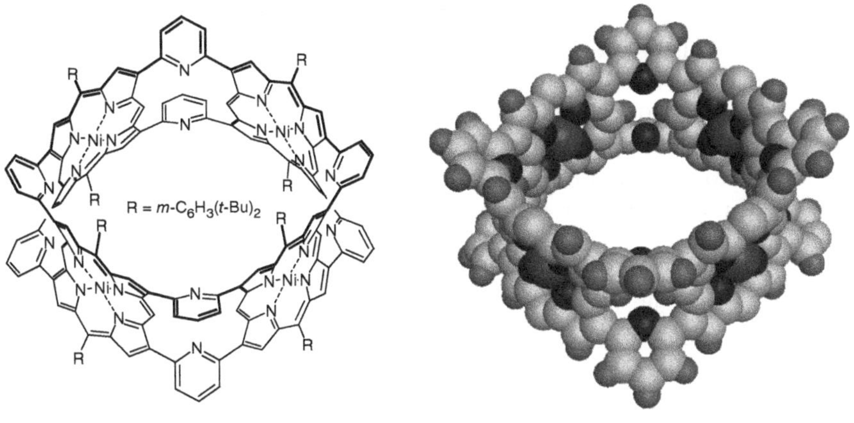

Figure 4.21 The porphyrin-based belt molecule (left) and its X-ray structure (right, 3,5-di-*tert*-butylphenyl groups were removed for clarity).

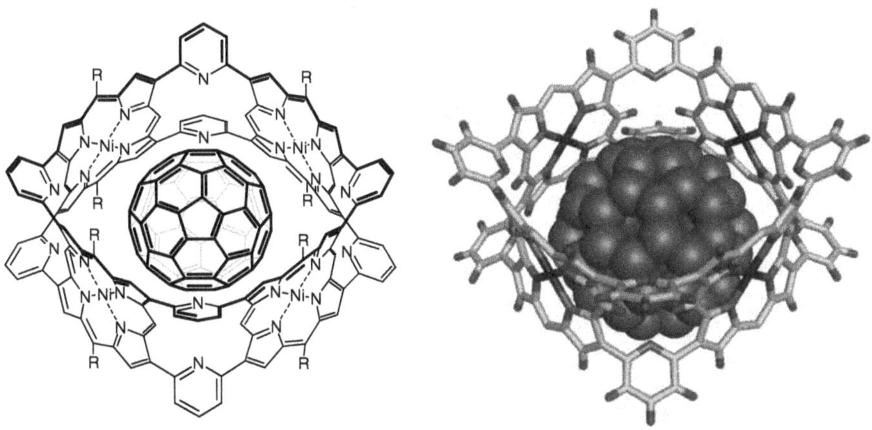

Figure 4.22 The porphyrin-based molecular belt complex with fullerene C_{60} (left, R = 3,5-$(t\text{-Bu})_2C_6H_3$) and its X-ray structure (right, 3,5-di-*tert*-butylphenyl groups were removed for clarity). Fullerene C_{60} guest molecule is partially disordered in the crystal.

The inner cavity of this porphyrin-based belt is large enough to accommodate bulky fullerene C_{60} guest molecule (Figure 4.22). This complex is quite strong ($K_a = 5.3 \times 10^5\ M^{-1}$ in toluene), even though an average distance between fullerene C_{60} and the belt sidewalls is relatively long (3.6 Å). Externally bound fullerene C_{60} molecules are also found in the crystal. They interact with the exterior concave porphyrin surfaces of the belt molecule [39].

Even larger molecular tubes can be constructed from porphyrin building blocks via template-directed synthesis (see also Chapter 1). The representative examples of these large porphyrin-based tubular assemblies include a molecular tube with 12 interconnected porphyrin units (Figure 4.23). The diameter of this large molecular tube is approximately 24 Å, the height of the core porphyrin-based sidewall is approximately 23 Å, and the internal cavity volume is about 4500 $Å^3$ [40].

Smaller cyclopentadienyl cobalt-stabilized cyclobutadiene rings can also be used for the construction of molecular belts. The cyclobutadiene ring has an obvious advantage over the traditional benzene ring since it can be constructed by dimerization of two acetylene fragments. An elegant one-pot synthesis of $[4.8]_3$beltene derivative utilizing this approach is presented in Figure 4.24. This molecular belt compound was prepared by irradiation of 5,6,11, 12-tetradehydrodibenzo[*a,e*]cyclooctene (Figure 4.24) in the presence of cyclopentadienyl $Co(CO)_2$. The substituted analogs of this molecule can be prepared using appropriately functionalized cyclopentadienyl cobalt derivatives. The highest yield (41%) was achieved with bulky pentamethylcyclopentadienyl $Co(CO)_2$. This synthesis is a rare example of controlled cyclo-oligomerization of an acetylene derivative leading to a belt-like macrocyclic compound [41].

The X-ray diffraction data show that the phenyl rings are bended outward (Figure 4.24). This bending might indicate a weak interaction between the cyclobutadiene rings and the benzene bridges. The calculated diameter of this $[4.8]_3$beltene is about 4.2 Å. Despite its relatively large size, no solvent molecules were observed inside the cavity [41].

Larger molecular belts with cyclobutadiene ring systems can be constructed via a somewhat different synthetic approach. For example, $[4.8]_4$beltene was prepared by Diels–Alder 2 + 2 cobalt-induced ring closure of a ribbon molecule with two terminal acetylene units (Figure 4.25). Even bigger $[4.8]_8$beltene with eight cobalt cyclopentadienyl units was also isolated from the reaction mixture. It was evidently produced by dimerization of the starting ribbon compound [42].

4.3 Pseudobelts

The preponderance of the true molecular belts described earlier has structural elements (e.g., aromatic rings) interconnected to each other with at least two linking bonds. This feature, while ensuring proper spatial arrangements of these building blocks within the molecular belt walls, complicates synthetic procedures and sometimes leads to low yields during the final macrocyclization step where at least two covalent bonds will have to be formed. Substantial simplification can be achieved if only one bridge connection will have to be closed during the final macrocyclization. These molecular belts can be named "pseudobelts", since some structural elements comprising the belt wall are interconnected with only one bond [43].

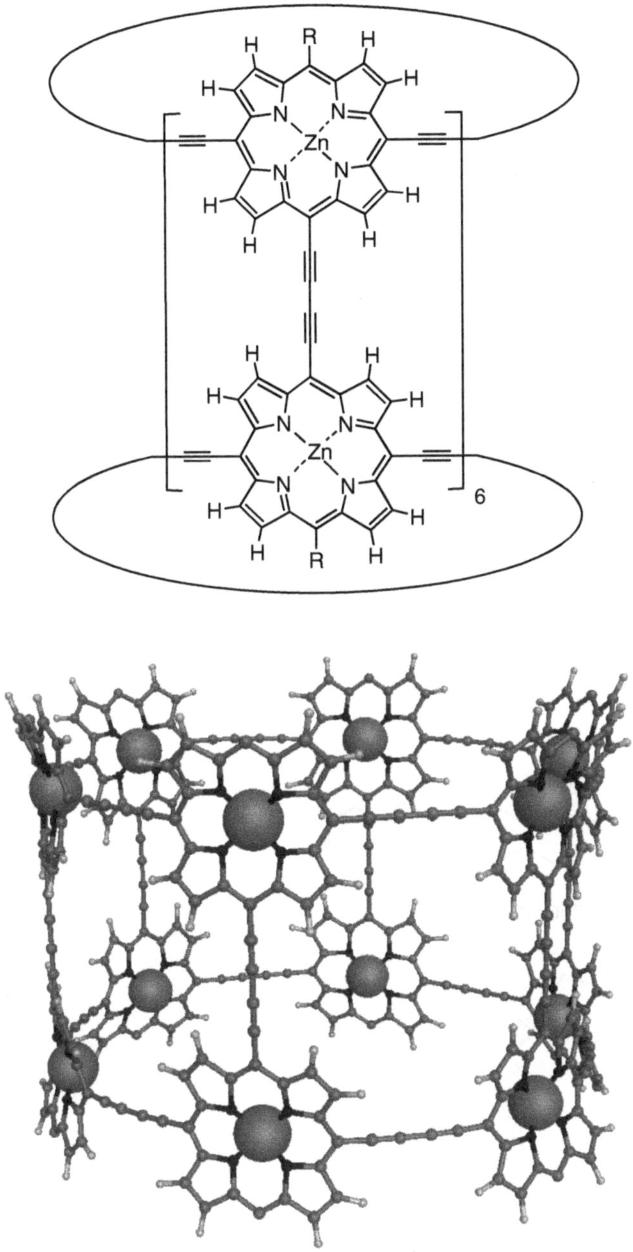

Figure 4.23 The porphyrin-based molecular tube (top, R = 3,5-[(*n*-$C_6H_{13})_3Si]_2C_6H_3$) and its energy-minimized 3D structure (bottom, 3,5-bis(tri-*n*-hexylsilyl)phenyl groups were removed for clarity).

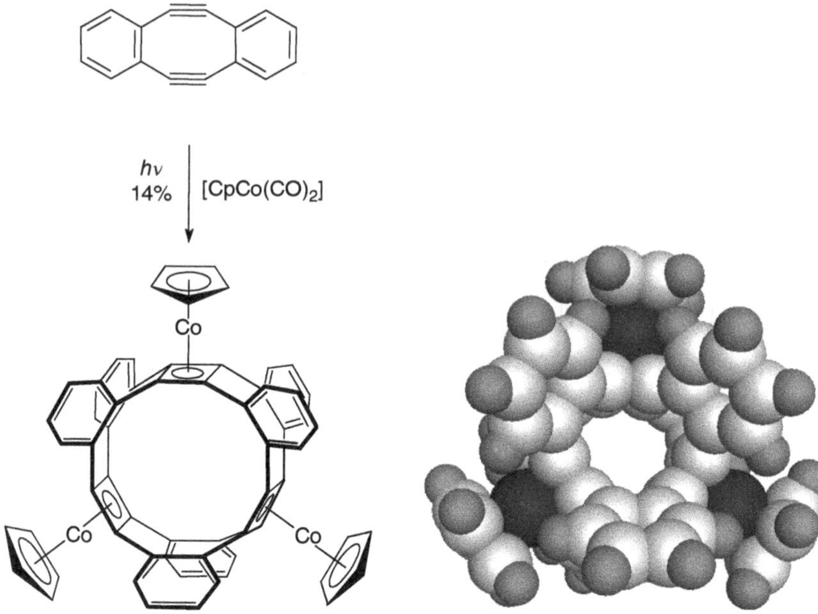

Figure 4.24 The synthesis (left) and the X-ray structure (right) of the $[4.8]_3$beltene cyclopentadienyl Co derivative.

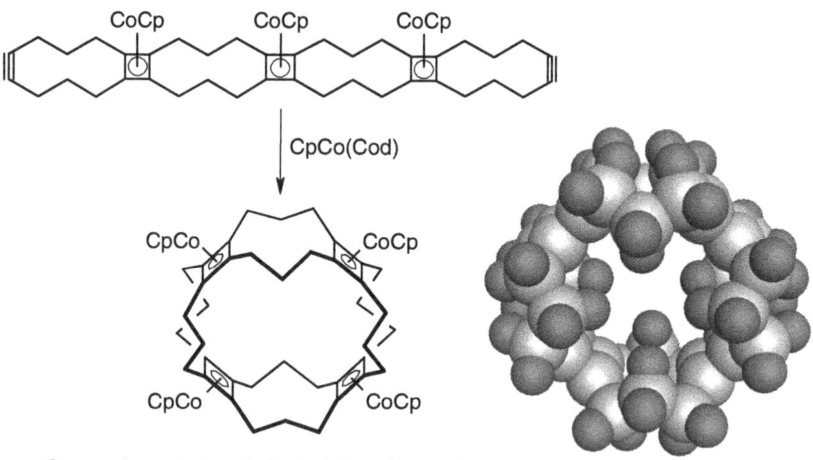

Cp - cyclopentadienyl; Cod - 1,5-cyclooctadiene

Figure 4.25 The synthesis (left) and the energy-minimized structure of $[4.8]_4$beltene (right, CoCp fragments were removed for clarity).

The major disadvantage of pseudobelts is higher degree of freedom attained by the structural elements interconnected with only one bond and less precise geometry of the molecular wall. Nevertheless, in many cases, proper spatial arrangements of these construction elements can be made even with only one interconnected bond by utilizing inherent strain of pseudobelts with small diameters.

Good examples of molecular pseudobelts are [*n*]cycloparaphenylenes, a family of macrocyclic belt molecules made of *n* benzene rings interconnected to each other in *para*-positions. Since benzene rings are planar, macrocyclization requires bending of these rings. This bending is forcing benzene ring alignment along the molecular belt axis, thus providing proper molecular wall configuration. Not surprisingly, the best spatial arrangements between the benzene rings are observed for smaller [*n*]cycloparaphenylenes, which have higher strain energy (Figure 4.26) [44]. The proper belt structure of large [*n*]cycloparaphenylenes can also be attained via complexation with round, spherical guests of appropriate sizes, such as fullerenes.

Although initial unsuccessful attempts to make [*n*]cycloparaphenylenes were reported more than 20 years ago [45], the real progress was achieved only recently using sophisticated synthetic strategies based on preformation and subsequent aromatization of mixed macrocycles [10, 46]. Direct synthesis of [*n*]cycloparaphenylenes by macrocyclization of the corresponding

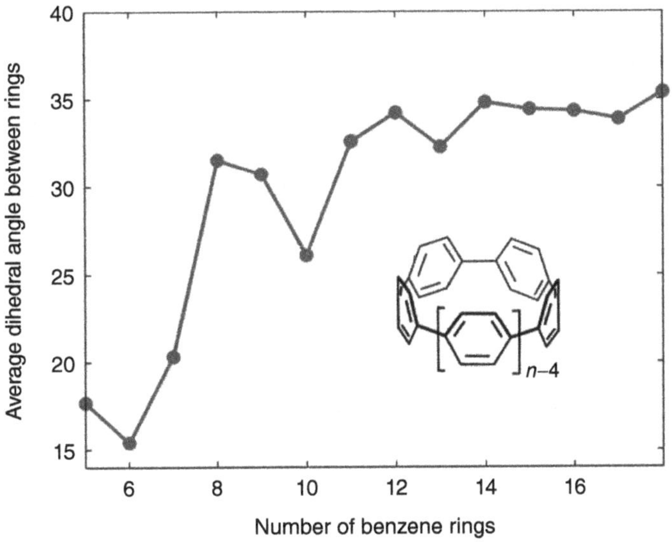

Figure 4.26 The calculated average dihedral angles between the adjacent benzene rings in energy-minimized molecular models of [*n*]cycloparaphenylenes. *Source*: Adapted from Wong 2009 [44]. Reproduced with permission of American Chemical Society.

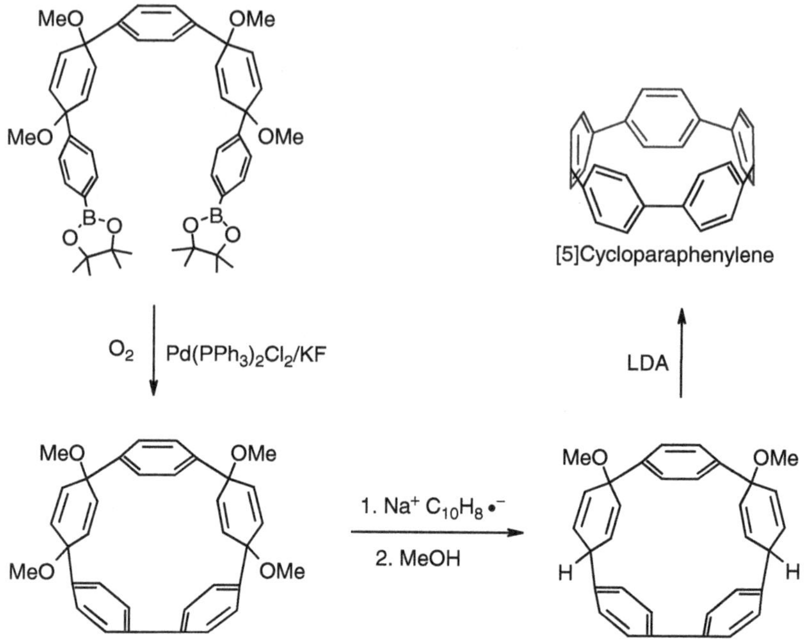

Scheme 4.18 The Synthesis of [5]cycloparaphenylene (LDA – lithium diisopropylamide).

linear [*n*]*para*phenylenes is not feasible due to unfavorable macrocyclic strain energy buildup leading to polymeric *para*phenylenes. Modern synthesis of [*n*]cycloparaphenylenes entails macrocyclization of curved mixed *para*phenylene–1,4-cyclohexylene oligomers followed by subsequent aromatization of cyclohexane (or 1,4-cyclohexadiene) moieties in already preformed macrocycles. The success of this approach could be attributed to energy gains during the aromatization offsetting the increase in the macrocyclic strain energy [10].

As an example of this approach, the synthesis of [5]cycloparaphenylene, the smallest known to date [*n*]cycloparaphenylene, is presented in Scheme 4.18 [47]. Aromatization of the corresponding mixed *para*phenylene–1,4-cyclohexadiene macrocycle was achieved in two steps underscoring the difficulties of the formation of highly strained [5]cycloparaphenylene having calculated macrocyclic strain energy of about 119 kcal/mol (23.8 kcal/mol per benzene ring) [47].

A convenient three-step synthesis of [5]cycloparaphenylene allows thorough characterization of this intriguing belt molecule. The compound is reported as a dark red [47] (or deep purple) [48] solid, which produces deep-ruby solutions [47]. It appears to be unstable in air [47]. However, other authors indicate good chemical stability [48]. The single-crystal X-ray analysis (Figure 4.27)

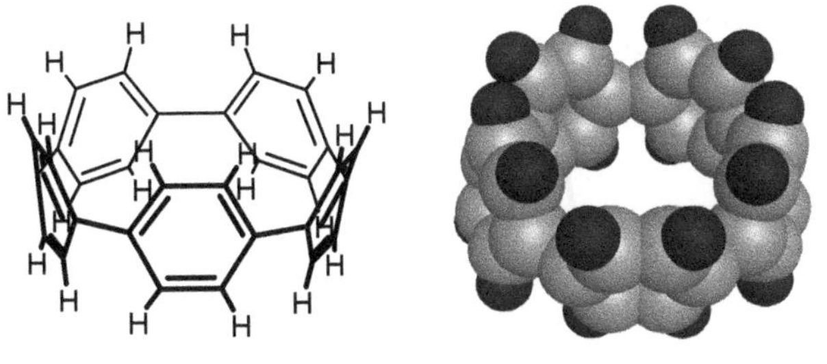

Figure 4.27 The X-ray crystal structure of [5]cycloparaphenylene [47].

confirms a round, belt-like alignment of benzene units with average dihedral angle between the adjacent rings of only 12.4°, which is even lower than the calculated 16.4°, and the smallest among other known [*n*]cycloparaphenylenes [47]. Macrocyclic strain is forcing benzene rings to adopt nonplanar, shallow boat conformation. As a result, *para*-carbons are bended out of the benzene planes by approximately 15.6°. Despite this significant nonplanarity, benzene rings in [5]cycloparaphenylene retain their aromatic character with shorter, almost equivalent in-ring carbon–carbon distances (1.38–1.40 Å) and longer macrocyclic out-of-ring *para*-connectors (1.49 Å). The internal cavity of [5]cycloparaphenylene has a diameter of about 6.7 Å, which is probably large enough to accommodate small guests, such as metal cations [47].

A different synthetic strategy was successfully used for the preparation of the next homolog, [6]cycloparaphenylene, having six benzene rings instead of five (Scheme 4.19) [49]. In this case, low-strain rectangle macrocycle preformation was achieved via Pt links having proper 90° bond geometry. Subsequent elimination of the Pt bridges was accomplished with XeF_2. The increase in macrocyclic strain energy associated with [6]cycloparaphenylene creation is compensated by the formation of four energetically favorable C—C bonds. This orange compound has somewhat smaller calculated macrocyclic strain energy of about 97 kcal/mol (16.2 kcal/mol per benzene ring) compared to [5]cycloparaphenylene [47] and is reported to be stable in air.

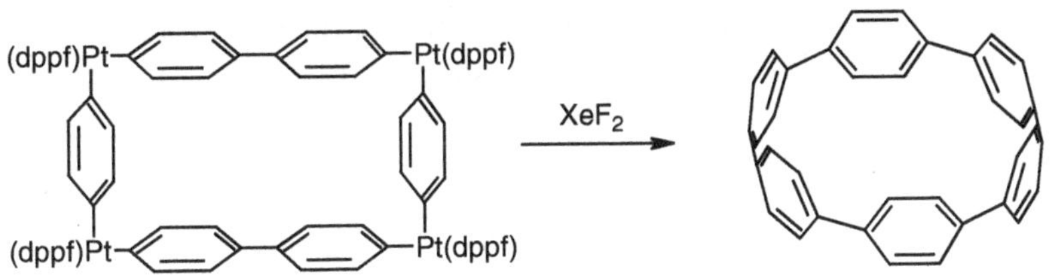

dppf – 1,1′-bis(Diphenylphosphino)ferrocene

[6]Cycloparaphenylene

Scheme 4.19 The synthesis of [6]cycloparaphenylene.

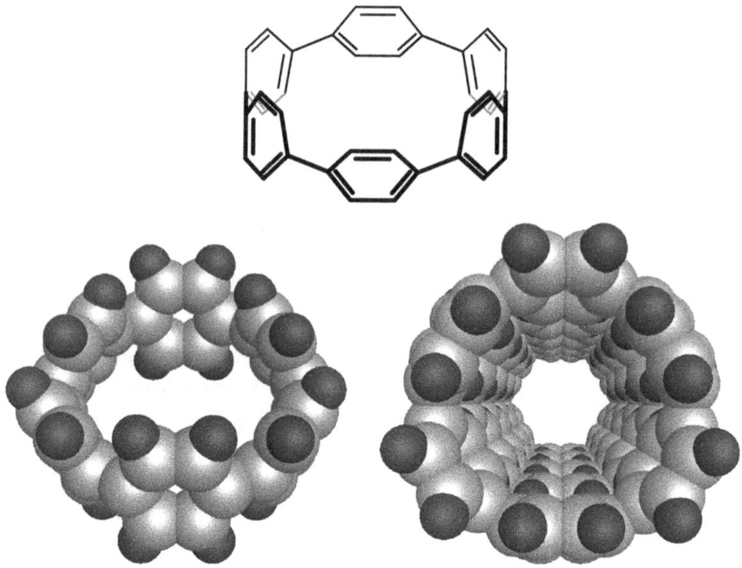

Figure 4.28 The X-ray crystal structure (bottom left) and packing (bottom right) of [6]cycloparaphenylene (top) [50].

The X-ray single-crystal data analysis shows that the dihedral angles between the benzene rings in [6]cycloparaphenylene are more uniform and larger on average (26.4°) than in [5]cycloparaphenylene (Figure 4.28) [50]. These experimentally determined angles are also substantially larger than predicted by semiempirical calculations [44]. Another interesting feature of this compound is the tube-like alignment in the crystal (Figure 4.28) [50]. This tubular alignment is probably responsible for the relatively large dihedral angles between the adjacent benzene rings, which allow for tight top-to-bottom molecular belt stacking and minimization of unfavorable nonbonding H—H interactions in the crystal. The inner diameter of this tube is about 8.1 Å, which might be adequate for the accommodation of some narrow, linear molecules [50].

Both [7]cycloparaphenylene and [8]cycloparaphenylene can be prepared from the same precursor (Scheme 4.20). The final oxidative aromatization was achieved by prolonged heating of the corresponding preformed cyclohexane-based mixed macrocycles in the presence of an acid catalyst ($NaHSO_4$) at 150 °C in nonairtight conditions [51].

Starting with [7]cycloparaphenylene, the internal cavities of [*n*]cycloparaphenylenes are big enough to capture various organic molecules, most notably organic solvents. For example, 1:1 adduct of [7]cycloparaphenylene with cyclohexane was recently examined by X-ray crystallography (Figure 4.29) [51]. This belt compound is more sterically relaxed compared to [5]cycloparaphenylene and [6]cycloparaphenylene discussed earlier, while

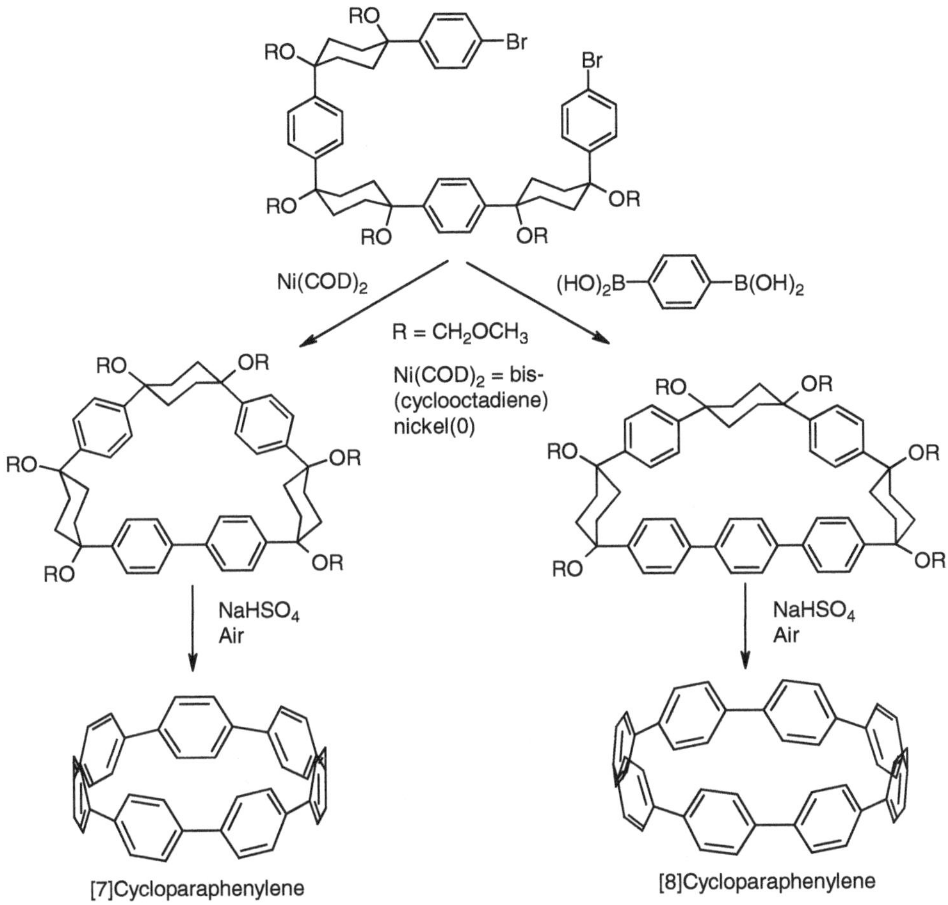

Scheme 4.20 The synthesis of [7]- and [8]cycloparaphenylenes.

retaining almost a perfect round shape. The calculated internal cavity size of [7]cycloparaphenylene is about 9.8 Å [52].

The crystal structure of [8]cycloparaphenylene appears to be more distorted. This belt molecule is slightly elliptical, with broad distribution of the dihedral angles between the adjacent benzene rings, from about 0° to almost 45° (Figure 4.30). The internal diameter of [8]cycloparaphenylene is approximately 10.9 Å [53]. The elliptical shape distortion of the [8]cycloparaphenylene belt is even more pronounced in deep-green tetra-K^+ salt of the corresponding tetra-anion (Figure 4.30), which can be easily produced by the action of potassium metal on [8]cycloparaphenylene in the presence of [18]crown-6 [54].

Even larger [*n*]cycloparaphenylenes, with up to 18 benzene rings, can be prepared using appropriate synthetic strategies [10]. An important feature of

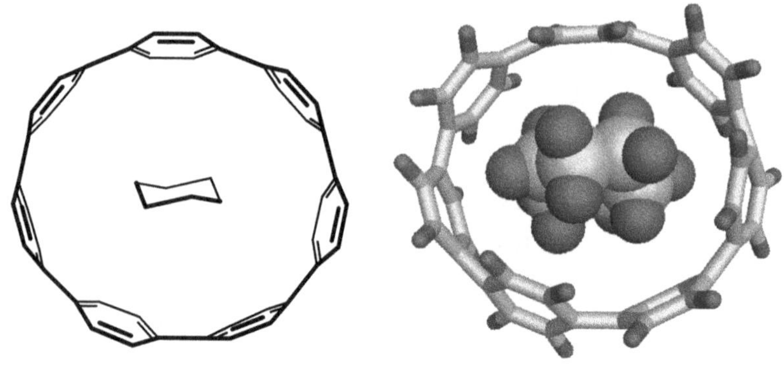

Figure 4.29 The X-ray crystal structure of [7]cycloparaphenylene complex with cyclohexane.

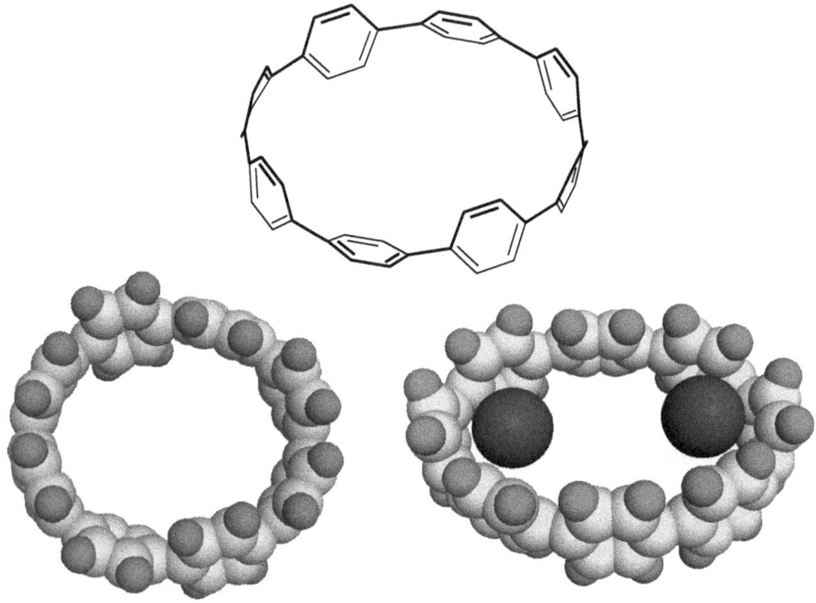

Figure 4.30 [8]Cycloparaphenylene (top), its X-ray crystal structure (bottom left), and the X-ray structure of the tetra-K^+ salt of [8]cycloparaphenylene tetra-anion (bottom right, [18]crown-6 molecules and two externally-bound K^+ cations were removed for clarity).

these large [*n*]cycloparaphenylenes is the ability to produce complexes with fullerenes and endohedral fullerenes, which can be useful for the separation of these spherical guest molecules [55]. Two [10]cycloparaphenylene complexes (with fullerene C_{60} and endohedral fullerene $Li^+@C_{60}$), together with "empty"

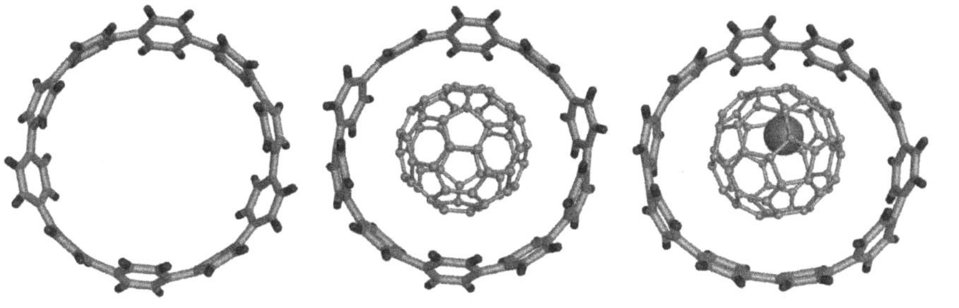

Figure 4.31 The X-ray crystal structures of empty [10]cycloparaphenylene (left), its complex with fullerene C_{60} (middle), and with endohedral fullerene $Li^+@C_{60}$ (right).

[10]cycloparaphenylene, are presented in Figure 4.31 [53, 56]. It is interesting to note that the expected flattening of the molecular belt wall was observed only in the case of [10]cycloparaphenylene complex with $Li^+@C_{60}$ but not with C_{60}, thus underscoring the importance of proper electronic and spatial arrangements for the construction of optimized [*n*]cycloparaphenylene host–guest systems. The average dihedral angle between the adjacent benzene rings in [10]cycloparaphenylene complex with $Li^+@C_{60}$ is about 20.4°, which is smaller than in the parent [10]cycloparaphenylene (27.3°), and in its complex with C_{60} (28.5°) [53]. Both complexes of [10]cycloparaphenylene with $Li^+@C_{60}$ ($K_a = 4.8 \times 10^4\,M^{-1}$, in CH_2Cl_2) and with C_{60} ($K_a = 6.0 \times 10^3\,M^{-1}$, in *o*-$C_6H_4Cl_2$) [57] are quite strong, with the distances between the guest fullerene molecules and the host walls close to optimal for π–π stacking (about 3.4 Å) [53, 56]. The internal diameter of [10]cycloparaphenylene is about 13.7 Å [53].

A simple extension of [*n*]cycloparaphenylene walls can be achieved using phenyl-substituted derivatives, which leads to the so-called π-extended polyphenylene cylinders. The synthesis of one of these π-extended cylinders is shown in Scheme 4.21. So far, all attempts to fuse the extended benzene rings and to construct a rigid molecular tube were unsuccessful. One of the possible explanations of this failure is out-of-plane ring distortions clearly observed in the X-ray crystal structures of these molecules [58]. An alternative approach entails the use of iterative reductive aromatization/ring-closing metathesis strategy [59].

Bridged, or 3D-interconnected, [*n*]cycloparaphenylene molecular belts are also known. They can be constructed by the incorporation of two *meta*-substituted benzene-1,3,5-triyl "bridgeheads" as three-way links [60, 61]. More complex [*n*]cycloparaphenylene-based nanocages containing four bridgeheads are also known [62]. The X-ray crystal structure of "[4.4.4]carbon nanocage" as a representative example of bridged [*n*]cycloparaphenylenes is shown in Figure 4.32. Similar to [8]cycloparaphenylene, the inner cavity of this

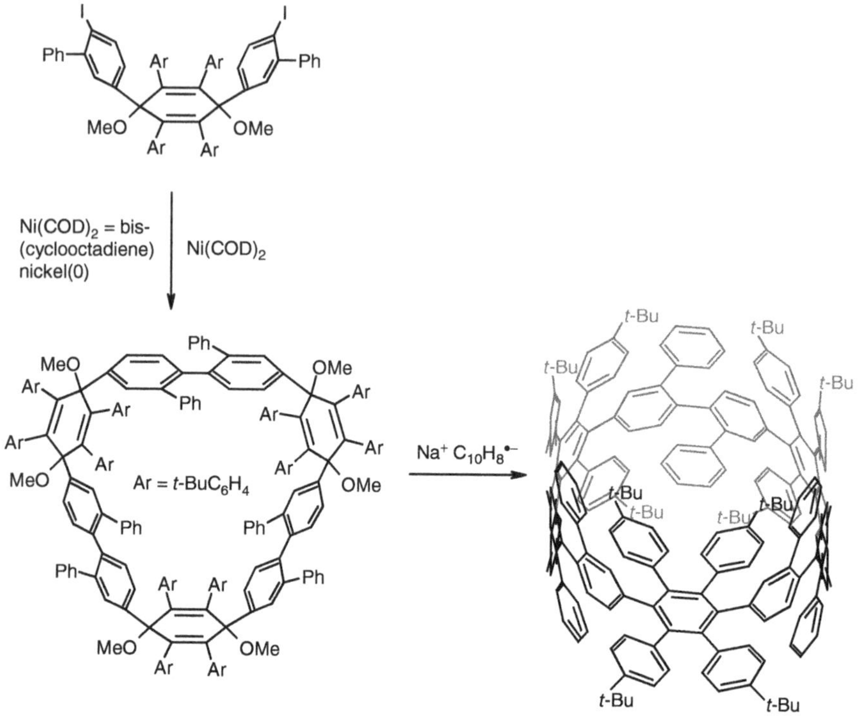

Scheme 4.21 The synthesis of the π-extended polyphenylene cylinder (Ar = 4-*t*-BuC$_6$H$_4$) [58].

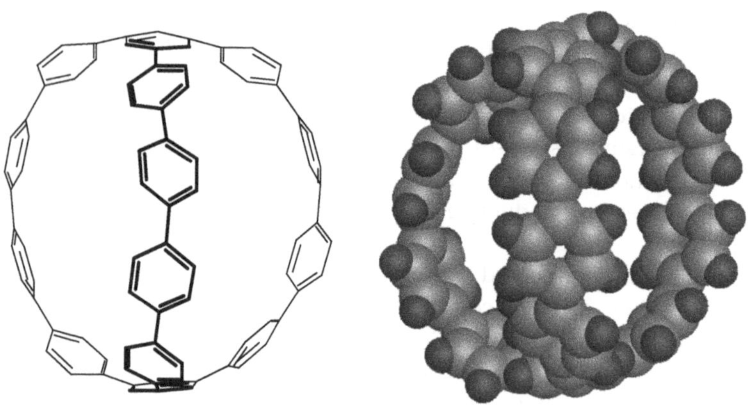

Figure 4.32 The X-ray structure of the [4.4.4]carbon nanocage [61].

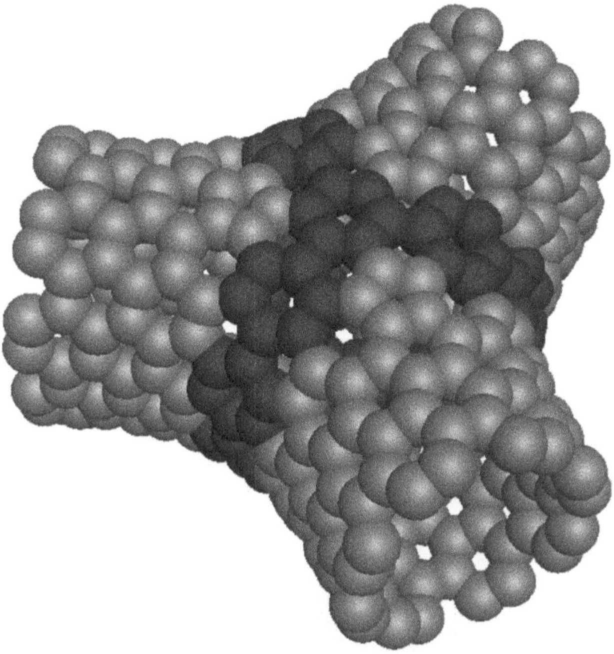

Figure 4.33 The energy-minimized structure of a hypothetical three-way junction between all-carbon molecular tubes based on [4.4.4]carbon nanocage substructure.

[4.4.4]carbon nanocage is ellipsoid, with significant dihedral angles between the adjacent benzene rings. The distance between the two bridgeheads is about 13.0 Å, and the estimated strain energy is approximately 96.0 kcal/mol [61].

Similar to [*n*]cycloparaphenylenes, which can be considered as structural units of straight all-carbon molecular tubes, [*n.n.n*]carbon nanocages can be envisioned as elements of three-way junctions between these tubes. A hypothetical example of such junctions is depicted in Figure 4.33 [60].

A different synthetic strategy was employed for the synthesis of the so-called picotube, which can be considered as a π-extended [4]cycloparaphenylene. It can also be classified as [4]cyclo-9,10-anthracenylene since it has four anthracene moieties interconnected in 9,10 positions. This belt compound can be prepared in three steps from 9-bromoanthracene via tetradehydrodianthracene (Scheme 4.22) [63].

The preliminary X-ray data indicate that the picotube has an internal diameter of approximately 5.4 Å and an overall length of 8.2 Å. Despite the presence of four double bonds in the [4]cycloparaphenylene substructure, this belt molecule is surprisingly chemically and thermally stable. It does not react with *m*-chloroperbenzoic acid and bromine in normal conditions and remains

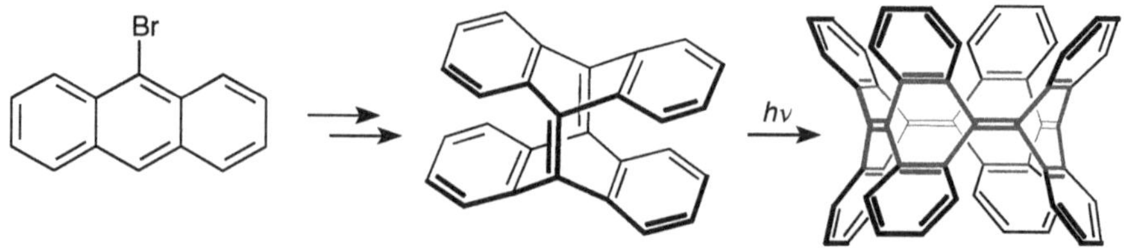

Scheme 4.22 The synthesis of the picotube.

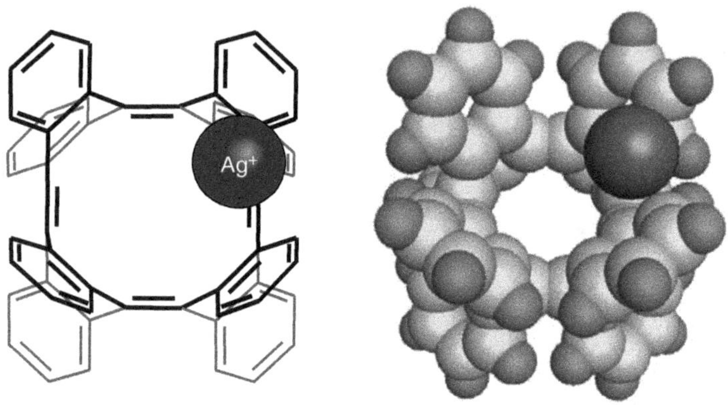

Figure 4.34 The X-ray crystal structure of the Ag^+-picotube complex.

intact at the temperatures up to 450 °C [63]. The reaction of the "picotube" with lithium metal leads to a stable tetraanion [64]. Attempts to prepare a small all-carbon [4.4] molecular tube fragment from the picotube by flash vacuum pyrolysis at 800 °C were unsuccessful [65].

The picotube produces strong Ag^+ complex with well-defined geometry, which is fully characterized by single-crystal X-ray crystallography (Figure 4.34). The Ag^+ cation is coordinated nonsymmetrically to two carbon atoms of the adjacent benzene rings with the average Ag^+—C distance of about 2.54 Å. The host belt molecule has a distinct square-type shape [66].

Alkylation of "picotube" with a large excess of *t*-BuCl and catalytic amounts of $AlCl_3$ leads primarily to two diastereomers with eight *t*-Bu group "tilted" uniformly along each rim (Figure 4.35). They can be separated by HPLC. One of these isomers (D_4, Figure 4.35) is chiral and can be further resolved into two enantiomers on a chiral HPLC column [67].

Pyrene rings instead of benzene rings can be successfully employed for the construction of molecular belts. A representative synthesis of pyrenylene-based molecular pseudobelt ([4]cyclo-2,7-pyrenylene) is presented in Scheme 4.23 [68]. The success of this approach is attributable to

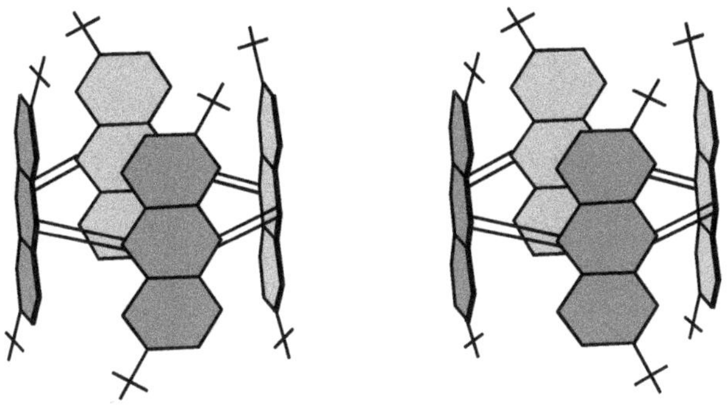

Figure 4.35 The chemical structures of the two *t*-Bu derivatives of the picotube.

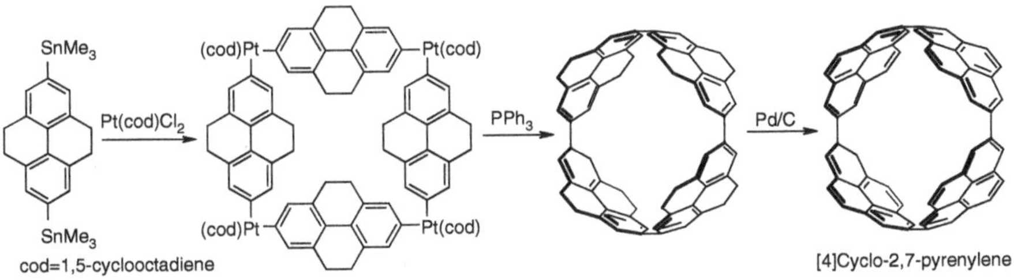

Scheme 4.23 The synthesis of [4]cyclo-2,7-pyrenylene.

step-by-step macrocyclic strain energy buildup strategy, which entails the use of more flexible octahydropyrene derivatives during the initial stages of the synthesis. Strain-inducing rearomatization was performed during the last step of the synthetic sequence.

Also known are [*n*]cycloparaphenylene analogs having anthanthren units instead of benzene units. The syntheses of (12,8)-[4]cyclo-2,8-anthanthrenylene derivatives were accomplished in several steps from commercially available pigment "Red 168" (Scheme 4.24) [69]. Due to a larger size of anthanthren building blocks, strain-related synthetic obstructions are weaker compared to pyrene building blocks, thus allowing a shorter preparative route without the need of a tedious step-by-step strain energy buildup sequence. The separation of the final products is complicated by the presence of several sets of enantiomers and diastereomers (a barrier of rotation of anthanthren units is calculated to be $\Delta H^\ddagger = +21$ kcal/mol and $\Delta S^\ddagger = -15$ cal/mol/K for *n*-hexyl derivatives) [69].

The structure of one of these belt molecules, 6,12-bis[(triisopropylsilyl) ethynyl]-(*P*)-(12,8)-[4]cyclo-2,8-anthanthrenylene, is slightly deviated from

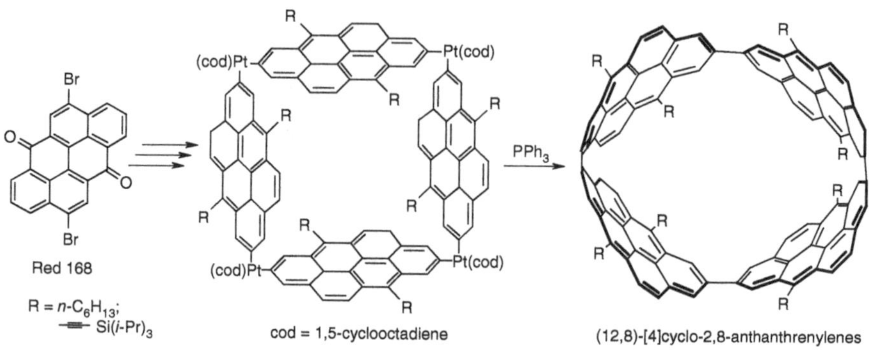

Scheme 4.24 The synthesis of (12,8)-[4]cyclo-2,8-anthanthrenylene derivatives.

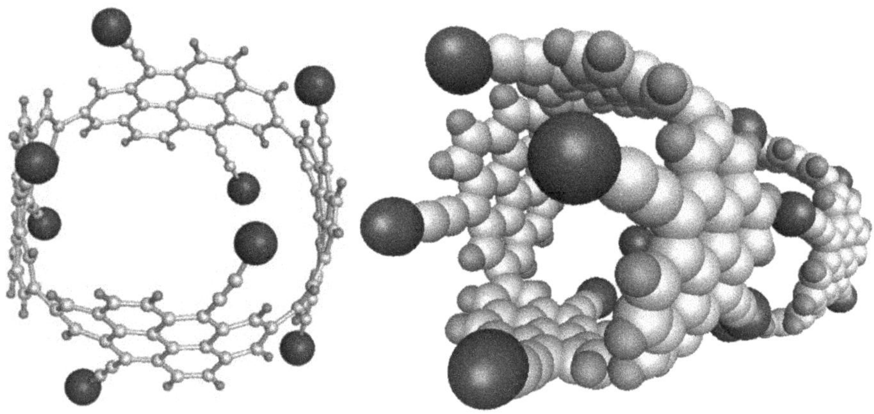

Figure 4.36 The X-ray crystal structure (left) and packing (right) of 6,12-bis[(triisopropylsilyl)ethynyl]-(12,8)-(*P*)-[4]cyclo-2,8-anthanthrenylene. The isopropyl groups were removed for clarity.

a perfectly round shape toward a square shape due to strain-related factors (Figure 4.36). The inner diameter of the belt is around 14.1 Å. In crystal, individual molecular belts are packed atop each other, thus producing infinite molecular tubes (Figure 4.36).

Due to the optimal size of the internal cavity and extended π-network, (12,8)-[4]cyclo-2,8-anthanthrenylenes are good host systems for some fullerenes. The complex of 6,12-bis(*n*-hexyl)-(12,8)-(*P*)-[4]cyclo-2,8-anthanthrenylene with fullerene C_{60} is depicted in Figure 4.37. This fullerene C_{60} complex and similar complex with fullerene C_{70} have very high association constants ($K_a = 5.0 \times 10^9\ M^{-1}$ and $4.0 \times 10^9\ M^{-1}$ in o-$C_6H_4Cl_2$, respectively) [70].

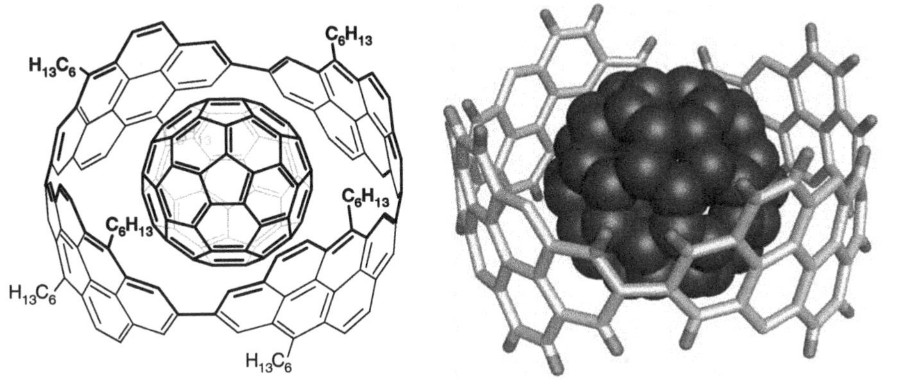

Figure 4.37 The X-ray crystal structure of 6,12-bis(*n*-hexyl)-(12,8)-(*P*)-[4]cyclo-2,8-anthanthrenylene complex with fullerene C_{60} (right, *n*-hexyl groups were removed for clarity). The fullerene C_{60} guest molecule is partially disordered in the crystal [70].

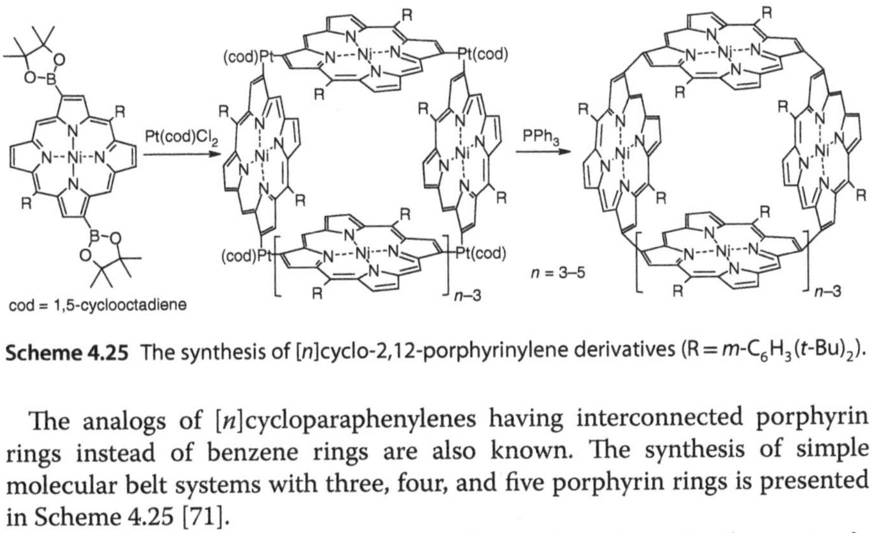

Scheme 4.25 The synthesis of [*n*]cyclo-2,12-porphyrinylene derivatives ($R = m\text{-}C_6H_3(t\text{-}Bu)_2$).

The analogs of [*n*]cycloparaphenylenes having interconnected porphyrin rings instead of benzene rings are also known. The synthesis of simple molecular belt systems with three, four, and five porphyrin rings is presented in Scheme 4.25 [71].

The macrocyclic strain causes significant distortions in these simple porphyrin-based molecular belts, most notably reflected in bending of the porphyrin rings (Figure 4.38). The distortions are more pronounced in the smallest molecular belt, the [3]cyclo-2,12-porphyrinylene derivative (Figure 4.38). The average diameters of the porphyrin-based molecular belts are 9.32 Å for $n=3$, 11.95 Å for $n=4$, and 15.60 Å for $n=5$. The strain energies of these [*n*]cyclo-2,12-porphyrinylenes are calculated to be about 77 kcal/mol for $n=3$, 57 kcal/mol for $n=4$, and 47 kcal/mol for $n=5$ [71].

Other heteroaromatic systems, such as carbazoles and benzocarbazoles, can be successfully used for the construction of pseudobelts. The best results

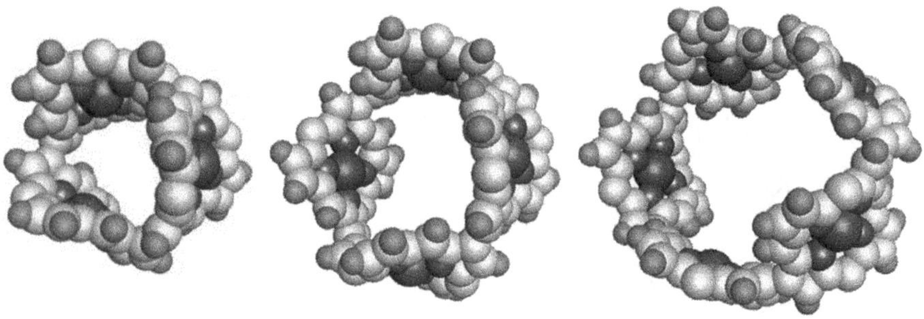

Figure 4.38 The X-ray crystal structures of the [*n*]cyclo-2,12-porphyrinylene derivatives; $n = 3$ (left), $n = 4$ (middle), and $n = 5$ (right). The 3,5-di-*tert*-butylphenyl groups were removed for clarity.

were achieved with end-capped heteroaromatic pseudobelt molecules, in which proper alignment of the heteroaromatic building blocks within the belt sidewall was secured via 3D cap stabilization discussed earlier. The effective one-step synthesis of carbazole- and benzo[*def*]carbazole-based capped molecular pseudobelts from preorganized "three-arms" precursors is shown in Scheme 4.26. The success of this approach underscores the utility of preorganized intermediates for the preparation of highly strained molecules. Similar methodology was successfully employed for the construction of calix[*n*]arene-capped [*n*]cyclometaphenylenes [72]. In the cases of carbazole- and benzo[*def*]carbazole-based capped molecular pseudobelts, the macrocyclic strain energy is estimated to be about 138 kcal/mol for the carbazole derivative and approximately 144 kcal/mol for the benzo[*def*]carbazole derivative [73].

The single-crystal X-ray diffraction analysis of the carbazole derivative reveals that this capped belt molecule has three bent carbazole subunits within the sidewall (Figure 4.39). These subunits are interconnected in 2,7 positions with three slightly elongated (1.513 Å on average) C—C bonds, thus forming [6]cycloparaphenylene substructure. Each of the six benzene rings in this [6]cycloparaphenylene substructure is tilted at approximately 77° relative to the cap plane. The dimensions of the internal cavity are about 4.6 Å (height) to 8.6 Å (belt diameter). The internal space of this capped molecular belt is filled with a 1,2-dichloroethane solvent molecule displaying weak C···Cl van der Waals interactions with the host molecule [73].

The benzo[*def*]carbazole derivative has a slightly longer and narrower cavity (height 5.9 Å, diameter 8.1 Å), deep enough to enclose a small dichloromethane solvent molecule. The X-ray crystal structures of this benzo[*def*]carbazole-based capped belt molecule and its complex with CH_2Cl_2 are shown in Figure 4.40 [73].

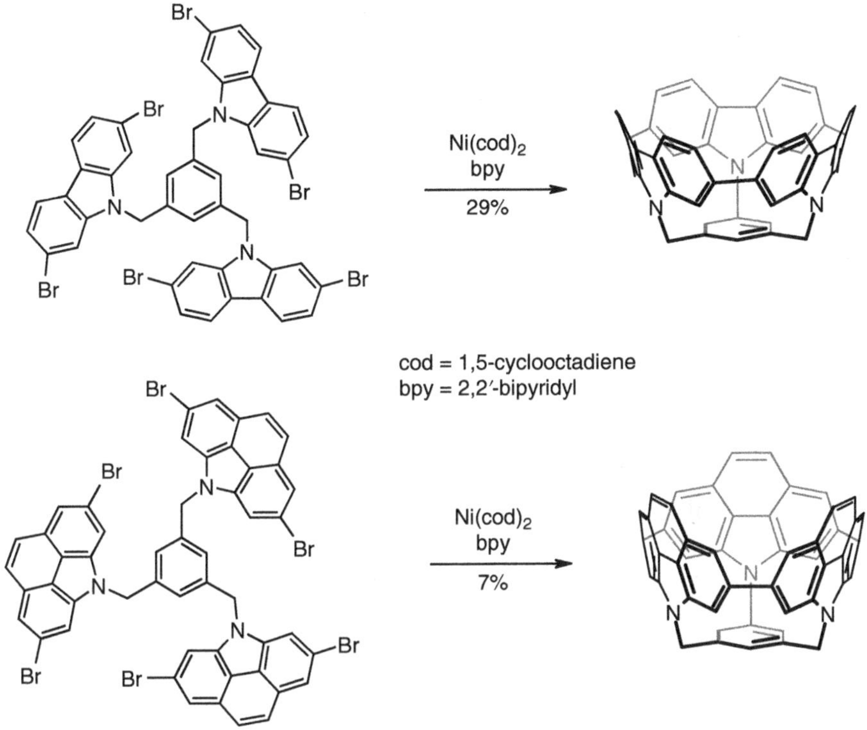

Scheme 4.26 The synthesis of the capped carbazole-based belt molecule (top) and the corresponding benzo[*def*]carbazole derivative (bottom).

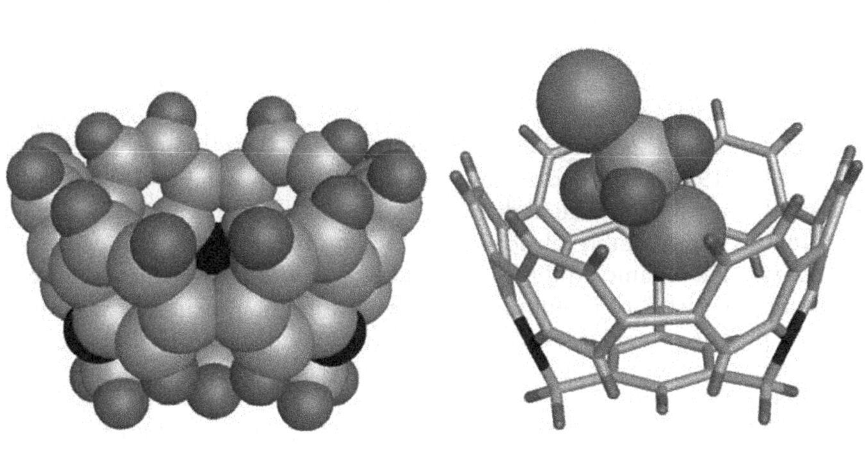

Figure 4.39 The X-ray crystal structures of the capped carbazole-based molecular belt (left) and its complex with 1,2-dichloroethane (right).

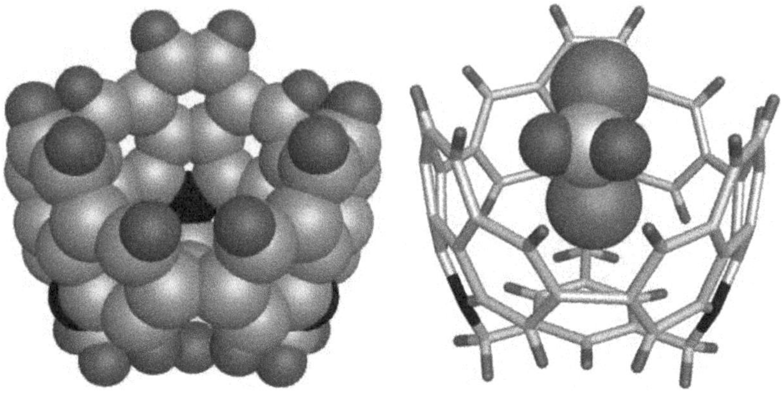

Figure 4.40 The X-ray crystal structures of the benzo[*def*]carbazole derivative (left) and its complex with CH_2Cl_2 (right).

4.4 Cucurbiturils

The synthesis of belt molecules is not always a complicated task. Sometimes nature plays a chemist's hand by providing unparalleled, easy access to seemingly very complicated structures via self-assembly pathway, usually executed in a form of self-templating repetitive synthesis (see Chapter 1). The serendipitous discovery of cucurbit[*n*]urils in 1905, which were prepared by protracted heating of glycoluril and formaldehyde in acidic media [74], is a clear demonstration of the creative power of this synthetic paradigm. The structure of the main product, cucurbit[6]uril (the systematic name of this compound is dodecahydro-1*H*,4*H*,14*H*,17*H*-2,16:3,15-dimethano-5*H*, 6*H*, 7*H*, 8*H*,9*H*,10*H*, 11*H*, 12*H*,13*H*,18*H*,19*H*,20*H*,21*H*,22*H*,23*H*,24*H*,25*H*,26*H*-2,3,4a,5a,6a,7a,8a, 9a,10a, 11a,12a, 13a,15,16,17a,18a,19a, 20a,21a,22a,23a,24a,25a,26a-tetracosa azabispentaleno[1‴, 6‴:5″,6″,7″]cycloocty[1″,2″,3″:3′,4′]pentaleno(1′,6′:5,6,7) cycloocta(1,2,3-*gh*:1′, 2′, 3′-*g*′*h*′)cycloocta(1,2,3-*cd*:5,6,7-*c*′*d*′)dipentalene-1,4,6, 8,10,12,14,17,19,21,23,25-dodecone), was finally established only in 1981, almost 80 years after the initial discovery (Scheme 4.27) [75]. In essence, this spontaneous self-assembly fabrication method outpaced the normal development of rational organic synthesis by decades, even though this inevitably led to "premature birth" of useful molecules lying essentially dormant until the scientific progress would eventually catch up with them. Because of easy access and versatility, cucurbit[*n*]urils play significant role in the applied sciences of molecular belts and tubes [76–83].

The most important member of cucurbit[*n*]uril family, cucurbit[6]uril, is a barrel-shaped molecular belt compound with reported effective diameter of the two portals of about 3.9–4.0 Å and the internal cavity of about 5.5–5.8 Å [75, 83]. The effective height of cucurbit[6]uril is about 9.1 Å, and the calculated

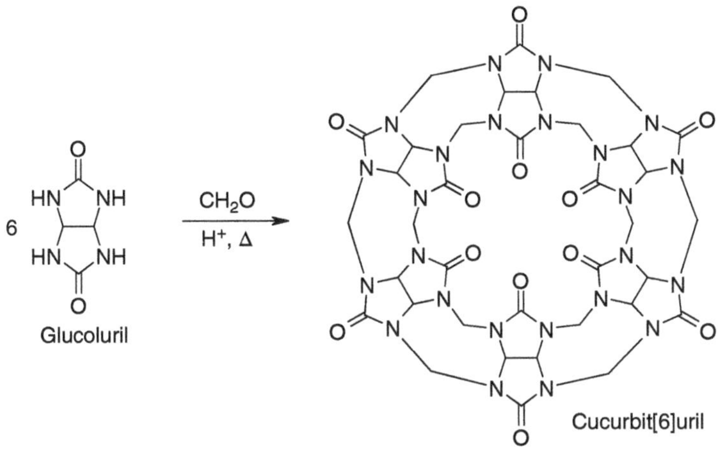

Scheme 4.27 The self-assembly synthesis of cucurbit[6]uril.

effective volume of inner cavity of is approximately 142 Å^3 [83]. The crystal structure of almost perfectly round cucurbit[6]uril molecular belt, taken from a recent publication, is presented in Figure 4.41 [84]. Based on this X-ray structure, the nominal diameter of the cucurbit[6]uril portals is about 6.9 Å (the average maximum distance between the periphery oxygen atoms), the internal cavity diameter is about 10.2 Å (measured between the carbon atoms in the middle of the belt), and the height is about 6.0 Å. The differences between the nominal dimensions and the effective dimensions can be attributed to van der Waals atomic radius adjustments.

The upper and lower rims of cucurbit[6]uril are made of polar, hydrophilic carbonyl groups, whereas the interior is less polar and more hydrophobic. Not surprisingly, this host compound produces strong complexes with certain cationic diammonium salts. In these complexes, two polar cationic centers are

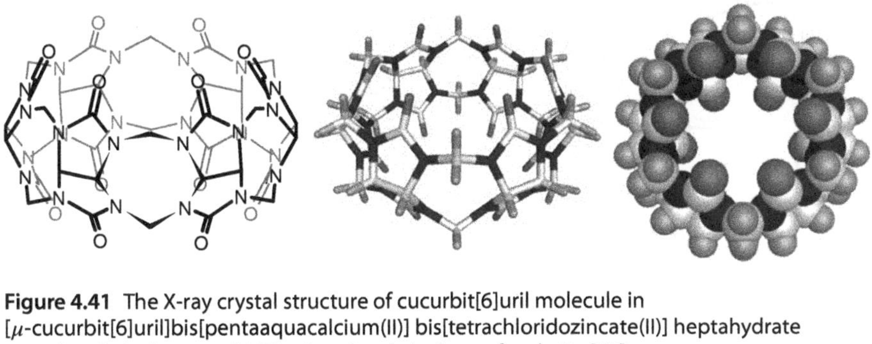

Figure 4.41 The X-ray crystal structure of cucurbit[6]uril molecule in [μ-cucurbit[6]uril]bis[pentaaquacalcium(II)] bis[tetrachloridozincate(II)] heptahydrate complex. Only the cucurbit[6]uril molecule is shown for clarity [84].

located near the polar rims, and less polar alkyl fragments reside deep inside the barrel. This type of binding underscores the importance of ion–dipole interactions and hydrogen bonding between the ammonium groups and the carbonyl oxygen atoms of the host as well as hydrophobic "water-exclusion" effect attributed to the hydrocarbon region of the guest. Ammonium salts with bulkier hydrocarbon moieties (i.e., adamantyl group) are discriminated since they do not fit well into the small cavity of cucurbit[6]uril.

As an example, the X-ray crystal structure of cucurbit[6]uril complex with *para*-phenylenediammonium dication is depicted in Figure 4.42. The cucurbit[6]uril molecular belt in this complex is elliptical to better accommodate the flat guest molecule. The ability of cucurbit[*n*]urils to adapt to the specific shapes of the guest molecules is an important feature of these host molecules, especially notable for larger and more flexible cucurbit[*n*]urils [85].

A detailed investigation of host–guest interactions between cucurbit[6]uril and various α-*n*-alkylammonium and α,ω-*n*-alkylenediammonium salts shows a clear preference toward diammonium salts with the proper spatial matching between the guest and the host molecules. These trends are depicted in Figure 4.43. The best results were achieved with $H_3N^+(CH_2)_6NH_3^+$ and $H_3N^+(CH_2)_5NH_3^+$ dications [86].

The previously described cucurbit[6]uril is the main, but not the only component of the mixture of cucurbit[*n*]urils produced by acid-promoted macrocyclization of glycoluril and formaldehyde. Other members of cucurbit[*n*]urils family can be isolated from this reaction mixture, most notably cucurbit[5]uril, cucurbit[7]uril, and cucurbit[8]uril. Their X-ray crystal structures are presented in Figure 4.44. All of them retain a characteristic barrel-type shape of cucurbit[6]uril, but with more flexibility for larger cucurbit[*n*]urils, as evident in the cases of slightly distorted cucurbit[7]uril and cucurbit[8]uril [87].

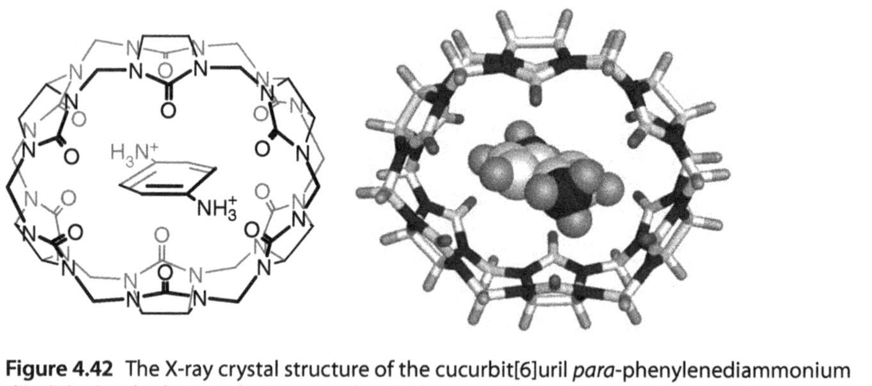

Figure 4.42 The X-ray crystal structure of the cucurbit[6]uril *para*-phenylenediammonium diiodide decahydrate inclusion complex. Only cucurbit[6]uril host molecule and *para*-phenylenediammonium guest molecule are shown for clarity.

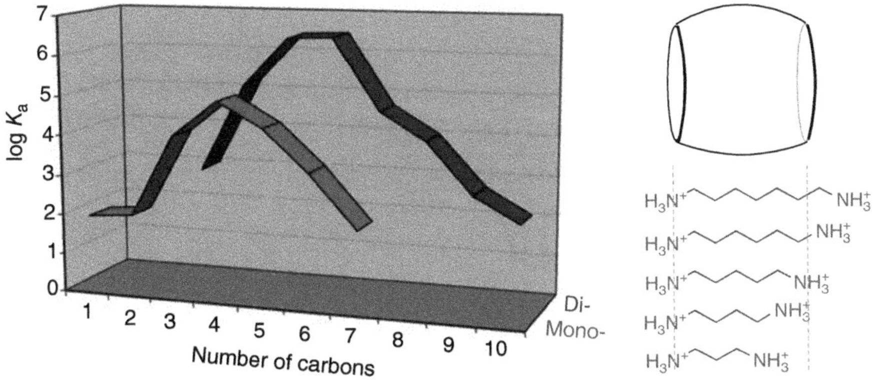

Figure 4.43 Trends in association constants (log K_a, M^{-1}) of complexes of aliphatic α-*n*-alkylammonium (mono-, $CH_3(CH_2)_{n-1}NH_3^+$, front curve) and α,ω-*n*-alkylenediammonium (di-, $H_3N^+(CH_2)_{n-1}NH_3^+$, back curve) salts with cucurbit[6]uril in 85% formic acid–water (1:1) at 40 °C.

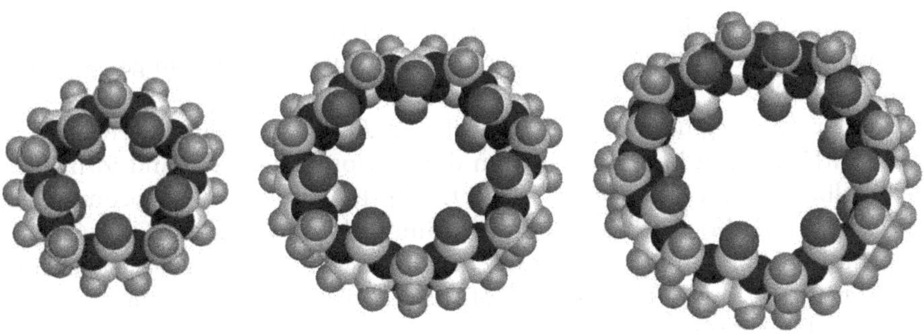

Figure 4.44 The X-ray crystal structures of cucurbit[5]uril (left), cucurbit[7]uril (middle), and cucurbit[8]uril (right).

Higher cucurbit[*n*]urils are also isolated from the reaction mixture (up to cucurbit[14]urils, see Möbius and other twisted belts in Section 4.5) [88], but the separation of pure compounds represents a formidable challenge due to self-aggregation of small and large cucurbit[*n*]urils in matryoshka-type complexes (see Chapter 1). For example, cucurbit[5]uril and cucurbit[10]uril bind together so strongly that they can be separated only via a sophisticated multistep competitive complexation–decomplexation procedure [89]. The internal cavity of cucurbit[10]uril is big enough to accommodate several small guest molecules. The X-ray crystal structure of this large molecular belt is presented in Figure 4.45 [80].

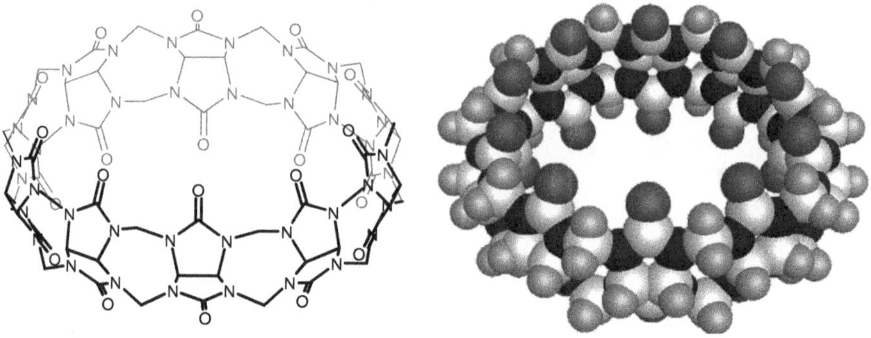

Figure 4.45 The X-ray crystal structure of cucurbit[10]uril.

Substituted cucurbit[*n*]urils are also known. They can be produced either by macrocyclization reactions of glycoluril derivatives or by chemical transformations of already formed cucurbit[*n*]uril molecular belts. For example, decamethylcucurbit[5]uril can be prepared using the first approach from dimethylglycoluril and formaldehyde (Figure 4.46) [89]. This compound was later found to be "capped" with two ammonium cations [90]. "Lid-free" decamethylcucurbit[5]uril with improved affinity toward certain gases is also available [90].

Similar to decamethylcucurbit[5]uril, 2,2-dimethylpropane derivative of cucurbit[5]uril can be prepared from dimethylpropanediurea and formaldehyde, albeit in a very low yield (1–5%) [91, 92]. The solubility of this compound in water (11.5 mM), methanol (5.6 mM), dimethyl sulfoxide (DMSO) (9.6 mM), and dimethylformamide (DMF) (2.2 mM) is higher than the parent cucurbit[5]uril [91]. The X-ray structure of this 2,2-dimethylpropane derivative is presented in Figure 4.47 [91].

Cycloalkanocucurbit[*n*]urils are of particular interest due to their enhanced solubility. They can be prepared by the standard macrocyclization reactions of

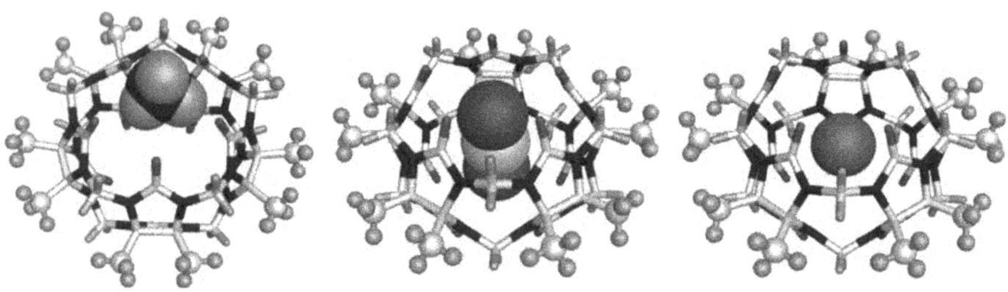

Figure 4.46 The X-ray crystal structures of capped decamethylcucurbit[5]uril (left, only one of the two NH_4^+ cationic lids is presented for clarity) and lid-free complexes of decamethylcucurbit[5]uril with CO_2 (middle) and Kr (right).

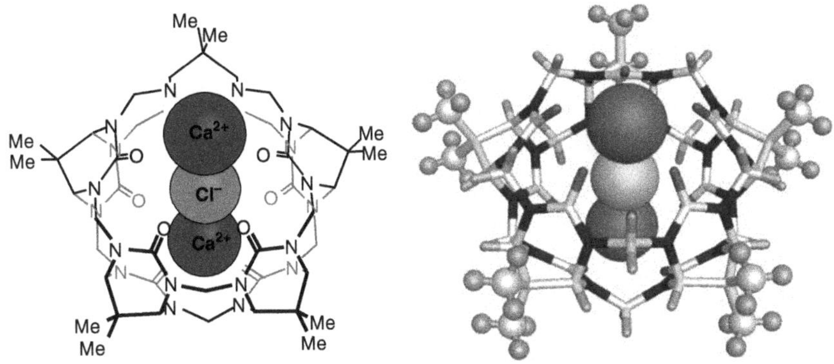

Figure 4.47 The X-ray crystal structure of cucurbit[5]dimethylpropanediuril in (μ-chloro)-(μ-cucurbit[5]dimethylpropanediuril)-diaqua-di-calcium trichloride hydrate complex (only one central Cl^- anion and two Ca^{2+} cations are shown for clarity).

cyclohexanoglycoluril or cyclopentanoglycoluril with formaldehyde in acidic conditions. For example, the reaction of cyclohexanoglycoluril with formaldehyde yields predominantly cyclohexanocucurbit[5]uril (16%), together with small amounts of cyclohexanocucurbit[6]uril (2%), and higher homologs. Both compounds have good solubility in neutral water (up to $\sim 2 \times 10^{-1}$ M) and DMSO (up to $\sim 3 \times 10^{-2}$ M). The X-ray structures of these cucurbit[*n*]uril derivatives are presented in Figure 4.48 [93].

Cyclopentanocucurbit[5]uril, cyclopentanocucurbit[6]uril, and cyclopentanocucurbit[7]uril were similarly prepared from cyclopentanoglycoluril and formaldehyde. In optimized conditions, cyclopentanocucurbit[6]uril is the

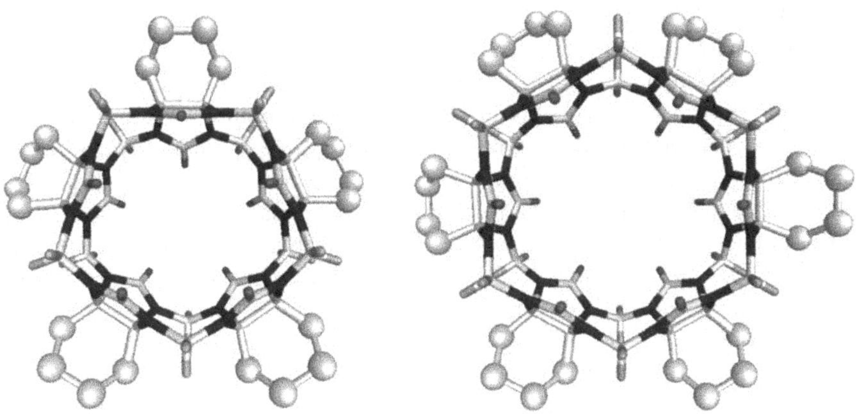

Figure 4.48 The X-ray crystal structures of cyclohexanocucurbit[5]uril (left) and cyclohexanocucurbit[6]uril (right). The hydrogen atoms of cyclohexane rings were omitted for clarity.

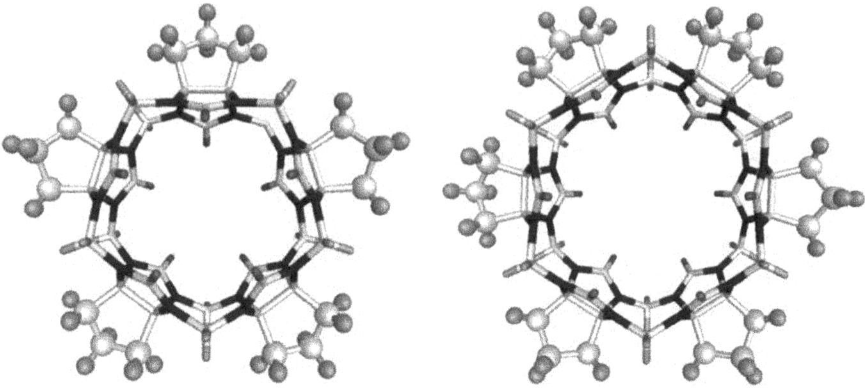

Figure 4.49 The X-ray crystal structures of cyclopentanocucurbit[5]uril (left) and cyclopentanocucurbit[6]uril (right).

main isolated product, followed by cyclopentanocucurbit[5]uril, and, finally, cyclopentanocucurbit[7]uril (this relative content in some reaction conditions can be shifted in favor of cyclopentanocucurbit[5]uril). Higher cyclopentanocucurbit[*n*]urils, such as cyclopentanocucurbit[8]uril, were also detected, but were not isolated. Analysis of the reaction mixtures and separation of the three cyclopentanocucurbit[*n*]uril homologs are facilitated by the use of appropriate guest molecules, such as 1,4-dioxane and 1-adamantylammonium salts, which have selective binding affinities toward cyclopentanocucurbit[5]uril and cyclopentanocucurbit[7]uril, respectively. The X-ray structures of cyclopentanocucurbit[5]uril and cyclopentanocucurbit[6]uril are shown in Figure 4.49 [94].

The second approach (chemical transformations of already formed cucurbit[*n*]uril molecular belts) was successfully employed for the synthesis of dodecahydroxycucurbit[6]uril (Figure 4.50). This compound was prepared by oxidative hydroxylation of cucurbit[6]uril with $K_2S_2O_8$. Unlike unsubstituted cucurbit[6]uril, the dodecahydroxy derivative is soluble in DMSO and moderately soluble in DMF and can be further derivatized via alkylation or acylation of the twelve hydroxy groups. Decahydroxycucurbit[5]uril can be prepared in similar fashion from cucurbit[5]uril, but the yields are much lower for the higher homologs of cucurbit[6]uril [95].

Incorporation of benzene rings into the walls of cucurbit[*n*]urils leads to belt molecules with elliptical internal cavity. This can be achieved by introduction of bis(phthalhydrazide) into the cucurbit[*n*]uril self-assembly synthetic pathway. Due to the presence of several electron-withdrawing groups, the final macrocyclization reaction between glycoluril oligomeric intermediates and bis(phthalhydrazide) requires strong acidic conditions (CH_3SO_2OH). The

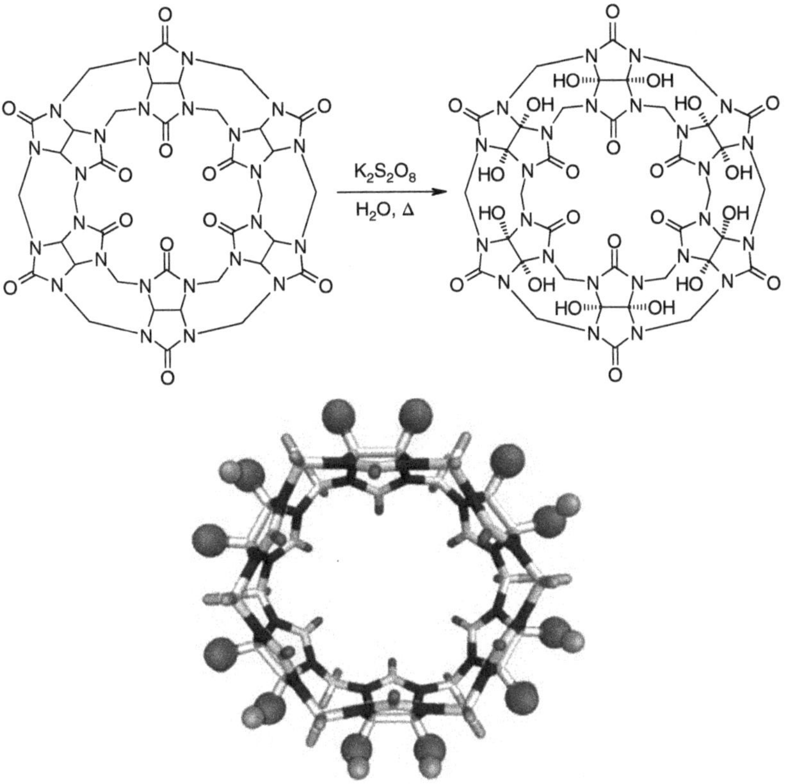

Figure 4.50 The synthesis (top) and the X-ray crystal structure (bottom) of dodecahydroxycucurbit[6]uril.

X-ray crystal structure of one of these cucurbit[*n*]uril analogs is shown in Figure 4.51 [96].

The hollow structure and high affinity toward bifunctional guest molecules make cucurbit[6]uril a valuable building block for the construction of complex interlocked structures, such as molecular necklaces, using self-assembly methodology (Figure 4.52). Successful fabrication of these supramolecular compounds requires a careful selection of appropriate metal cations with optimal coordination geometry suitable for the construction of bridgeheads, counter-anions with desired metal complexation parameters, as well as appropriate flexible links and other assembly elements [97].

The simplest molecular necklace can be constructed using infinite polymeric "string" composed of short individual metal-interconnected bifunctional ligands passing through the cucurbit[6]uril molecular beads. One of the possible synthetic strategies employing this design idea is depicted in Scheme 4.28. With

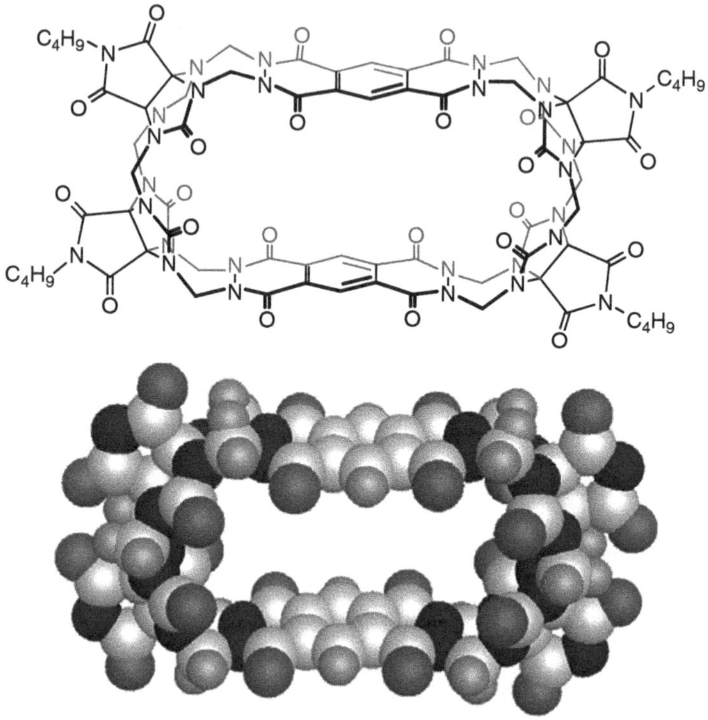

Figure 4.51 The bis(phthalhydrazide)-based cucurbituril analog (top) and its X-ray crystal structure (bottom). The partially disordered *n*-butyl fragments were removed for clarity (bottom).

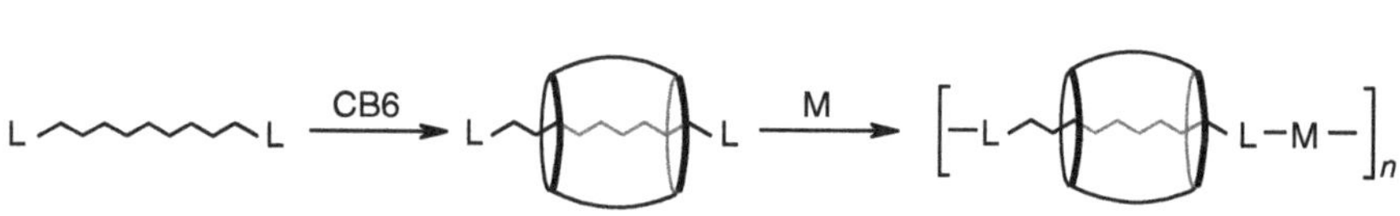

Scheme 4.28 The synthetic strategy for the construction of an infinite molecular necklaces (CB6 – cucurbit[6]uril; M – a metal cation; L – a ligand).

proper execution, this approach is capable of producing polymeric materials with high structural regularity [98].

As a practical example, an infinite molecular necklace was prepared in one step by threading cucurbit[6]uril molecules with *N,N′*-bis(4-pyridylmethyl)-1, 4-diaminobutane dihydrochloride and then adding $Cu(NO_3)_2$ to produce a linear coordination polymer (Figure 4.53). The success of this approach can be attributed to high affinity of cucurbit[6]uril toward 1,4-diaminobutane salts as well as site-selective coordination of exactly two pyridine ligands with copper linking units. This is the first example of a coordination polymer with a

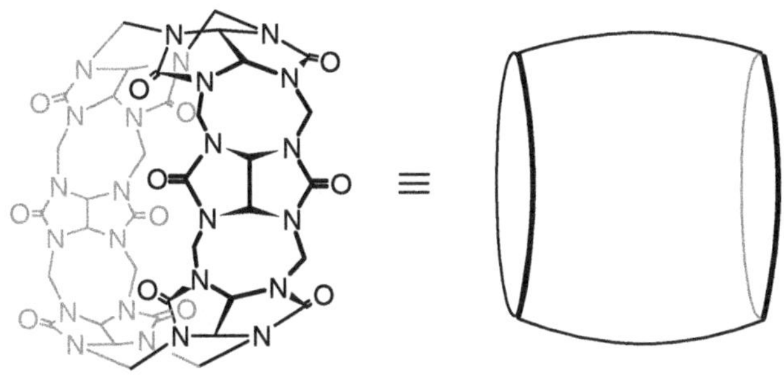

Figure 4.52 A symbolic representation of cucurbit[6]uril molecule as a hollow individual "bead" for the construction of molecular necklaces.

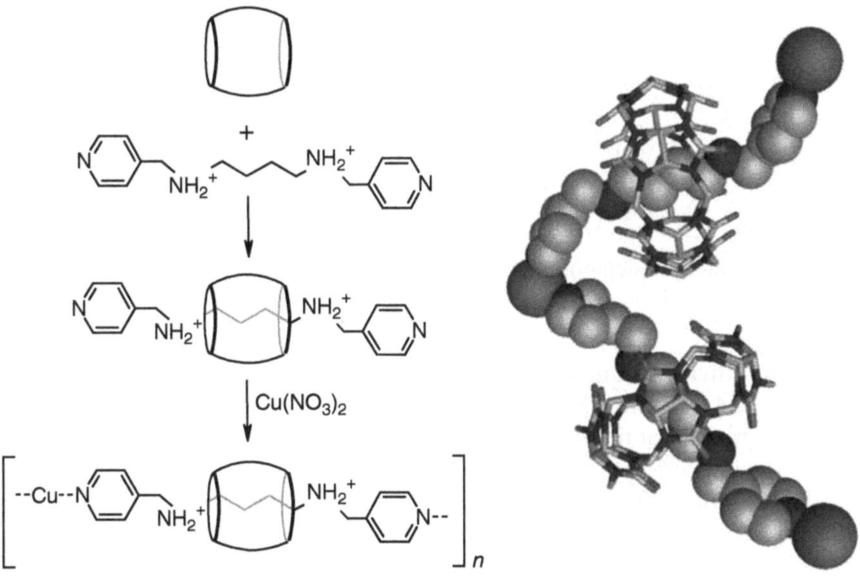

Figure 4.53 The synthesis of the an infinite molecular necklace based on *N,N′*-bis(4-pyridylmethyl)-1,4-diaminobutane (left) and the X-ray structure of a necklace chain fragment (right).

cyclic component at every repeating unit, which is structurally characterized by single-crystal X-ray crystallography (Figure 4.53) [98].

Cyclic molecular necklaces with cucurbit[6]uril beads are also known. The smallest necklace was prepared in one-pot, self-assembly fashion from nine building blocks including three molecular beads, three strings, and three "angle connectors". The synthetic procedure is quite trivial and entails a simple stirring of a 1:1:1 mixture of cucurbit[6]uril (a bead), *N,N′*-bis(4-pyridylmethyl)-1,4-diaminobutane dihydronitrate (a string), and

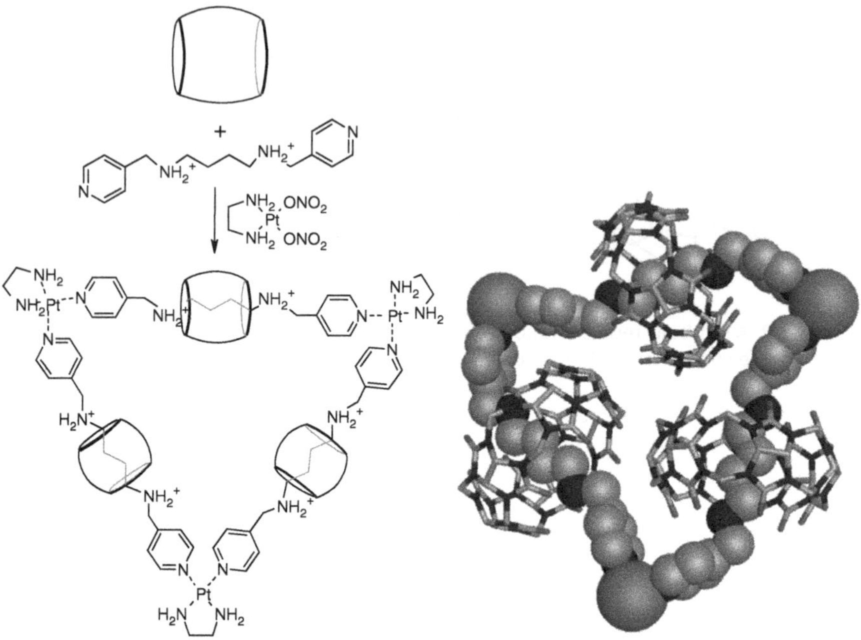

Figure 4.54 The synthesis (left) and the X-ray structure of the smallest molecular necklace (right, ethylenediamine ligands and hydrogen atoms were omitted for clarity).

$Pt(NO_3)_2$–ethylenediamine complex (an angle connector) in boiling water for 24 h (Figure 4.54). The X-ray crystal structure analysis of this cyclic molecular necklace reveals that three cucurbit[6]uril beads are threaded onto a triangular pendant with distinctive triskelion shape. Each corner of the triangle is occupied by the Pt–ethylenediamine linking unit holding the whole molecular necklace assembly together (Figure 4.54) [99].

A slightly different "2 + 2 approach" was used to construct a molecular necklace with four cucurbit[6]uril beads. The key distinctive feature of this synthetic strategy is the use of preorganized, L-shaped molecules with two molecular beads. These L-shaped molecules can then be joined together via appropriate metal angle connectors to produce the molecular necklace with four beads. A practical realization of this strategy is presented in Scheme 4.29 [100].

The X-ray crystal structure analysis of this molecular necklace with four cucurbit[6]uril beads reveals that four molecular beads are threaded over a large (19 × 17 Å) rectangle (Figure 4.55). The cucurbit[6]uril beads are held in place by a network of hydrogen bonds, predominantly between their carbonyl oxygen atoms and the ammonium cationic groups of the necklace pendant, with some additional coordination to the copper metal centers. The large inner cavity of this assembly is filled with water molecules and nitrate anions [100].

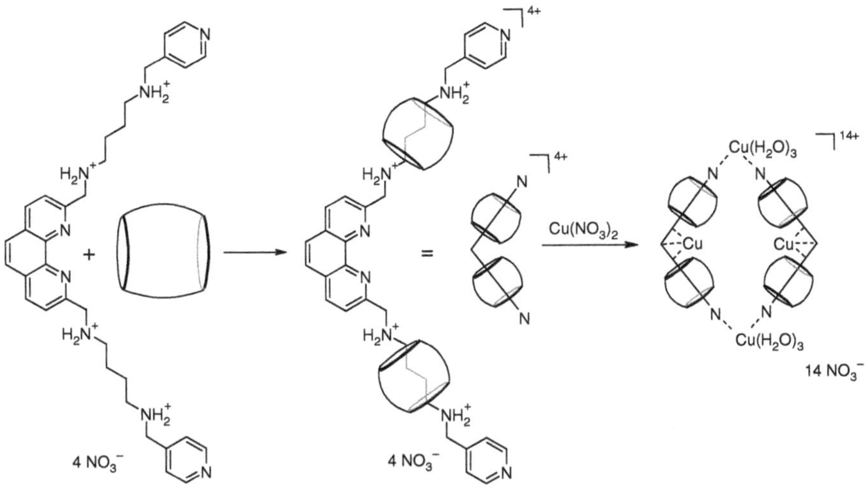

Scheme 4.29 The synthesis of the molecular necklace with four cucurbit[6]uril beads.

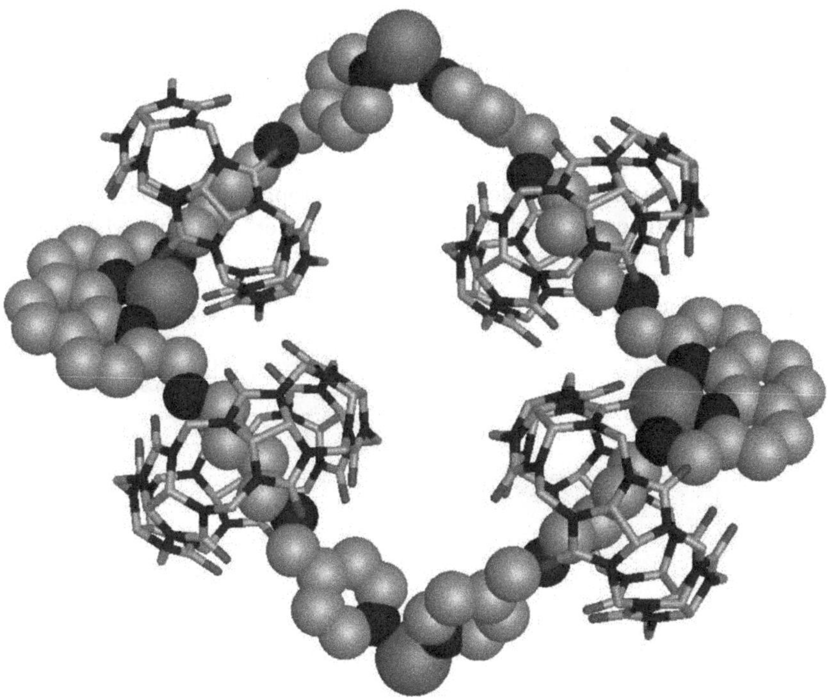

Figure 4.55 The X-ray structure of the molecular necklace with four cucurbit[6]uril beads. The water molecules, nitrate anions, and hydrogen atoms were omitted for clarity.

4.5 Möbius and Other Twisted Molecular Belts

Möbius and other twisted belts are topological variants of "traditional" belts, even though in scientific papers they are frequently named as strips or bands, not belts, to reflect their unique spatial features. Mathematically, Möbius strip is an example of nonorientable, one-sided two-dimensional surface with a simple closed-curve boundary, and, therefore, its edge is homomorphic (topologically equivalent) to a circle. Möbius band is a special chiral object: a half-twist clockwise or counterclockwise produces *right-* or *left-*handed structures, while maintaining the abovementioned homomorphism. Half-twisted (Möbius) belts have found sporadic applications in conventional technology. Examples include abrasive belts with even side distribution of wear, continuous-loop recording tapes with double play time, and typewriter ribbons with twice the effective area of utility. They were also proposed as noninductive resistors, as compact resonators with half of the resonance frequency, and as other electronic components. Möbius themes have been artistically explored in fashion industry. Some examples of conventional Möbius-type devices and fashion items are presented in Figure 4.56.

Molecular twisted belts are rare – no general synthetic approach has been developed so far [101, 102]. Some hypothetical [16]cyclacene-based twisted belts are depicted in Figure 4.57. Obviously, they would have even higher macrocyclic strain energy than the hitherto unknown [16]cyclacene (Figure 4.57), and therefore might not be easily accessible via common synthetic methodologies.

First Möbius half-twisted molecular belt was reported in 1982 [103]. It was prepared by macrocyclization of a "rope-ladder" diol-ditosylate with

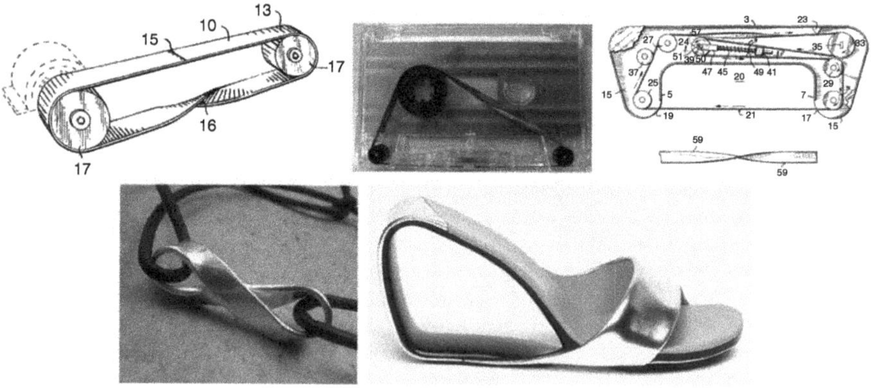

Figure 4.56 The conventional Möbius-type devices and fashion items. *Source*: Courtesy of Harris Owen, US Patent 2308309; Barrie Sutcliffe; Kondur Nicholas, US Patent 3621968; Amy DeLorenzo, PartsbyNC; Rem D. Koolhaas, United Nude.

Figure 4.57 The energy-minimized hypothetical structures of "normal" [16]cyclacene (left), half-twisted (Möbius) [16]cyclacene (middle), and full-twisted [16]cyclacene (right).

three ethylene units interconnected with polyether links from both sides. Under high dilution conditions, this diol-ditosylate yields three distinct belt molecules: one "traditional" untwisted belt and two enantiomeric half-twisted Möbius-type belts (Scheme 4.30).

As expected, ozonolysis of these topological isomers produces distinctively different molecules. Traditional, untwisted belt (Scheme 4.31) yields two molecules of a macrocyclic triketone, whereas both Möbius-twisted enantiomers produce the same double-sized macrocyclic hexaketone (Scheme 4.31) [104].

Another notable example of a twisted belt molecule is giant cucurbit[14]uril. This is a prized compound produced via the self-assembly pathway since it is not only the largest cucurbituril ever documented but also a cucurbituril molecule with a full twist. The compound was isolated in low yield (1.2%) from a reaction mixture produced by the standard macrocyclization reaction between glycoluril and formaldehyde in acidic conditions. The twisted belt nature of cucurbit[14]uril is confirmed by X-ray crystallography of its Eu^{3+} complex (Figure 4.58). Due to its twisted shape, cucurbit[14]uril is fairly soluble in water (up to 4.1×10^{-2} M) and DMSO (up to 1.8×10^{-2} M) [105].

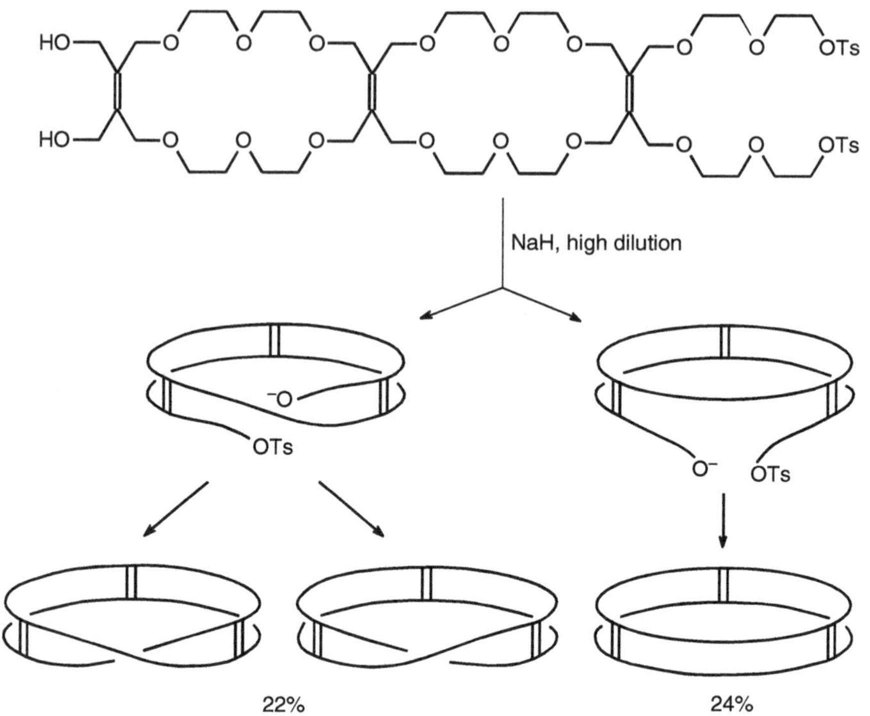

Scheme 4.30 The synthesis of the two enantiomeric Möbius-type half-twisted belts (bottom left and middle) and the untwisted "traditional" belt (bottom right).

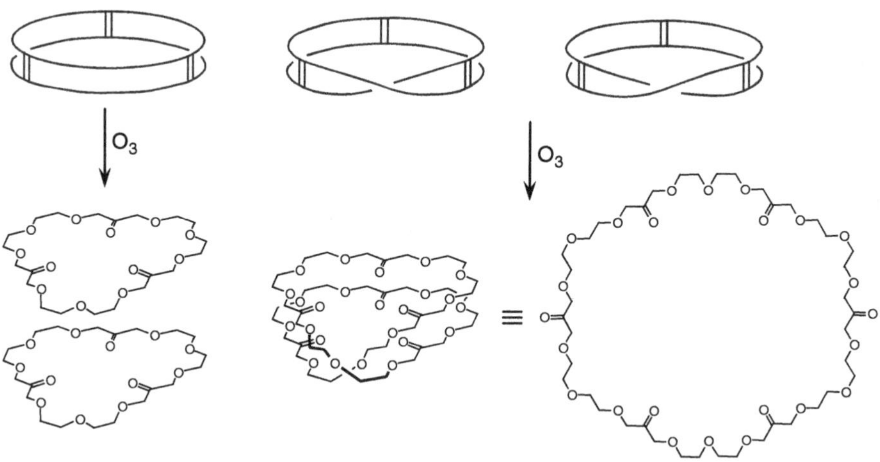

Scheme 4.31 Ozonolysis of the untwisted traditional molecular belt (left) and the two enantiomeric Möbius belts (middle, right).

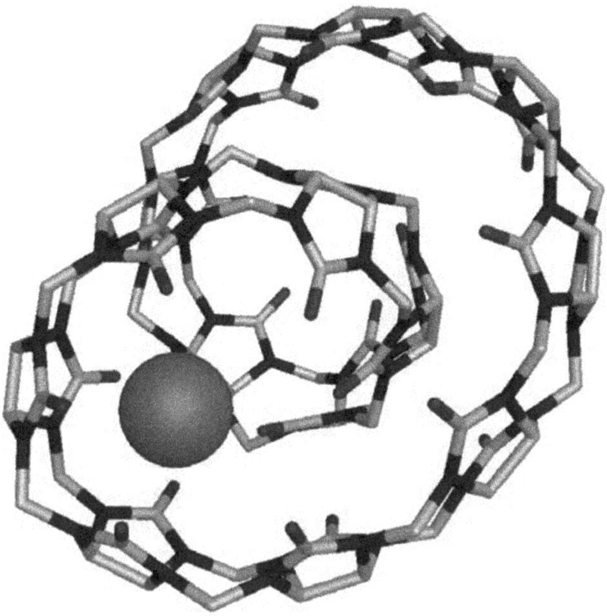

Figure 4.58 The X-ray crystal structure of the Eu^{3+} complex of twisted cucurbit[14]uril. The hydrogen atoms were removed for clarity.

Twisted and half-twisted expanded porphyrin pseudobelts are of theoretical interest with respect to the so-called Möbius-type aromaticity, which, unlike "normal", Hückel aromaticity, requires $4n$ cyclic π-electrons instead of $4n+2$ π-electrons [106, 107]. Due to the inherent conformational flexibility of pseudobelts, some expanded porphyrins display a unique dynamics between Möbius-aromatic and Hückel-antiaromatic states. Crystal structures of a half-twisted Möbius-aromatic porphyrin with 28 π-electrons [106] and a fully twisted "figure-eight" Hückel-antiaromatic porphyrin with 32 π-electrons [107] are presented in Figure 4.59. The former compound undergoes rapid exchange between Möbius-aromatic and Hückel-antiaromatic conformations in solution at elevated temperatures [106].

4.6 Applications

Molecular belts and tubes were proposed for numerous applications, primarily in the biomedical field and material sciences. Other active areas of applied research include sensors, chemical separations, and catalysis. Early examples include an attempt to use [12]cyclacene derivatives attached to the surface of a piezoelectric quartz crystal as chemical sensors. In the presence of an

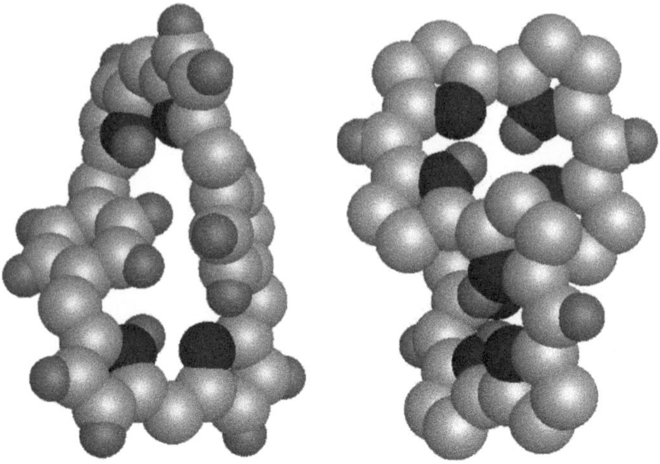

Figure 4.59 The X-ray crystal structures of the half-twisted Möbius porphyrin (left) and the fully twisted "figure-eight" porphyrin with 32 π-electrons (right). Only core porphyrin-based macrocycles are shown for clarity.

appropriate guest, the oscillation frequency of the crystal changes due to mass increase. A high sensitivity toward nitrobenzene was demonstrated with the use of a hexaepoxyoctacosahydro[12]cyclacene derivative [108].

Some large [*n*]cycloparaphenylenes ($n = 8–13$) exhibit strong fluorescence with emission maxima ranging from about 530 nm (for $n = 8$) to about 425 nm (for $n = 13$) and can be used as molecular electronic components and censoring elements (Figure 4.60) [109]. It is interesting to note that the adsorption maximum at around 340 nm remains almost the same regardless of the ring size for these large [*n*]cycloparaphenylenes with *n* above 7 [109], but it starts shifting to the lower wavelengths with *n* below 7, particularly for the smallest [5]cycloparaphenylene with $\lambda_{max} = 335$ nm [46]. This shift is also associated with almost complete loss of fluorescence for small [*n*]cycloparaphenylenes ($n = 6$ to 5), presumably due to bending of the benzene subunits [46].

There was also a proposal to use [*n*]cycloparaphenylenes for the separation of various fullerene molecules, particularly endohedral fullerenes. The idea is supported by experimental data indicating that, with the proper size matching, some [*n*]cycloparaphenylenes can differentiate fullerene molecules based on the size and electron density. For example, solubility differences between the complexes of various endohedral metallofullerenes and [11]cycloparaphenylene were successfully employed for the selective precipitation of $Gd@C_{82}$-C_{2v} (Figure 4.61) [55].

The preponderance of practical applications of molecular tubes is actually associated with the use of the oldest class of technomimetic molecules – cucurbiturils [79, 83, 90, 110, 111]. One of the opportunities in

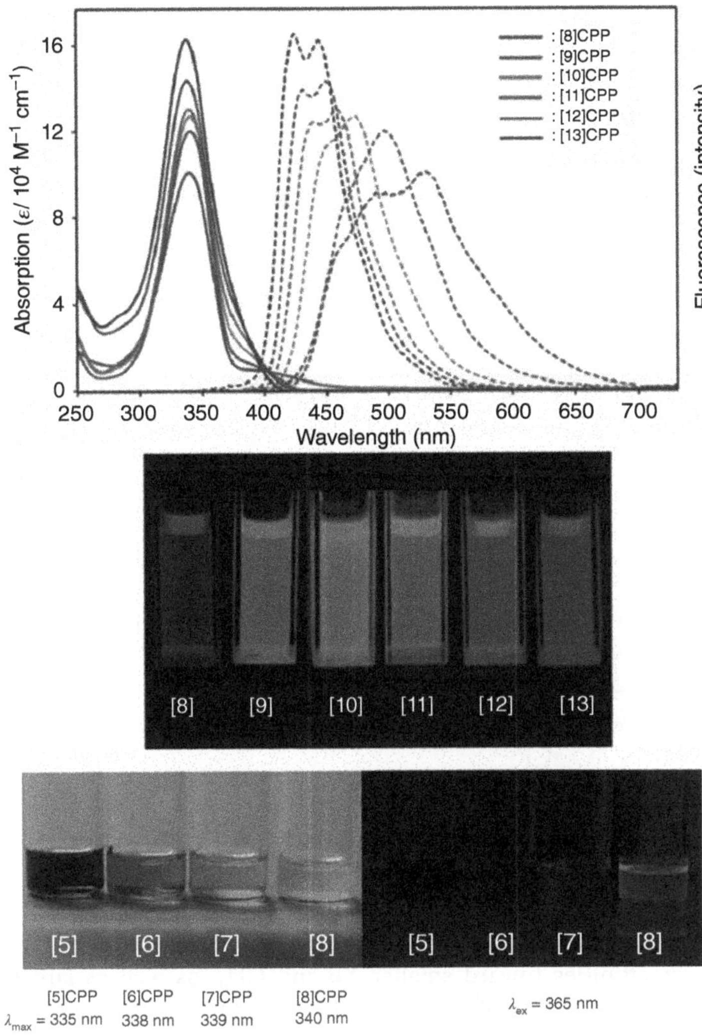

Figure 4.60 The fluorescent and adsorption properties of [*n*]cycloparaphenylenes. *Source*: Adapted with permission from Iwamoto *et al.* 2011 (top) [109], and Golder and Jasti 2015 (bottom) [46]. Copyright 2011 and 2015 American Chemical Society.

applied science of cucurbiturils is related to the presence of well-defined, rigid hydrophobic cavities in these molecular hosts. This, in turn, translates in selective binding of some gases – a valuable feature for separation technologies. The best association constants can be achieved with size matching between the host cucurbituril molecules and the guest gas molecules. For example, cucurbit[6]uril has high binding affinities toward SF_6, C_3H_8, and C_4H_{10}, but

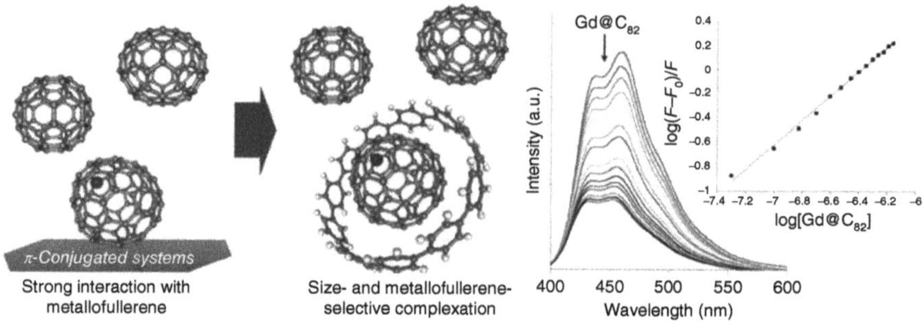

Figure 4.61 The proposed strategy for selective complexation and subsequent separation of metallofullerenes (left) and the fluorescence spectra of [11]cycloparaphenylene (λ_{exc} = 370 nm) in toluene in the presence of Gd@C_{82}-C_{2v} (right, $K_a = 1.8 \times 10^6$ M^{-1}). *Source:* Adapted from Nakanishi *et al.* 2014 [55] and Lewis 2015 [10]. Reproduced with permission of The Royal Society of Chemistry.

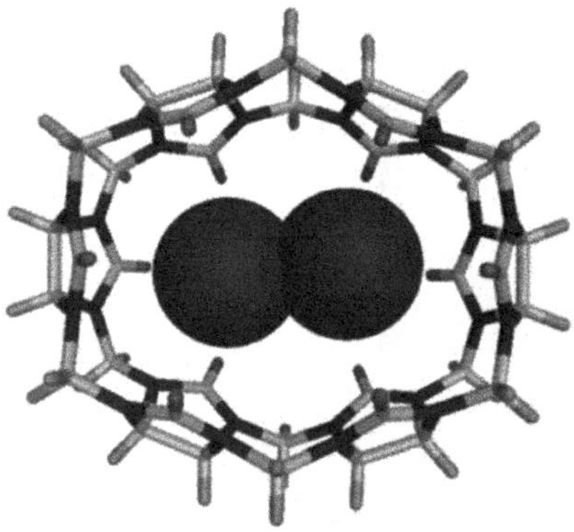

Figure 4.62 The X-ray crystal structure of the cucurbit[6]uril complex with Br_2 [112].

substantially lower affinities toward smaller Xe and CH_4, as well as larger C_6H_{14} [110]. The X-ray structure of a rare complex of cucurbit[6]uril with the reactive Br_2 guest molecule is presented in Figure 4.62 [112]. Cucurbiturils are known to form extremely strong complexes with properly designed diammonium guest molecules. For example, the complexes of diamantane bis(trimethylammonium) salts with cucurbit[7]uril have association constants in excess of 10^{15} M^{-1} in water-based solutions, which is on par with the biotin–avidin pair frequently used in biomedicine as a part of various detection and purification systems. The X-ray structure of one of these complexes is presented in Figure 4.63 [113].

Good binding affinities of cucurbiturils toward some drug molecules make them attractive drug delivery carriers [111]. Drug delivery applications require larger internal cavity volumes, so larger cucurbit[7]uril and cucurbit[8]uril are

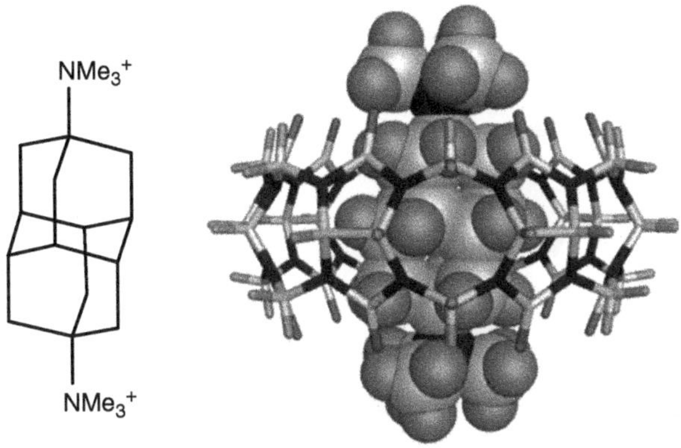

Figure 4.63 Diamantane bis(trimethylammonium) bis(iodide) (left) and the X-ray structure of its complex with cucurbit[7]uril (right). The iodide counteranions were removed for clarity.

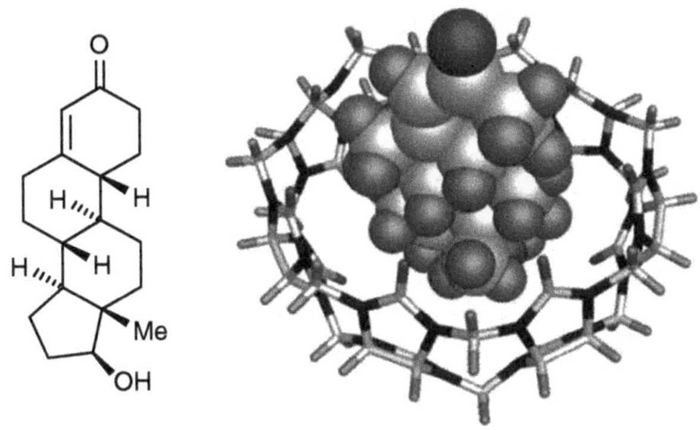

Figure 4.64 Nandrolone (left) and the X-ray crystal structure of its complex with cucurbit[7]uril (right) [114].

the most frequently used molecular hosts. The use of these cucurbit[*n*]urils can substantially improve the bioavailability of drugs while at the same time provide better photostability and protection against chemical degradation [111]. Recent examples include nanomolar binding of steroids to cucurbit[*n*]urils (Figure 4.64) [114]. In many cases, cucurbituril encapsulation of steroid molecules results in improvements of their solubility and chemical stability [114].

There are also several reports outlining the ability of cucurbiturils to form strong complexes with medically important regulatory proteins, such as insulin. The X-ray structure of human insulin complex with cucurbit[7]uril is presented in Figure 4.65 [115]. In this complex, the aromatic side chain of the N-terminal phenylalanine residue is encapsulated in the cavity of cucurbit[7]uril. The complex is relatively strong, with reported $K_a = 1.5 \times 10^6\ M^{-1}$.

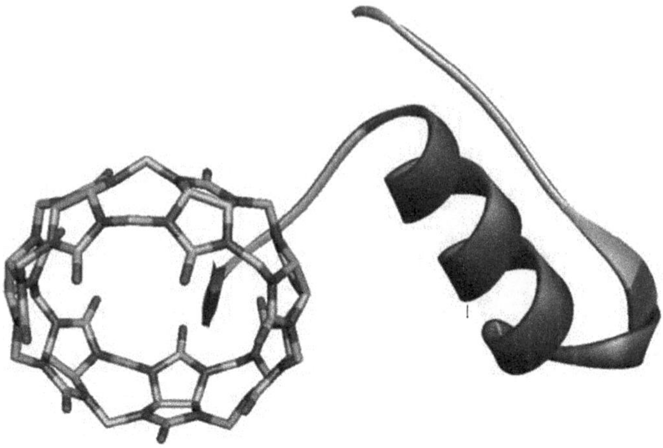

Figure 4.65 The X-ray crystal structure of the human insulin complex with cucurbit[7]uril. The hydrogen atoms of the cucurbit[7]uril moiety were removed for clarity.

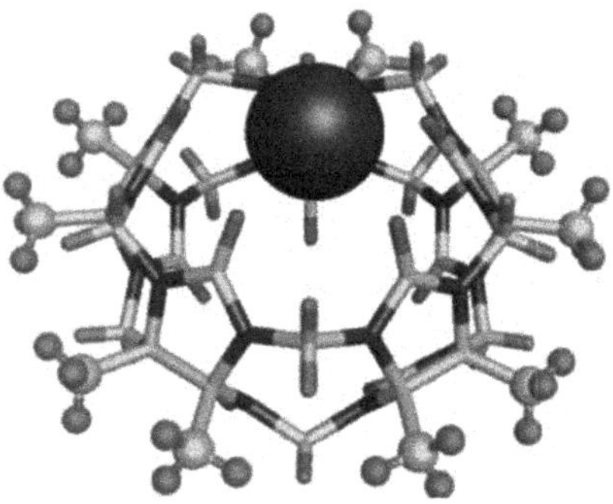

Figure 4.66 The X-ray crystal structure of the $AgNO_3$ complex with decamethylcucurbit [5]uril. Only one Ag^+ cation is presented for clarity [116].

Perhaps the most promising application field of cucurbiturils is related to catalysis. In the simplest case, the metal catalytic nanoparticles can be prepared from the complexes of cucurbiturils with appropriate metal cations. The X-ray structure of one of these complexes with Ag^+ is presented in Figure 4.66. The subsequent reduction of this complex with H_2 provides an easy access to Ag^0 nanoparticles with excellent catalytic activity in the $NaBH_4$ reduction reaction of industrially important nitrophenols [116]. Similar Pd^0 catalytic nanoparticles were also prepared from the complex of $PdCl_2$ with cucurbit[6]uril [117].

The presence of two open orifices in cucurbiturils presents an opportunity to use them as enzyme-type catalytic reactors. In essence, two properly functionalized reactants can reach simultaneously the internal cavity of these molecular belts from both ends. This internal cavity will then provide

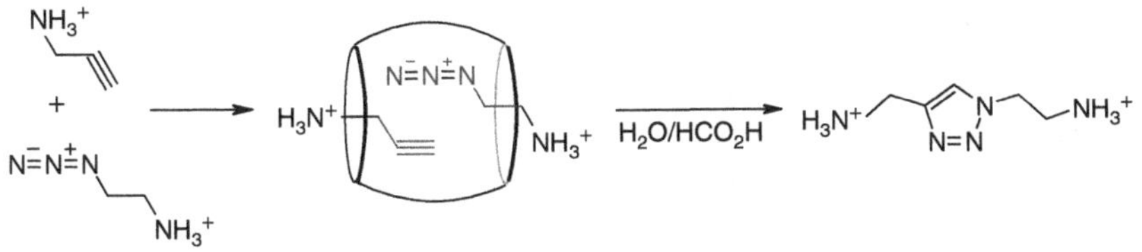

Scheme 4.32 The cycloaddition [2 + 3] click reaction catalyzed by cucurbit[6]uril [118].

a proper closed environment to improve the speed of the desired chemical transformation. This enzyme-type catalytic methodology was successfully demonstrated by employing two alkylammonium reagents equipped with diene and dienophile reactive sites (Scheme 4.32) [118]. The pericyclic reaction is markedly accelerated by a factor of 55,000 in the presence of cucurbit[6]uril. This catalytic reaction displays a number of enzyme-like features. Thus, the reaction displays saturation behavior: the rate of the reaction becomes independent of substrate concentrations with sufficient amounts of the diene and the dienophile reagents. Also, high concentrations of the diene slow down the cycloaddition (substrate inhibition), and the release of 1,2,3-triazole product from its complex with cucurbit[6]uril is found to be the rate-determining step. This cucurbit[6]uril-catalyzed "click" reaction was subsequently used for the synthesis of various molecular switches and valves [119, 120]. Other notable examples of cucurbituril catalysis include photochemical reactions inside the cavity [121–126], hydrolysis [127], oxidation [128, 129], and other reactions [130].

In summary, molecular belts and tubes, prepared by rational organic synthesis methodologies, represent a valuable class of technomimetic molecules with a broad appeal in various areas of applied research, including separation technologies, biomedical applications, sensors, and catalysis. While their applied science is only in early, exploratory stages of development, the presented examples clearly demonstrate a possibility of attractive commercial opportunities in the near future. Many belt and tube molecules are also of academic interest, including topological aspects of aromaticity as well as organic reactions in confined space. The recent rapid flow of research papers reflects continuing attention to this vibrant area of research.

References

1 Schröder, A., Mekelburger, H.-B., and Vögtle, F. (1994) Belt-, ball-, and tube-shaped molecules, in *Cyclophanes*, Topics in Current Chemistry, vol. **172** (ed. E. Weber), Springer-Verlag, Berlin/Heidelberg, pp. 179–201.

2 Tahara, K. and Tobe, Y. (2006) Molecular loops and belts. *Chem. Rev.*, **106**, 5274–5290.
3 Gleiter, R., Hellbach, B., Gath, S., and Schaller, R.J. (2006) From superphanes to beltenes. *Pure Appl. Chem.*, **78**, 699–706.
4 Schaller, G.R. and Herges, R. (2011) Aromatic belts as sections of nanotubes, in *Fragments of Fullerenes and Carbon Nanotubes: Designed Synthesis, Unusual Reactions, and Coordination Chemistry* (eds I.A. Petrukhina and L.T. Scott), John Wiley & Sons, pp. 259–289.
5 Eisenberg, D., Shenhar, R., and Rabinovitz, M. (2010) Synthetic approaches to aromatic belts: building up strain in macrocyclic polyarenes. *Chem. Soc. Rev.*, **39**, 2879–2890.
6 Gleiter, R., Esser, B., and Kornmayer, S.C. (2009) Cyclacenes: hoop-shaped systems composed of conjugated rings. *Acc. Chem. Res.*, **42**, 1108–1116.
7 Iyoda, M., Yamakawa, J., and Rahman, M.J. (2011) Conjugated macrocycles: concepts and applications. *Angew. Chem. Int. Ed.*, **50**, 10522–10553.
8 Herges, R. (2004) Fully conjugated beltenes (belt-like and tubular aromatics), in *Modern Cyclophane Chemistry* (eds R. Gleiter and H. Hopf), Wiley-VCH, pp. 337–358.
9 Bunz, U.H.F., Menning, S., and Martín, N. (2012) *para*-Connected cyclophenylenes and hemispherical polyarenes: building blocks for single-walled carbon nanotubes? *Angew. Chem. Int. Ed.*, **51**, 7094–7101.
10 Lewis, S.E. (2015) Cycloparaphenylenes and related nanohoops. *Chem. Soc. Rev.*, **44**, 2221–2304.
11 Evans, P.J. and Jasti, R. (2012) Molecular belts, in *Polyarenes I*, Topics in Current Chemistry, vol. **349** (eds J.S. Siegel and Y.-T. Wu), Springer, Berlin/Heidelberg, pp. 249–290.
12 Feuerbacher, N. and Vögtle, F. (1998) Iterative synthesis in organic chemistry, in *Dendrimers*, Topics in Current Chemistry, vol. **197** (ed. F. Vögtle), Springer, Berlin/Heidelberg, pp. 1–18.
13 Kohnke, F.H. and Stoddart, J.F. (1989) The evolution of molecular belts and collars. *Pure Appl. Chem.*, **61**, 1581–1586.
14 Mathias, J.P. and Stoddart, J.F. (1992) Constructing a molecular LEGO set. *Chem. Soc. Rev.*, **21**, 215–225.
15 Kohnke, F.H., Slawin, A.M.Z., Stoddart, J.F., and Williams, D.J. (1987) Molecular belts and collars in the making: a hexaepoxyoctacosahydro[12]cyclacene derivative. *Angew. Chem. Int. Ed.*, **26**, 892–894.
16 Ashton, P.R., Brown, G.R., Isaacs, N.S. *et al.* (1992) Molecular LEGO. 1. Substrate-directed synthesis via stereoregular Diels–Alder oligomerizations. *J. Am. Chem. Soc.*, **114**, 6330–6353.
17 Ashton, P.R., Isaacs, N.S., Kohnke, F.H. *et al.* (1988) Towards the making of [12]collarene. *Angew. Chem. Int. Ed.*, **27**, 966–969.
18 Ashton, P.R., Isaacs, N.S., Kohnke, F.H. *et al.* (1989) Trinacrene – a product of structure-directed synthesis. *Angew. Chem. Int. Ed.*, **28**, 1261–1263.

19 Ashton, P.R., Girreser, U., Giuffrida, D. *et al.* (1993) Molecular belts. 2. Substrate-directed syntheses of belt-type and cage-type structures. *J. Am. Chem. Soc.*, **115**, 5422–5429.

20 Benkhoff, J., Boese, R., Klärner, F.-G., and Wigger, A.E. (1994) Synthesis of sterically rigid macrocycles by the use of pressure-induced repetitive Diels–Alder reactions. *Tetrahedron Lett.*, **35**, 73–76.

21 Godt, A., Enkelmann, V., and Schlüter, A. (1989) Double-stranded molecules: a [6]beltene derivative and the corresponding open-chain polymer. *Angew. Chem. Int. Ed.*, **28**, 1680–1682.

22 Cory, R.M., McPhail, C.L., Dikmans, A.J., and Vittal, J.J. (1996) Macrocyclic cyclophane belts via double Diels–Alder cycloadditions: macroannulation of bisdienes by bisdienophiles. Synthesis of a key precursor to an [8]cyclacene. *Tetrahedron Lett*, **37**, 1983–1986.

23 Cory, R.M. and McPhail, C.L. (1996) Transformations of a macrocyclic cyclophane belt into advanced [8]cyclacene and [8]cyclacene triquinone precursors. *Tetrahedron Lett.*, **37**, 1987–1990.

24 Lou, K., Prior, A.M., Wiredu, B. *et al.* (2010) Synthesis of cyclododeciptycene quinones. *J. Am. Chem. Soc.*, **132**, 17635–17641.

25 Neudorff, W.D., Lentz, D., Anibarro, M., and Schlüter, A.D. (2003) The carbon skeleton of the belt region of fullerene C_{84} (D_2). *Chem. Eur. J.*, **9**, 2745–2757.

26 Stuparu, M., Gramlich, V., Stanger, A., and Schlüter, A.D. (2006) Double-stranded cycles: toward C_{84}'s belt region. *J. Org. Chem.*, **72**, 424–430.

27 Denekamp, C., Etinger, A., Amrein, W. *et al.* (2008) Towards a fully conjugated, double-stranded cycle: a mass spectrometric and theoretical study. *Chem. Eur. J.*, **14**, 1628–1637.

28 Standera, M., Häfliger, R., Gershoni-Poranne, R. *et al.* (2011) Evidence for fully conjugated double-stranded cycles. *Chem. Eur. J.*, **17**, 12163–12174.

29 Nakamura, E., Tahara, K., Matsuo, Y., and Sawamura, M. (2003) Synthesis, structure, and aromaticity of a hoop-shaped cyclic benzenoid [10]cyclophenacene. *J. Am. Chem. Soc.*, **125**, 2834–2835.

30 Scott, L.T., Jackson, E.A., Zhang, Q. *et al.* (2011) A short, rigid, structurally pure carbon nanotube by stepwise chemical synthesis. *J. Am. Chem. Soc.*, **134**, 107–110.

31 Yang, H., Beavers, C.M., Wang, Z. *et al.* (2010) Isolation of a small carbon nanotube: the surprising appearance of D_{5h}(1)-C_{90}. *Angew. Chem. Int. Ed.*, **49**, 886–890.

32 Esser, B., Rominger, F., and Gleiter, R. (2008) Synthesis of $[6.8]_3$cyclacene: conjugated belt and model for an unusual type of carbon nanotube. *J. Am. Chem. Soc.*, **130**, 6716–6717.

33 Esser, B., Bandyopadhyay, A., Rominger, F., and Gleiter, R. (2009) From metacyclophanes to cyclacenes: synthesis and properties of $[6.8]_3$cyclacene. *Chem. Eur. J*, **15**, 3368–3379.
34 Kang, H.C., Hanson, A.W., Eaton, B., and Boekelheide, V. (1985) $[2_6]$(1,2,4,5)Cyclophane (deltaphane) and related compounds. Simultaneous π-electron interaction among three benzene rings. *J. Am. Chem. Soc.*, **107**, 1979–1985.
35 Vögtle, F., Schröder, A., and Karbach, D. (1991) Strategy for the synthesis of tube-shaped molecules. *Angew. Chem. Int. Ed. Engl.*, **30**, 575–577.
36 Josten, W., Karbach, D., Nieger, M. *et al.* (1994) Gürtelförmige moleküle mittels repetitiver synthesestrategie. *Chem. Ber.*, **127**, 767–777.
37 Josten, W., Neumann, S., Vögtle, F. *et al.* (1994) Ein neues, deltaphan und bandförmige cyclophane. *Chem. Ber.*, **127**, 2089–2096.
38 Aratani, N. and Osuka, A. (2015) Exploration of giant functional porphyrin arrays. *Bull. Chem. Soc. Jpn.*, **88**, 1–27.
39 Song, J., Aratani, N., Shinokubo, H., and Osuka, A. (2010) A porphyrin nanobarrel that encapsulates C_{60}. *J. Am. Chem. Soc.*, **132**, 16356–16357.
40 Neuhaus, P., Cnossen, A., Gong, J.Q. *et al.* (2015) A molecular nanotube with three-dimensional π-conjugation. *Angew. Chem. Int. Ed.*, **54**, 7344–7348.
41 Hellbach, B., Rominger, F., and Gleiter, R. (2004) Synthesis of beltenes by reactions of 5,6,11,12-tetradehydrodibenzo[*a*,*e*]cyclooctene with $[CpCo(CO)_2]$ derivatives. *Angew. Chem. Int. Ed.*, **43**, 5846–5849.
42 Schaller, R.J., Gleiter, R., Hofmann, J., and Rominger, F. (2002) Beltlike macrocycles with four and eight (cyclopentadienyl)(cyclobutadienyl)cobalt building blocks, formed by stepwise oligomerization of 1,6-cyclodecadiyne units. *Angew. Chem. Int. Ed.*, **41**, 1181–1183.
43 Schwierz, H. and Vögtle, F. (1999) Linear and macrocyclic molecular thread, ribbon and belt assemblies of nanometric scale. *Synthesis*, **1999**, 295–305.
44 Wong, B.M. (2009) Optoelectronic properties of carbon nanorings: excitonic effects from time-dependent density functional theory. *J. Phys. Chem. C*, **113**, 21921–21927.
45 Friederich, R., Nieger, M., and Vögtle, F. (1993) Auf dem weg zu makrocyclischen *para*-phenylenen. *Chem. Ber.*, **126**, 1723–1732.
46 Golder, M.R. and Jasti, R. (2015) Syntheses of the smallest carbon nanohoops and the emergence of unique physical phenomena. *Acc. Chem. Res.*, **48**, 557–566.
47 Evans, P.J., Darzi, E.R., and Jasti, R. (2014) Efficient room-temperature synthesis of a highly strained carbon nanohoop fragment of buckminsterfullerene. *Nat. Chem.*, **6**, 404–408.
48 Kayahara, E., Patel, V.K., and Yamago, S. (2014) Synthesis and characterization of [5]cycloparaphenylene. *J. Am. Chem. Soc.*, **136**, 2284–2287.

49 Kayahara, E., Iwamoto, T., Suzuki, T., and Yamago, S. (2013) Selective synthesis of [6]-, [8]-, and [10]cycloparaphenylenes. *Chem. Lett.*, **42**, 621–623.
50 Xia, J. and Jasti, R. (2012) Synthesis, characterization, and crystal structure of [6]cycloparaphenylene. *Angew. Chem. Int. Ed.*, **51**, 2474–2476.
51 Sibbel, F., Matsui, K., Segawa, Y. *et al.* (2013) Selective synthesis of [7]- and [8]cycloparaphenylenes. *Chem. Commun.*, **50**, 954–956.
52 Sisto, T.J., Golder, M.R., Hirst, E.S., and Jasti, R. (2011) Selective synthesis of strained [7]cycloparaphenylene: an orange-emitting fluorophore. *J. Am. Chem. Soc.*, **133**, 15800–15802.
53 Xia, J., Bacon, J.W., and Jasti, R. (2012) Gram-scale synthesis and crystal structures of [8]- and [10]CPP, and the solid-state structure of C_{60}@[10]CPP. *Chem. Sci.*, **3**, 3018–3021.
54 Zabula, A.V., Filatov, A.S., Xia, J. *et al.* (2013) Tightening of the nanobelt upon multielectron reduction. *Angew. Chem. Int. Ed.*, **52**, 5033–5036.
55 Nakanishi, Y., Omachi, H., Matsuura, S. *et al.* (2014) Size-selective complexation and extraction of endohedral metallofullerenes with cycloparaphenylene. *Angew. Chem. Int. Ed.*, **53**, 3102–3106.
56 Ueno, H., Nishihara, T., Segawa, Y., and Itami, K. (2015) Cycloparaphenylene-based ionic donor–acceptor supramolecule: isolation and characterization of Li^+@C_{60} [10]CPP. *Angew. Chem. Int. Ed.*, **54**, 3707–3711.
57 Iwamoto, T., Watanabe, Y., Sadahiro, T. *et al.* (2011) Size-selective encapsulation of C_{60} by [10]cycloparaphenylene: formation of the shortest fullerene-peapod. *Angew. Chem. Int. Ed.*, **50**, 8342–8344.
58 Golling, F.E., Quernheim, M., Wagner, M. *et al.* (2014) Concise synthesis of 3D π-extended polyphenylene cylinders. *Angew. Chem. Int. Ed.*, **53**, 1525–1528.
59 Golder, M.R., Colwell, C.E., Wong, B.M. *et al.* (2016) Iterative reductive aromatization/ring-closing metathesis strategy toward the synthesis of strained aromatic belts. *J. Am. Chem. Soc.*, **138**, 6577–6582.
60 Matsui, K., Segawa, Y., Namikawa, T. *et al.* (2012) Synthesis and properties of all-benzene carbon nanocages: a junction unit of branched carbon nanotubes. *Chem. Sci.*, **4**, 84–88.
61 Matsui, K., Segawa, Y., and Itami, K. (2014) All-benzene carbon nanocages: size-selective synthesis, photophysical properties, and crystal structure. *J. Am. Chem. Soc.*, **136**, 16452–16458.
62 Kayahara, E., Iwamoto, T., Takaya, H. *et al.* (2013) Synthesis and physical properties of a ball-like three-dimensional π-conjugated molecule. *Nat. Commun.*, **4**, 2694.
63 Kammermeier, S., Jones, P.G., and Herges, R. (1996) Ring-expanding metathesis of tetradehydro-anthracene – synthesis and structure of a tube-like, fully conjugated hydrocarbon. *Angew. Chem. Int. Ed.*, **35**, 2669–2671.

64 Treitel, N., Deichmann, M., Sternfeld, T. *et al.* (2003) Picotube tetraanion: a novel lithiated tubular system. *Angew. Chem. Int. Ed.*, **42**, 1172–1176.
65 Deichmann, M., Näther, C., and Herges, R. (2003) Pyrolysis of a tubular aromatic compound. *Org. Lett.*, **5**, 1269–1271.
66 Kammermeier, S., Jones, P.G., Dix, I., and Herges, R. (1998) A silver(I) complex of a tube-shaped hydrocarbon. *Acta Crystallogr. Sect. C*, **54**, 1078–1081.
67 Herges, R., Deichmann, M., Wakita, T., and Okamoto, Y. (2003) Synthesis of a chiral tube. *Angew. Chem. Int. Ed.*, **42**, 1170–1172.
68 Iwamoto, T., Kayahara, E., Yasuda, N. *et al.* (2014) Synthesis, characterization, and properties of [4]cyclo-2,7-pyrenylene: effects of cyclic structure on the electronic properties of pyrene oligomers. *Angew. Chem. Int. Ed.*, **53**, 6430–6434.
69 Matsuno, T., Kamata, S., Hitosugi, S., and Isobe, H. (2013) Bottom-up synthesis and structures of π-lengthened tubular macrocycles. *Chem. Sci.*, **4**, 3179–3183.
70 Matsuno, T., Sato, S., Iizuka, R., and Isobe, H. (2015) Molecular recognition in curved π-systems: effects of π-lengthening of tubular molecules on thermodynamics and structures. *Chem. Sci.*, **6**, 909–916.
71 Jiang, H.-W., Tanaka, T., Mori, H. *et al.* (2015) Cyclic 2,12-porphyrinylene nanorings as a porphyrin analogue of cycloparaphenylenes. *J. Am. Chem. Soc.*, **137**, 2219–2222.
72 André, E., Boutonnet, B., Charles, P. *et al.* (2016) A new, simple and versatile strategy for the synthesis of short segments of zigzag-type carbon nanotubes. *Chem. Eur. J.*, **22**, 3105–3114.
73 Myśliwiec, D., Kondratowicz, M., Lis, T. *et al.* (2015) Highly strained nonclassical nanotube end-caps. A single-step solution synthesis from strain-free, non-macrocyclic precursors. *J. Am. Chem. Soc.*, **137**, 1643–1649.
74 Behrend, R., Meyer, E., and Rusche, F.I. (1905) Ueber condensationsproducte aus glycoluril und formaldehyd. *Justus Liebigs Ann. Chem.*, **339**, 1–37.
75 Freeman, W.A., Mock, W.L., and Shih, N.Y. (1981) Cucurbituril. *J. Am. Chem. Soc.*, **103**, 7367–7368.
76 Cintas, P. (1994) Cucurbituril: supramolecular perspectives for an old ligand. *J. Incl. Phenom. Mol. Recognit. Chem.*, **17**, 205–220.
77 Mock, W. (1995) Cucurbituril, in *Supramolecular Chemistry II—Host Design and Molecular Recognition*, Topics in Current Chemistry, vol. **175** (ed. E. Weber), Springer, Berlin/Heidelberg, pp. 1–24.
78 Lagona, J., Mukhopadhyay, P., Chakrabarti, S., and Isaacs, L. (2005) The cucurbit[*n*]uril family. *Angew. Chem. Int. Ed.*, **44**, 4844–4870.
79 Kim, K., Selvapalam, N., Ko, Y.H. *et al.* (2007) Functionalized cucurbiturils and their applications. *Chem. Soc. Rev.*, **36**, 267–279.

80 Isaacs, L. (2009) Cucurbit[*n*]urils: from mechanism to structure and function. *Chem. Commun.*, 619–629.

81 Davis, F. and Higson, S. (2011) Cucurbiturils, in *Macrocycles: Construction, Chemistry and Nanotechnology Applications*, John Wiley & Sons, pp. 325–380.

82 Masson, E., Ling, X., Joseph, R. *et al.* (2012) Cucurbituril chemistry: a tale of supramolecular success. *RSC Adv.*, **2**, 1213–1247.

83 Assaf, K.I. and Nau, W.M. (2015) Cucurbiturils: from synthesis to high-affinity binding and catalysis. *Chem. Soc. Rev.*, **44**, 394–418.

84 Shao, Y., Li, Y.-Z., Shi, J.-P., and Lu, G.-Y. (2007) [μ-Cucurbit[6]uril(2^-)]bis[pentaaquacalcium(II)] bis[tetrachloridozincate(II)] heptahydrate. *Acta Crystallogr. Sect. E*, **63**, m1480–m1481.

85 Huang, W.-H., Zavalij, P.Y., and Isaacs, L. (2007) Cucurbit[6]uril *p*-phenylenediammonium diiodide decahydrate inclusion complex. *Acta Crystallogr. Sect. E*, **63**, o1060–o1062.

86 Mock, W.L. and Shih, N.Y. (1986) Structure and selectivity in host–guest complexes of cucurbituril. *J. Org. Chem.*, **51**, 4440–4446.

87 Kim, J., Jung, I.-S., Kim, S.-Y. *et al.* (2000) New cucurbituril homologues: syntheses, isolation, characterization, and X-ray crystal structures of cucurbit[*n*]uril (n = 5, 7, and 8). *J. Am. Chem. Soc.*, **122**, 540–541.

88 Day, A., Arnold, A.P., Blanch, R.J., and Snushall, B. (2001) Controlling factors in the synthesis of cucurbituril and its homologues. *J. Org. Chem.*, **66**, 8094–8100.

89 Liu, S., Zavalij, P.Y., and Isaacs, L. (2005) Cucurbit[10]uril. *J. Am. Chem. Soc.*, **127**, 16798–16799.

90 Miyahara, Y., Abe, K., and Inazu, T. (2002) "Molecular" molecular sieves: lid-free decamethylcucurbit[5]uril absorbs and desorbs gases selectively. *Angew. Chem. Int. Ed.*, **41**, 3020–3023.

91 Jiang, X., Yao, X., Huang, X. *et al.* (2015) A cucurbit[5]uril analogue from dimethylpropanediurea–formaldehyde condensation. *Chem. Commun.*, **51**, 2890–2892.

92 Ustrnul, L., Kulhanek, P., Lizal, T., and Sindelar, V. (2015) Pressocucurbit[5]uril. *Org. Lett.*, **17**, 1022–1025.

93 Zhao, J., Kim, H.-J., Oh, J. *et al.* (2001) Cucurbit[*n*]uril derivatives soluble in water and organic solvents. *Angew. Chem. Int. Ed.*, **40**, 4233–4235.

94 Wu, F., Wu, L.-H., Xiao, X. *et al.* (2012) Locating the cyclopentano cousins of the cucurbit[*n*]uril family. *J. Org. Chem.*, 77, 606–611.

95 Jon, S.Y., Selvapalam, N., Oh, D.H. *et al.* (2003) Facile synthesis of cucurbit[*n*]uril derivatives via direct functionalization: expanding utilization of cucurbit[*n*]uril. *J. Am. Chem. Soc.*, **125**, 10186–10187.

96 Lagona, J., Fettinger, J.C., and Isaacs, L. (2003) Cucurbit[*n*]uril analogues. *Org. Lett.*, **5**, 3745–3747.

97 Park, K.-M., Heo, J., Roh, S.-G. *et al.* (1999) Self-assembly of interlocked structures: rotaxanes, polyrotaxanes and molecular necklaces. *Mol. Cryst. Liq. Cryst. Sci. Technol. Sect. A*, **327**, 65–70.
98 Whang, D., Jeon, Y.-M., Heo, J., and Kim, K. (1996) Self-assembly of a polyrotaxane containing a cyclic "bead" in every structural unit in the solid state: cucurbituril molecules threaded on a one-dimensional coordination polymer. *J. Am. Chem. Soc.*, **118**, 11333–11334.
99 Whang, D., Park, K.-M., Heo, J. *et al.* (1998) Molecular necklace: quantitative self-assembly of a cyclic oligorotaxane from nine molecules. *J. Am. Chem. Soc.*, **120**, 4899–4900.
100 Roh, S., Park, K., Park, G. *et al.* (1999) Synthesis of a five-membered molecular necklace: a 2+2 approach. *Angew. Chem. Int. Ed.*, **38**, 638–641.
101 Herges, R. (2006) Topology in chemistry: designing Möbius molecules. *Chem. Rev.*, **106**, 4820–4842.
102 Schaller, G.R. and Herges, R. (2013) Möbius molecules with twists and writhes. *Chem. Commun.*, **49**, 1254–1260.
103 Walba, D.M., Richards, R.M., and Haltiwanger, R.C. (1982) Total synthesis of the first molecular Möbius strip. *J. Am. Chem. Soc.*, **104**, 3219–3221.
104 Walba, D.M., Homan, T.C., Richards, R.M., and Haltiwanger, R.C. (1993) Topological stereochemistry. 9. Synthesis and cutting "in half" of a molecular Möbius strip. *New J. Chem.*, **17**, 661–681.
105 Cheng, X.-J., Liang, L.-L., Chen, K. *et al.* (2013) Twisted cucurbit[14]uril. *Angew. Chem. Int. Ed.*, **52**, 7252–7255.
106 Stępień, M., Latos-Grażyński, L., Sprutta, N. *et al.* (2007) Expanded porphyrin with a split personality: a Hückel–Möbius aromaticity switch. *Angew. Chem. Int. Ed.*, **46**, 7869–7873.
107 Bröring, M., Jendrny, J., Zander, L. *et al.* (1995) Octaphyrin-(1.0.1.0.1.0.1.0). *Angew. Chem. Int. Ed.*, **34**, 2515–2517.
108 Elmosalamy, M.A.F., Moody, G.J., Thomas, J.D.R. *et al.* (1989) Studies on two epoxyoctacosahydro[12]cyclacene derivatives as sensor coatings on quartz piezoelectric crystals for detecting aromatic vapours. *Anal. Proc.*, **26**, 15–16.
109 Iwamoto, T., Watanabe, Y., Sakamoto, Y. *et al.* (2011) Selective and random syntheses of [*n*]cycloparaphenylenes (n = 8–13) and size dependence of their electronic properties. *J. Am. Chem. Soc.*, **133**, 8354–8361.
110 Barrow, S.J., Kasera, S., Rowland, M.J. *et al.* (2015) Cucurbituril-based molecular recognition. *Chem. Rev.*, **115**, 12320–12406.
111 Saleh, N., Ghosh, I., and Nau, W.M. (2013) Cucurbiturils in drug delivery and for biomedical applications, in *Supramolecular Systems in Biomedical Fields, Monographs in Supramolecular Chemistry* (ed. H.-J. Schneider), Royal Society of Chemistry, Cambridge, pp. 164–212.
112 El-Sheshtawy, H.S., Bassil, B.S., Assaf, K.I. *et al.* (2012) Halogen bonding inside a molecular container. *J. Am. Chem. Soc.*, **134**, 19935–19941.

113 Cao, L., Šekutor, M., Zavalij, P.Y. *et al.* (2014) Cucurbit[7]uril•guest pair with an attomolar dissociation constant. *Angew. Chem. Int. Ed.*, **53**, 988–993.
114 Lazar, A.I., Biedermann, F., Mustafina, K.R. *et al.* (2016) Nanomolar binding of steroids to cucurbit[*n*]urils: selectivity and applications. *J. Am. Chem. Soc.*, **138**, 13022–13029.
115 Chinai, J.M., Taylor, A.B., Ryno, L.M. *et al.* (2011) Molecular recognition of insulin by a synthetic receptor. *J. Am. Chem. Soc.*, **133**, 8810–8813.
116 Li, H.-F., Lü, J., Lin, J.-X., and Cao, R. (2014) Monodispersed Ag nanoparticles as catalyst: preparation based on crystalline supramolecular hybrid of decamethylcucurbit[5]uril and silver ions. *Inorg. Chem.*, **53**, 5692–5697.
117 Cao, M., Lin, J., Yang, H., and Cao, R. (2010) Facile synthesis of palladium nanoparticles with high chemical activity using cucurbit[6]uril as protecting agent. *Chem. Commun.*, **46**, 5088–5090.
118 Mock, W.L., Irra, T.A., Wepsiec, J.P., and Manimaran, T.L. (1983) Cycloaddition induced by cucurbituril. A case of Pauling principle catalysis. *J. Org. Chem.*, **48**, 3619–3620.
119 Tuncel, D., Özsar, Ö., Tiftik, H.B., and Salih, B. (2007) Molecular switch based on a cucurbit[6]uril containing bistable [3]rotaxane. *Chem. Commun.*, 1369–1371.
120 Tuncel, D., Ünal, Ö., and Artar, M. (2011) Supramolecular assemblies constructed by cucurbituril-catalyzed click reaction. *Isr. J. Chem.*, **51**, 525–532.
121 Jon, S.Y., Ko, Y.H., Park, S.H. *et al.* (2001) A facile, stereoselective [2 + 2] photoreaction mediated by cucurbit[8]uril. *Chem. Commun.*, 1938–1939.
122 Wu, X.-L., Luo, L., Lei, L. *et al.* (2008) Highly efficient cucurbit[8]uril-templated intramolecular photocycloaddition of 2-naphthalene-labeled poly(ethylene glycol) in aqueous solution. *J. Org. Chem.*, **73**, 491–494.
123 Lei, L., Luo, L., Wu, X.-L. *et al.* (2008) Cucurbit[8]uril-mediated photodimerization of alkyl 2-naphthoate in aqueous solution. *Tetrahedron Lett.*, **49**, 1502–1505.
124 Barooah, N., Pemberton, B.C., and Sivaguru, J. (2008) Manipulating photochemical reactivity of coumarins within cucurbituril nanocavities. *Org. Lett.*, **10**, 3339–3342.
125 Pemberton, B.C., Barooah, N., Srivatsava, D.K., and Sivaguru, J. (2009) Supramolecular photocatalysis by confinement-photodimerization of coumarins within cucurbit[8]urils. *Chem. Commun.*, **46**, 225–227.
126 Biedermann, F., Ross, I., and Scherman, O.A. (2014) Host–guest accelerated photodimerisation of anthracene-labeled macromolecules in water. *Polym. Chem.*, **5**, 5375–5382.
127 Klöck, C., Dsouza, R.N., and Nau, W.M. (2009) Cucurbituril-mediated supramolecular acid catalysis. *Org. Lett.*, **11**, 2595–2598.

128 Wang, Y.-H., Cong, H., Zhao, F.-F. *et al.* (2011) Selective catalysis for the oxidation of alcohols to aldehydes in the presence of cucurbit[8]uril. *Catal. Commun.*, **12**, 1127–1130.

129 de Lima, S.M., Gómez, J.A., Barros, V.P. *et al.* (2010) A new oxovanadium(IV)–cucurbit[6]uril complex: properties and potential for confined heterogeneous catalytic oxidation reactions. *Polyhedron*, **29**, 3008–3013.

130 Lu, X. and Masson, E. (2010) Silver-promoted desilylation catalyzed by ortho- and allosteric cucurbiturils. *Org. Lett.*, **12**, 2310–2313.

5

Molecular Tweezers

CHAPTER MENU

5.1 Introduction

Conventional tweezers are simple mechanical devices usually constructed from two pincers interconnected with a semiflexible link. This link is capable of changing the distance between these pincers with the retention of their relative face-to-face position. The major function of this instrument is related to its ability to control hold-and-release operations. Tweezers belong to the extended family of simple hand-operated mechanical devices, such as pliers, clamps, and vises. The same principle is utilized for the construction of various heavy machinery components, such as mechanical claw grabbers, and others (Figure 5.1).

In the molecular domain, very similar mechanical principles can be used for the construction of molecular devices that are capable of holding certain guest molecules between a pair of properly aligned interconnected flat molecular moieties. Given their mode of actions, these technomimetic molecules can be named as molecular tweezers. While molecular tweezers represent a distinct category of technomimetic molecules, other similar molecular systems are often assigned to the same class due to minor differences in their structure and functionality [1–5]. For example, various molecular clefts [6–8], clips [9, 10], and jaws [11] can be considered as members of the extended family of molecular tweezers. There are other types of molecular tweezer-like compounds, such as Venus flytraps [12]. Some of them are discussed in Section 5.5.

Molecular Devices: An Introduction to Technomimetics and Its Biological Applications,
First Edition. Andrei A. Gakh.

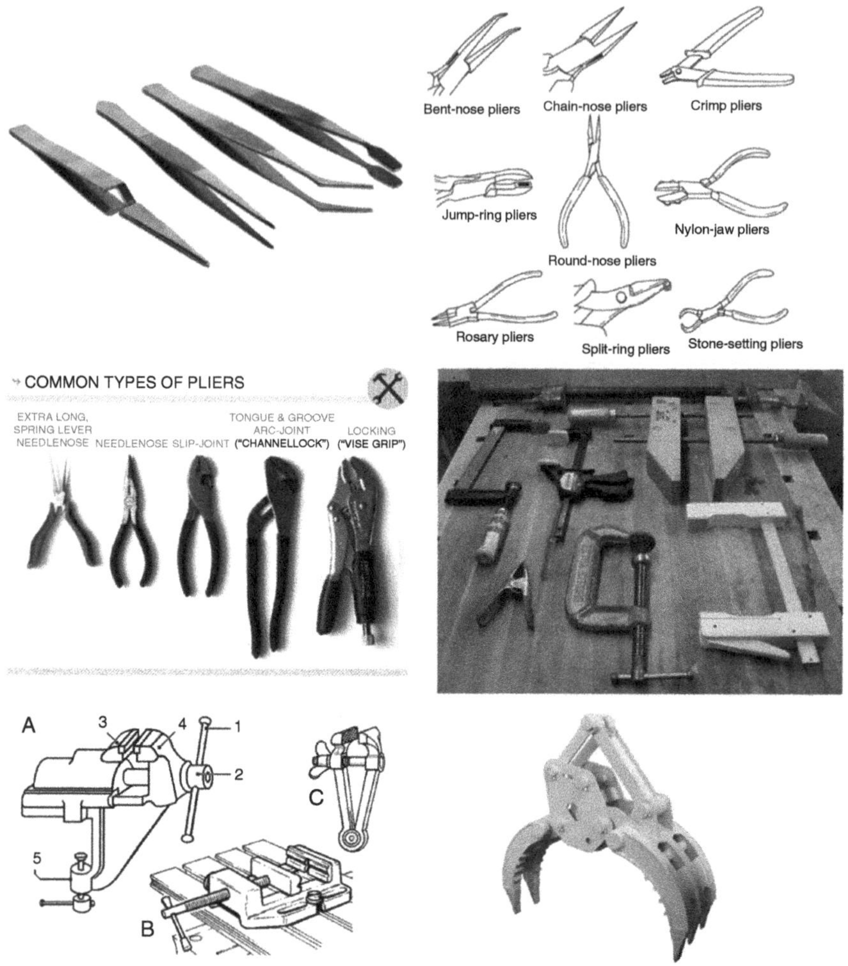

Figure 5.1 Various conventional tweezer-like mechanical devices. *Credits*: Gale Force 9, Primer Magazine, Luigi Zanasi, Jiangxi Nante Construction Machinery Co., Ltd.

The term "molecular tweezers" was originally introduced more than 30 years ago [13] to describe a specific host system equipped with two aromatic "pincers" linked together by a "spacer" (Figure 5.2). These pincers (also sometimes called "tips" or "chromophores") are usually flat but can also be concave to accommodate spherical molecular objects, such as fullerenes [14].

The principal mode of action of molecular tweezers is a direct mechanical analogy with conventional tweezers: two pincer elements hold an appropriate guest molecule. However, unlike conventional tweezers, whose hold-and-release mechanism of action involves an external force, molecular

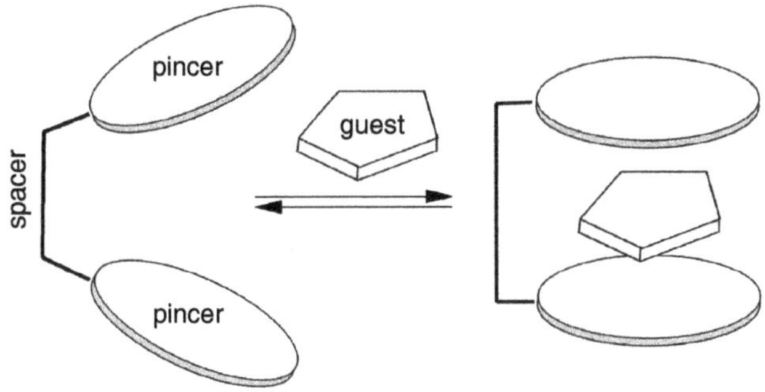

Figure 5.2 A typical "hold-and-release" mode of action of molecular tweezers.

tweezers usually rely on internal interactions between the guest molecule and the pincers. This "gabbing" action in classic molecular tweezers is primarily due to π-stacking interactions between a guest molecule and two π-rich surfaces of a host molecule, which adopts essentially a sandwich configuration [15]. Since neutral π–π interactions, even in the most favorable cases, are too weak for practical purposes, the preponderance of effective tweezer systems employs enhanced donor–acceptor π–π interactions [16, 17]. They can also be supplemented by hydrogen bonding and hydrophobic attraction [16, 18].

Simple molecular tweezers lack guest differentiation, which in conventional tweezers is typically achieved by intelligent actions of an automaton or a human being. Partial "intelligent" selectivity can be achieved in more complex molecular tweezer systems by incorporating additional functional groups (e.g., amino groups or carboxyl groups) attached to pincers or a spacer, thereby improving binding selectivity via the so-called multiple-site interactions. The use of multiple-site interactions can also improve the binding-affinity parameters beyond simple sum of all interactions, referred to as a "complex additivity in binding energies" [17]. Successful examples of this "enhancement-by-design" include systems that are capable of producing strong complexes ($K_a > 10^4\ M^{-1}$) with nitrogen heterocycles of biological origin, such as adenine derivatives, via both hydrogen bonding and π-stacking [19].

External control of molecular tweezers is an evolving area of research. Promising examples include photochemically [20–24] or electrochemically [25–27] controlled molecular tweezer systems. It is also possible to control the geometry of the internal cavity (and hence the binding affinity of molecular tweezers) using metal cation coordination with the pincers outside the cavity [28]. A variant of this approach entails the use of flexible molecular tweezers with extended cavities, which can incorporate both a metal cation and a guest molecule [29]. However, this is a less attractive design because of the potential interference between the guest molecules and the metal cations employed as the control elements.

One of the distinctive features of flexible molecular tweezers is a delicate balance of preorganization and self-organization, which allows to achieve the best "induced-fit" to a particular guest molecule. Not surprisingly, many molecular tweezers are designed with some level of flexibility, which can be accomplished using a flexible (usually single) bond connection of pincers to a spacer unit [17]. An alternative approach includes the use of semiflexible spacers.

The binding affinity of molecular tweezers varies significantly even for structurally related compounds. Performance improvement by a rational design can be achieved by applying computational analysis for the determination of the best complementary fit between the guest and the host molecules. This approach is very similar to the preproduction design of conventional mechanical devices with computer-aided design (CAD) software. One reported case includes the computation of the most favorable geometries, as well as binding energies, of the complexes between dimethylene- and tetramethylene-bridged tweezers as host molecules and various aliphatic and aromatic substrates as guest molecules [30]. Another publication describes *in silico* optimization of fluorinated molecular tweezers for iodine anion capture [31]. Better results in CAD of molecular tweezers can be attained with specialized software packages. For example, CAVEAT molecular design program was successfully employed for the construction of water-soluble molecular tweezers having optimal distances and angles between the aromatic pincers [32].

The spacer unit is an integral part of molecular tweezer systems, which acts as a basic construction element by defining the overall three-dimensional (3D) geometry of the host molecule. It may carry additional functional groups to facilitate guest bonding. The main role of the spacer is to provide a structural support for the pincers by preventing the collapse of the sandwich complexes. This requires a certain level of rigidity to achieve the desired spatial features [13, 17]. However, there are several successful examples of molecular tweezers with completely flexible aliphatic chain spacers [33–35]. Such flexible tweezers rely solely on self-organization.

The distance between the pincers is a critical parameter with regard to molecular tweezer functionality. The best results for simple planar aromatic guests are achieved with a pincer-to-pincer gap of about 6–8 Å [4, 17]. Unfortunately, this optimal configuration is also ideal for self-complexion, when a pincer of one tweezer molecule acts as a guest for another tweezer molecule [36–39]. This self-association is less pronounced in solutions, except for several specific cases, including unsymmetrical molecular tweezers containing both π-acceptor and π-donor pincers [17].

5.2 Flexible Molecular Tweezers

One of the first molecular tweezer systems [13] was a simple model constructed from two caffeine units interconnected by a diacetylene spacer (Figure 5.3, please note that in some review papers two nitrogen atoms were missed in the

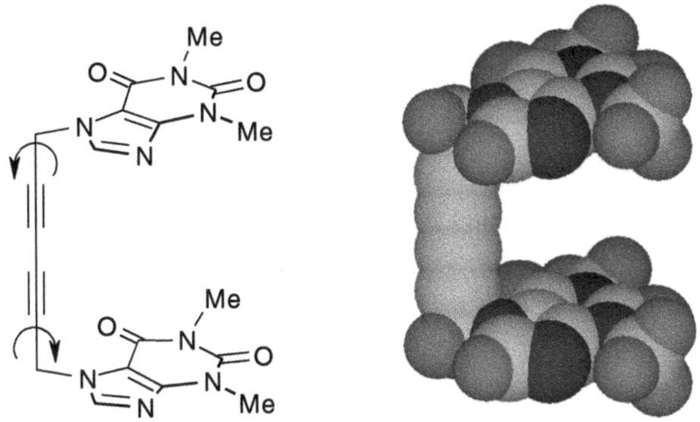

Figure 5.3 One of the first reported examples of molecular tweezers (left) and their energy-minimized 3D structure (right) [13].

structure drawings, and this omission was propagated into several subsequent publications). This system only partially obeys the specific tweezer design features outlined earlier. Specifically, the diacetylene spacer allows the free rotation of the two caffeine pincers out of the preferred *syn*-conformation. In addition, these two pincers are not parallel to each other due to the presence of the two methylene connectors with tetragonal 109.5° angle geometry. As a result, certain bending of the spacer would be required to achieve tight stacking with an aromatic guest. Nevertheless, even this nonoptimized molecular tweezer system, which was designed primarily to evaluate a feasibility of aligned sandwich-type complexes, demonstrated enhanced affinities toward some aromatic guests due to the formation of sandwich complexes stabilized by double π–π interactions. Hydrophobic interactions [16, 18] may also play a role in stabilizing the complexes with aromatic 2,6-dihydroxybenzoate guest in water [13].

Contrary to earlier pessimistic assumptions [17], a recent research clearly demonstrates that functional molecular tweezers can be successfully constructed using completely flexible spacers made of simple aliphatic or aliphatic/aromatic chains. For example, electron-deficient bipyridinium pincers interconnected with straight $(CH_2)_n$ aliphatic chains can be considered as a crude molecular tweezer device with a reasonable binding affinity toward electron-rich aromatic 2-naphthol (Figure 5.4) [33]. It is interesting to note that these flexible molecular tweezers have a very distinct correlation pattern between the association constants and the number of CH_2 linkages between the two bipyridinium pincers: the best results were achieved with the $(CH_2)_7$ ($n = 7$) spacer. It might be speculated that smaller spacers produce unfavorable out-of-plane bending of the bipyridinium pincers (Figure 5.4), whereas longer chains cannot hold them together. Other flexible spacers, which are discussed later, have similar spatial requirements and usually comprise 6–8 C—C or C—N bonds between the pincer units.

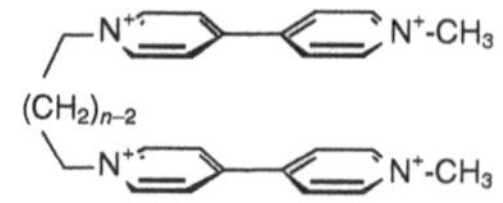

n	3	4	5	6	7	8
K_a, M^{-1}	32	50	130	110	210	96

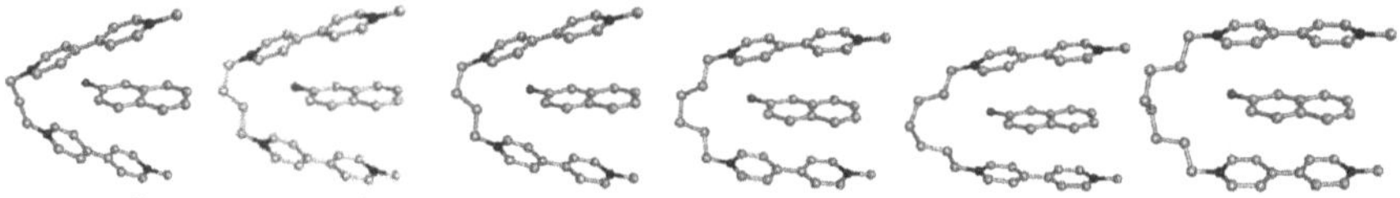

Figure 5.4 The association constants of 2-naphthol complexes with flexible bipyridinium molecular tweezers (top) and their energy-minimized structures (bottom, hydrogen atoms were removed for clarity).

Better guest-binding affinity and selectivity can be achieved by incorporating additional functional elements in flexible aliphatic spacers. One of such molecular tweezer systems was constructed on the basis of a diethylenetriamine spacer. Despite the flexible nature of the spacer, this host is capable of producing well-defined sandwich complexes with some dicarboxylic acids, such as phthalic acid or maleic acid (Figure 5.5). These complexes are held together by a network of H···O hydrogen bonds [35].

Crown ether backbone was successfully employed for the construction of flexible molecular tweezers (Figure 5.6). In this particular case, triptycene fragments were used as pincers. They provide adequate π–π binding interactions for certain cationic heteroaromatic guests despite their nonplanar structure. The best results ($K_a = 4080\,M^{-1}$) were achieved for compact bipyridinium dications, such as paraquat (*N,N′*-dimethyl-4,4′-bipyridinium dication). Single-crystal X-ray analysis indicates the involvement of crown ether moiety in hydrogen bonding with the paraquat guest molecule in addition to "classic" π–π-stacking interactions between the triptycene units and the bipyridinium ring system with approximately 3.3 Å distances [40].

Mixed aromatic–aliphatic spacers are frequently employed for the construction of flexible molecular tweezers. A representative example of these tweezers

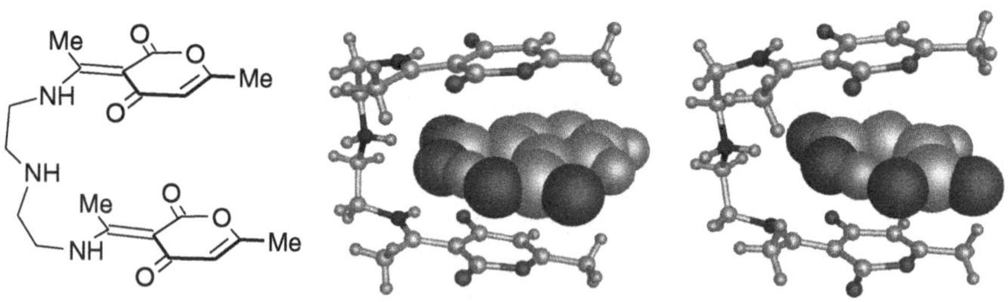

Figure 5.5 The flexible polyamine molecular tweezers (left) and the X-ray structures of their sandwich complexes with phthalic acid (middle) and maleic acid (right).

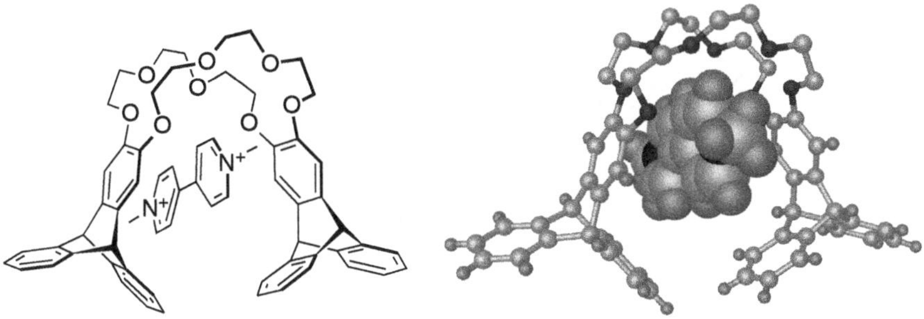

Figure 5.6 The complex of crown ether molecular tweezers with paraquat (left) and its X-ray structure (right, hydrogen atoms of the crown ether spacer were removed for clarity).

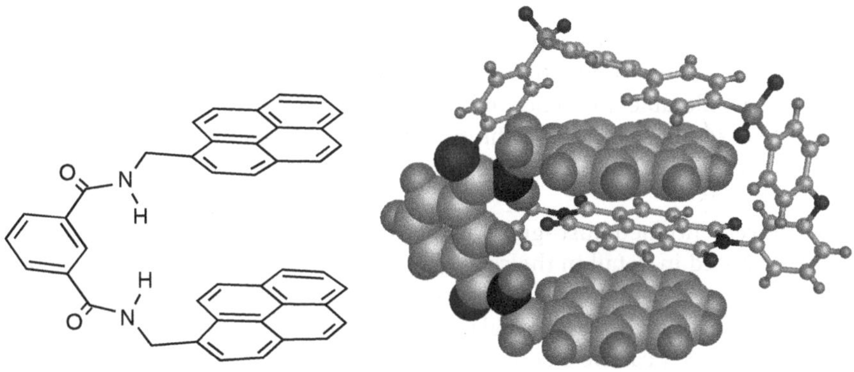

Figure 5.7 The flexible molecular tweezers with the 1,3-bis(carbonyl)benzene spacer and pyrene pincers (left) and the X-ray structure of their complementary complex with the macrocyclic guest molecule (right).

with a 1,3-bis(carbonyl)benzene spacer is shown in Figure 5.7. This molecular tweezer host produces a strong complex with a large macrocyclic guest molecule containing complementary electron-deficient naphthalene diimide (NDI) residue ($K_a = 24{,}300\ M^{-1}$). The enhanced stability of the complex can be attributed to additional "reverse" donor–acceptor π–π interactions between the electron-rich 3-aminophenoxy moiety in the guest molecule and the electron-deficient isophthaloyl residue of the tweezer spacer unit [41].

Similar molecular tweezer system with the 2,6-bis(carbonyl)pyridine spacer is presented in Figure 5.8. In this particular case, the pyridine unit and the amide links of the functionalized spacer provide supplementary binding of the nitroaromatic guests (such as 1,3,5-trinitrobenzene or 2,4,6-trinitrotoluene), in addition to the primary π–π charge–transfer interactions. The molecular tweezers with the pyridine-based spacer and amide links also demonstrate strong self-association in solutions, which is disrupted in the presence

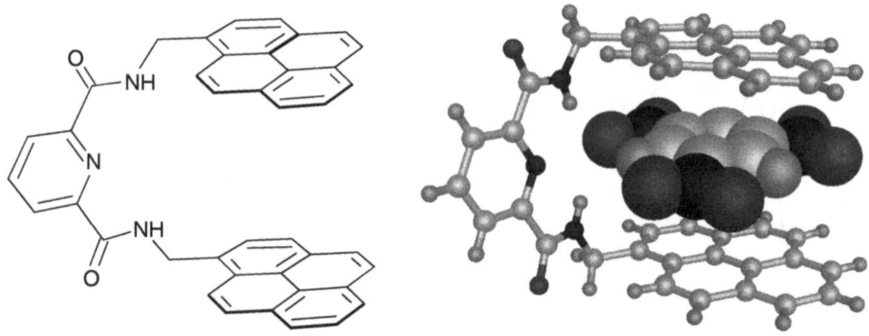

Figure 5.8 The flexible molecular tweezers with the 2,6-bis(carbonyl)pyridine spacer (left) and the X-ray structure of their 1,3,5-trinitrobenzene complex (right).

of the abovementioned nitroaromatic compounds ($K_a = 73{,}200\,M^{-1}$ for 1,3,5-trinitrobenzene and $769\,M^{-1}$ for 2,4,6-trinitrotoluene). The K_a data indicate considerable influence of the nature of the nitroaromatic guests on the tweezer host-binding affinity, whereby the addition of one methyl group reduces the strength of the sandwich complex by almost two orders of magnitude [42].

The abovementioned host–guest selectivity toward the nitroaromatic guests was investigated in detail in the case of molecular tweezers with a flexible tricyclic spacer. These molecular tweezers utilize extended heteropolycyclic pincer elements and have an additional heteroatom in the tricyclic spacer assembly (Figure 5.9), which can be a contributing factor for their relatively high selectivity. The best affinity was observed for the sandwich complexes of the molecular tweezer host with 2,4,7-trinitrofluorenone (Figure 5.9) [43]. This polynitroaromatic compound emerged as an important probe for the investigation of charge–transfer π–π interactions between the electron-deficient guests and the π-electron-rich pincers in the molecular tweezer systems.

The relation between the nature of the pincer elements in flexible molecular tweezers and binding affinity was further investigated using the tropylium cation as a versatile electron-deficient guest molecule probe (Figure 5.10) [44]. Similar to 2,4,7-trinitrofluorenone mentioned earlier, the tropylium cation is capable of producing strong charge–transfer sandwich complexes with flexible molecular tweezers having polyaromatic pincers. The formation of these π–π complexes can be easily monitored by color changes associated with the formation of a new broad UV–Vis absorption band in the 450–650 nm region [45]. Tropylium cation also serves as a template for the proper alignment of pincers to achieve the best energy-favored 3D configuration of the sandwich complexes. As noted previously, the flexible spacers cannot fix the pincers in proper positions, and successful operation of these technomimetic molecules relies primarily on self-organization.

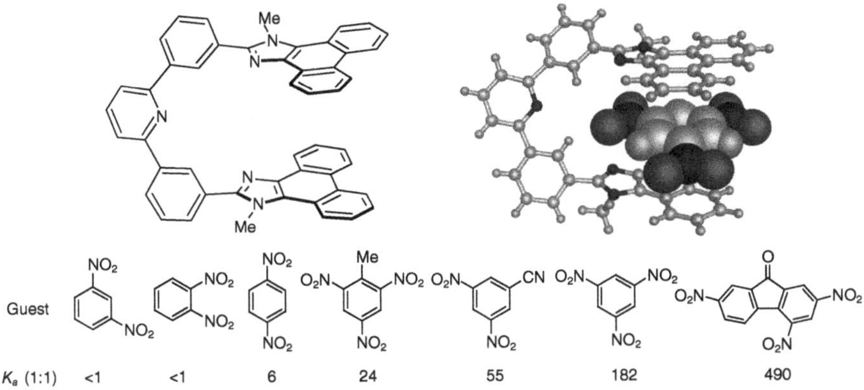

Figure 5.9 The flexible molecular tweezers with the tricyclic spacer (top left), the single-crystal X-ray structure of their complex with 1,3,5-trinitrobenzene (top right), and the association constants (K_a, M^{-1}) with various polynitroaromatic guest molecules in $CDCl_3$ (bottom).

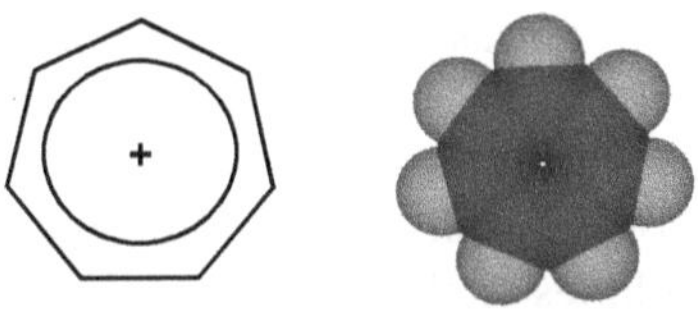

Figure 5.10 Tropylium cation – a versatile probe for investigation of π–π charge-transfer interactions in flexible molecular tweezer systems with polyaromatic pincers.

The influence of pincer structures on association constants is quite evident in the complexes of the tropylium cation with flexible molecular tweezers bearing benzene-, naphthalene-, and anthracene-based pincer units (Figure 5.11) [45]. The enlargement of aromatic pincers markedly improves the stability of the complexes. The molecular tweezers with benzene-based pincers form only weak charge–transfer complexes with the tropylium cation. The association

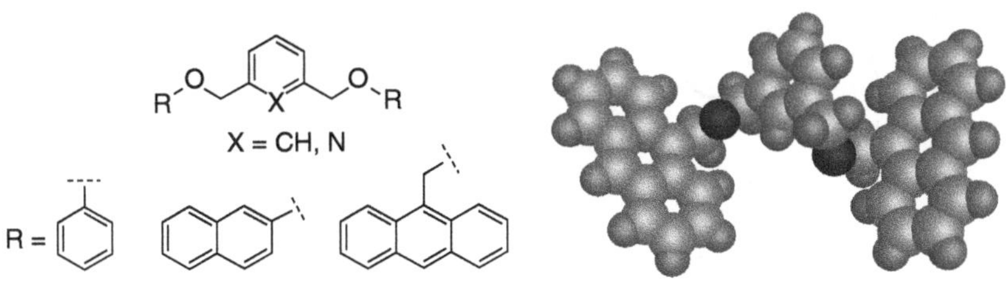

Figure 5.11 The flexible molecular tweezers with benzene- and pyridine-based spacers (left) and the single-crystal X-ray structure of the molecular tweezers with anthracene pincers and benzene-based spacer (X = CH, right).

constants of these complexes are comparable to the non-sandwich complexes of benzene with the tropylium cation. The complexes of naphthalene-based systems are stronger and have K_a values of 6.6 and 28 M^{-1} for X = CH and X = N, respectively (Figure 5.11). The bis(anthracene) molecules are of particular interest since these compounds display superior affinity toward the tropylium cation via charge–transfer, π–π, and cation–π interactions (K_a = 41 and 57 M^{-1}) [45].

The nature of an aromatic ring in a spacer unit is also important. Pyridine-based molecular tweezers with anthracene pincers have higher association constant (K_a = 57 M^{-1}; X = N, Figure 5.11) than benzene-based molecular tweezers (K_a = 41 M^{-1}; X = CH, Figure 5.11). The observed difference can be attributed partially to the presence of a nitrogen atom in the pyridine ring, which provides additional stabilization of the complexes via hydrogen bonding with the tropylium cation. These interactions can bring the tropylium cation closer to the host molecule [45].

Some molecular tweezers with polynitrogen electron-deficient heterocyclic spacers are prone for self-dimerization in the solid state. For example, molecular tweezers with a 2,4,6-triphenyl-1,3,5-triazine spacer dimerize in the solid state (Figure 5.12), but at the same time produce a reasonably strong complex (K_a = 80 M^{-1}) with 2,4,7-trinitrofluorenone in $CDCl_3$. In these dimers, the 1,3,5-triazine π-acceptor spacer subunit of one tweezer molecule is sandwiched between the anthracene π-donor pincers of another molecule. The distance between the anthracene and the 1,3,5-triazine planes is about 3.5 Å, indicating the presence of favorable π–π-stacking interactions in the solid state. No detectable self-assembly is observed in solutions [38].

Alicyclic spacers (such as 1,2-diaminocyclohexane) usually produce modest results due to a substantial angle between the pincer connectors preventing proper *syn*-orientation. This is quite evident in the case of a chiral molecular tweezer system constructed from (+)-usnic acid units as pincers and (1*R*,2*R*)-(−)-1,2-diaminocyclohexane as a spacer (Figure 5.13) [46–48]. The

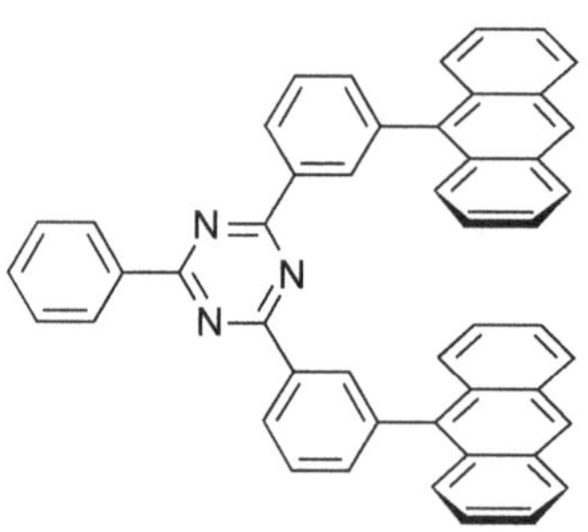

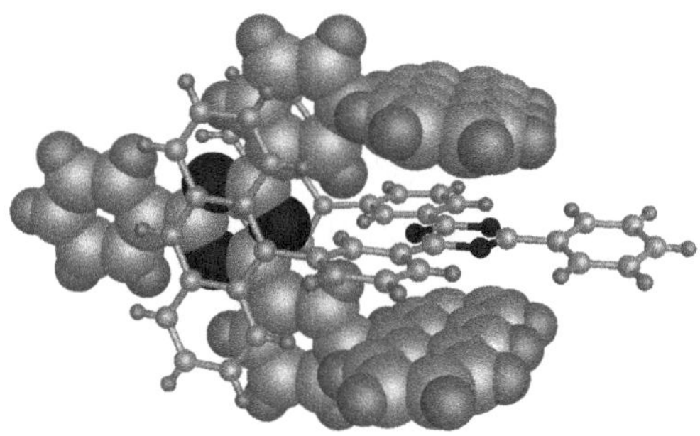

Figure 5.12 The flexible molecular tweezers with the 2,4,6-triphenyl-1,3,5-triazine spacer (left) and the X-ray structure of their self-associated dimer in the solid state (right).

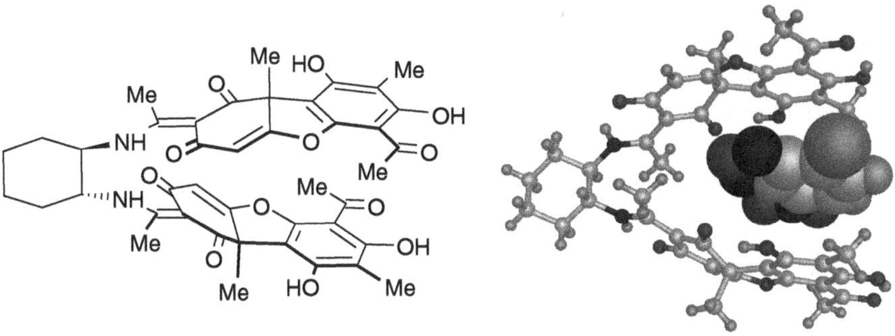

Figure 5.13 The chiral molecular tweezers constructed from (+)-usnic acid moieties as pincers and 1,2-diaminocyclohexane as a spacer (left) and the X-ray structure of their complex with 4-chloro-7-nitrobenzofurazan (right) [47].

highest binding affinity was observed for 2,4,7-trinitrofluorenone, one of the best electron-deficient polynitroaromatic guests. Even with this guest, the association constant is almost an order of magnitude lower ($K_a = 53\,M^{-1}$, NMR data) [47] than those of other types of molecular tweezers described earlier. Replacement of (+)-usnic acid pincers with phenanthrene pincers further reduces the association constant ($K_a = 24\,M^{-1}$ in CD_2Cl_2, NMR data) [47] and even favors external complexation [48]. Self-association is also notable in some of these 1,2-diaminocyclohexane-based molecular tweezers [46], especially with two different pincer elements [48].

Significantly better results were achieved with "argentivorous" molecular tweezers having a flexible macrocyclic spacer and polycyclic aromatic pincers [12]. One of the examples of argentivorous tweezers and their complex with Ag^+ are presented in Figure 5.14. Only complexes with Ag^+ demonstrate the

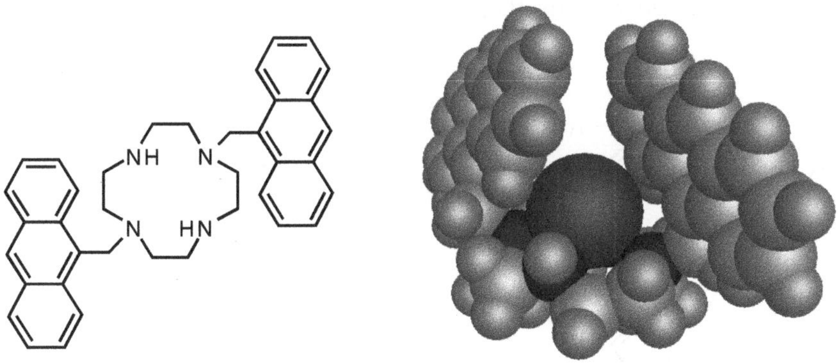

Figure 5.14 The argentivorous tweezer host with the 1,4,7,10-tetraazacyclododecane spacer and the anthracene pincer elements (left) and the X-ray structure of its complex with Ag^+ (right).

specific tweezer action mode in which Ag^+ cations are completely covered by anthracene pincers. No pincer cover of the metal cations in the host cavities is observed for Cu^{2+} and Hg^{2+} cations. Argentivorous molecules with extended pyrene polycyclic aromatic pincers also exhibit distinctive CH–π interactions [12].

Rigid polycyclic spacers in combination with conformationally mobile linking elements are also suitable for the construction of flexible molecular tweezers. A representative example of such systems is presented in Figure 5.15. In this particular case, anthracene moieties are employed for the design of both pincers and spacer elements [49, 50]. Anthracene is not an ideal scaffold for the construction of spacer units due to inadequate distance between the pincer attachment points (only about 5.1 Å). This problem was alleviated by the use of two conformationally flexible oxymethylene links. Single-crystal X-ray study shows that this design provides an optimal spatial separation of the pincer elements with an average interplanar distance between the anthracene pincers of approximately 7.1 Å [49]. This tweezer host system demonstrates an adequate binding affinity toward electron-deficient guest molecules such as 2,4,7-trinitrofluorenone ($K_a = 565\ M^{-1}$, $R = C_{12}H_{25}$) and tetracyanoquinodimethane (TCNQ) ($K_a = 4810\ M^{-1}$, $R = C_{12}H_{25}$, Figure 5.15) [49, 50].

Rigid steroid scaffolds can be used as spacer elements for the construction of chiral molecular tweezers. A representative example of flexible chiral tweezers with bile acid backbone is presented in Figure 5.16 [51–53]. These tweezer compounds bind electron-deficient aromatic guest molecules with reasonable association constants. The best results were achieved with 2,4,7-trinitrofluorenone ($K_a = 220\ M^{-1}$), while polyfluorinated compounds demonstrated lower binding affinities ($K_a = 19\ M^{-1}$ for perfluoronaphthalene) in $CDCl_3$ [52].

An interesting example of the molecular tweezer system with a ferrocene spacer was reported recently (Figure 5.17). These molecular tweezers have

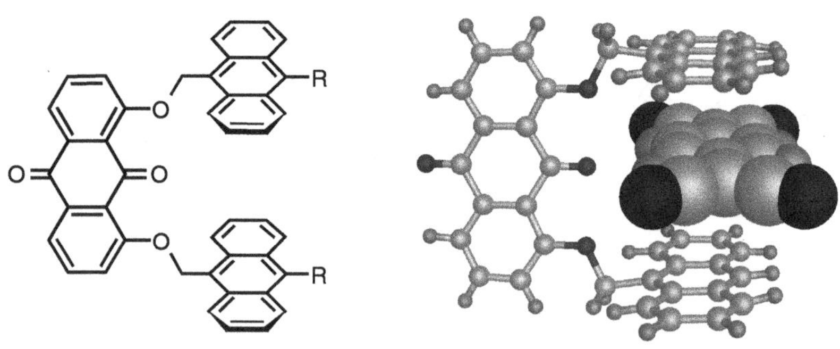

Figure 5.15 Flexible molecular tweezers with the 9,10-anthraquinone spacer (left, $R = H$, C_4H_9, $C_{12}H_{25}$) and the X-ray structure of their sandwich complex with TCNQ (right, $R = H$).

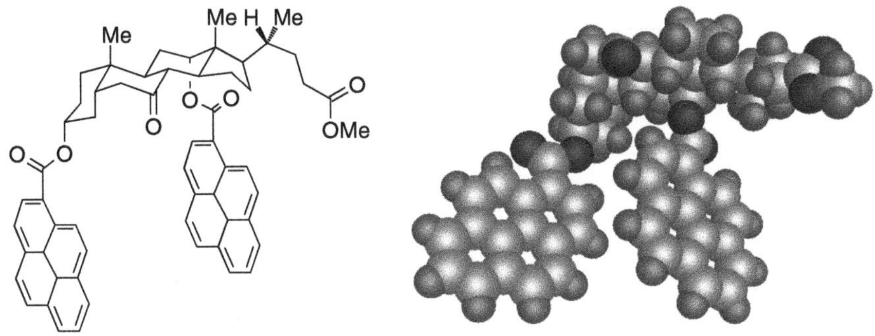

Figure 5.16 A representative example of flexible molecular tweezers with the bile acid spacer unit (left) and their X-ray structure (right) [52].

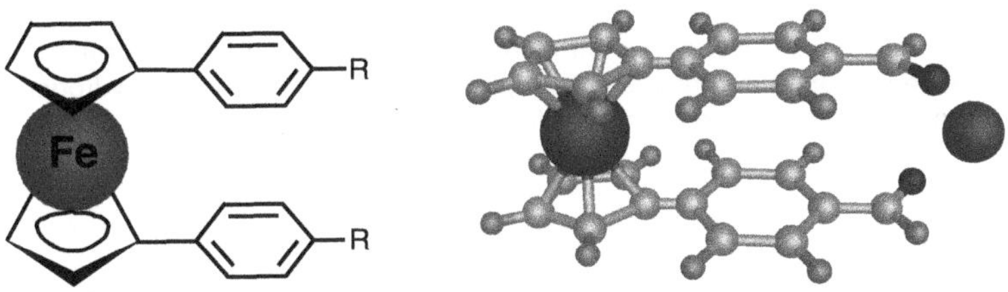

Figure 5.17 The flexible diarylferrocene molecular tweezers (left, R = H, Ph, OMe, CHO) and the energy-minimized 3D structure of their complex with Li^+ (right, R = CHO).

unusual dynamics due to rotation around the iron atom as a ball bearing element in a molecular scissor fashion. The best binding affinity under electrospray ionization mass spectrometry (ESI-MS) conditions is observed for small metal cations such as Li^+ [54].

Incorporation of crown ether functionality into the pincer elements of flexible molecular tweezers allows for efficient capture of some large spherical molecules, such as fullerenes. These functionalized pincers exploit the so-called cooperative n–π and π–π effects. One example of such systems is presented in Figure 5.18. These molecular tweezers are constructed on the basis of 9,10-bis(1,3-dithiol-2-ylidene)-9,10-dihydroanthracene-2,6-diol (exTTF), and they demonstrate very good binding affinity toward fullerene C_{60} ($K_a = 5.0 \times 10^6\ M^{-1}$ in benzonitrile). Even better binding is observed in the case of more electron-deficient fullerene C_{70} ($K_a = 2.57 \times 10^7\ M^{-1}$ in benzonitrile) [55].

Anionic binding sites can also be incorporated in pincer units. One of the practical implementations of this design idea is presented in Figure 5.19. A variant of CAD (CAVEAT) was used to achieve optimal (about 7.2 Å) distance between the linkage points of the spacer unit. This dianionic molecular tweezer

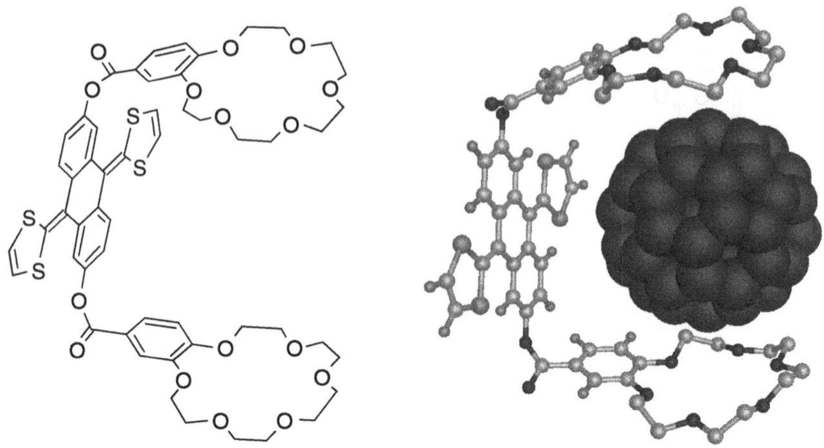

Figure 5.18 The flexible molecular tweezers with functionalized crown ether pincers (left) and the energy-minimized 3D structure of their complex with fullerene C_{60} (right, hydrogen atoms of the crown ether moieties were removed for clarity).

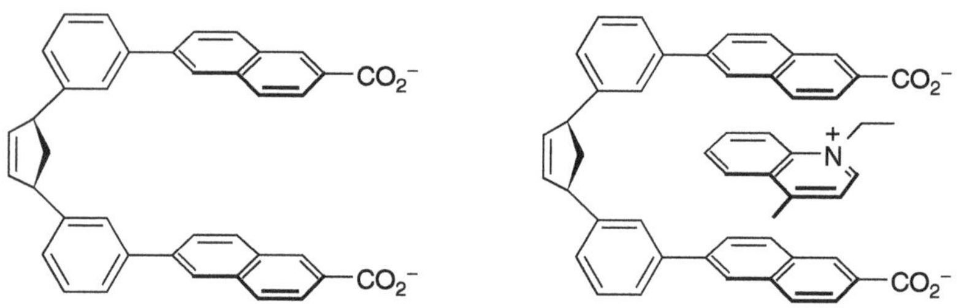

Figure 5.19 The flexible molecular tweezers with anionic pincers (left) and their sandwich complex with 1-ethyl-4-methylquinolinium cation (right).

system has reasonable binding affinity toward certain cationic guest molecules, such as 1-ethyl-4-methylquinolinium cation ($K_a = 950\,M^{-1}$) [32].

"Click" chemistry provides easy access to electroactive flexible molecular tweezers incorporating two tetrathiafulvalene units (Scheme 5.1). The distance between the tetrathiafulvalene pincers is only about 5 Å. The tweezer-type 3D structure is supported by weak hydrogen bonding between the pincer units and the pyridine nitrogen atom. This host molecule retains tweezer-type conformation in complexes with some divalent metals, such as Cd^{2+}, even though the metal cations are coordinated outside the narrow tweezer cavity [56].

Flexible molecular tweezers with anthracene- and pyrene-based pincers, also prepared via "click" chemistry, are capable of capturing large fullerene molecules (Figure 5.20). The most valuable feature of molecular tweezers with anthracene-based pincers is their ability to differentiate between spherical

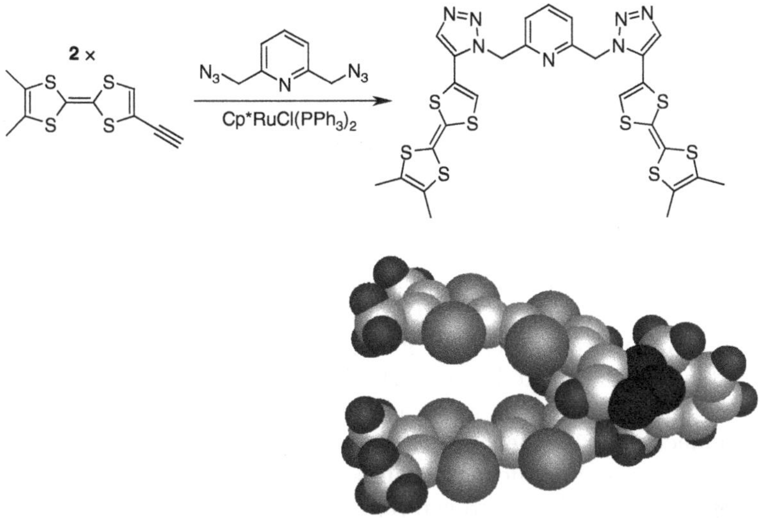

Scheme 5.1 Click synthesis of electroactive molecular tweezers incorporating two tetrathiafulvalene units (left) and their X-ray structure (bottom).

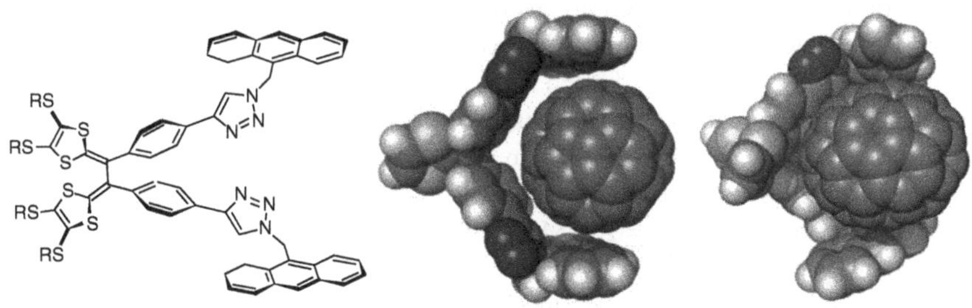

Figure 5.20 The flexible molecular tweezers with the tetrathiafulvalene vinylogue spacer and the anthracene pincers (left, $R = C_{10}H_{21}$) and the energy-minimized 3D structures of their complexes with fullerene C_{60} (middle) and fullerene C_{70} (right). $C_{10}H_{21}$ groups were replaced with hydrogen atoms in the energy-minimized structures to speed-up computations. *Source*: Adapted from Mulla *et al.* 2013 [57]. Reproduced with permission of American Chemical Society.

fullerene C_{60} and ellipsoidal fullerene C_{70}. Ellipsoidal fullerene C_{70} has a substantially higher association constant presumably due to better π-contacts with both anthracene pincers and the tetrathiafulvalene vinylogue spacer ($K_a = 79{,}000\,M^{-1}$ for C_{70}). It is interesting to note that this C_{60}/C_{70} selectivity is lost in the case of molecular tweezers with larger pyrene-based pincers [57].

The presence of nitrogen atoms in pyridine subunits of flexible spacers allows for the construction of controllable (also known as dynamic) molecular tweezers, which can be engaged or disengaged by the presence of additional control elements in the environment, such as Zn^{2+} or Cu^{+} metal cations. Some of these pyridine-based molecular tweezers were designed to switch between U and W shapes upon cation coordination, thus acting as dynamic devices capable of ion-induced allosteric conformational changes affecting guest-binding affinity. The presence of a bound guest molecule can be easily monitored via the appearance of colored charge–transfer sandwich complexes between the electron-rich pincers and the electron-deficient guests, such as 2,4,7-trinitrofluorenone and TCNQ [29].

An example of terpyridine-based molecular tweezers, which can be activated by Zn^{2+}, is presented in Figure 5.21. In this tweezer system, the Zn^{2+} cation serves as a template for preorganization of the host molecule into U shape by creating a rigid metal-coordinated spacer with the proper face-to-face orientation of two acridine pincers. It also serves as an additional coordination center for the guest molecule – 1,10-phenanthroline. In the double complex, the Zn^{2+} cation holds the guest and the host molecules together by providing binding with five nitrogen atoms – three nitrogen atoms of the host molecule and two nitrogen atoms of the guest molecule [29].

The proximity of the Zn^{2+} cation to the guest molecule and the pincer units in the abovementioned double complex precludes proper binding of the other electron-deficient guests, which do not have additional Zn^{2+} coordination centers. This obstacle can be overcome by extending the links between the pincers and the spacer. An example of such dynamic host with longer amide links is presented in Figure 5.22 [29]. This dynamic host system can be activated by the addition of Zn^{2+}. In the activated form, the host molecule is preorganized as molecular tweezers suitable for binding with electron-deficient guests, such as 2,4,7-trinitrofluorenone ($K_a = 130\,M^{-1}$) [29].

With proper modifications, dynamic host systems described earlier are capable of *releasing* guest molecules upon the activation by a metal cation controller. One of such systems is based on a pyrimidine-bis(pyridine) spacer (Figure 5.23). The replacement of the central pyridine subunit in

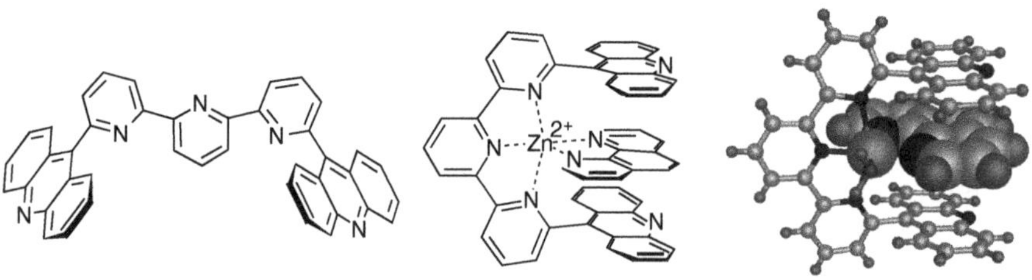

Figure 5.21 The tripyridine-based dynamic host (left), its double tweezer complex with 1,10-phenanthroline and Zn^{2+} (middle), and the X-ray structure of this complex (right, as monohydrate).

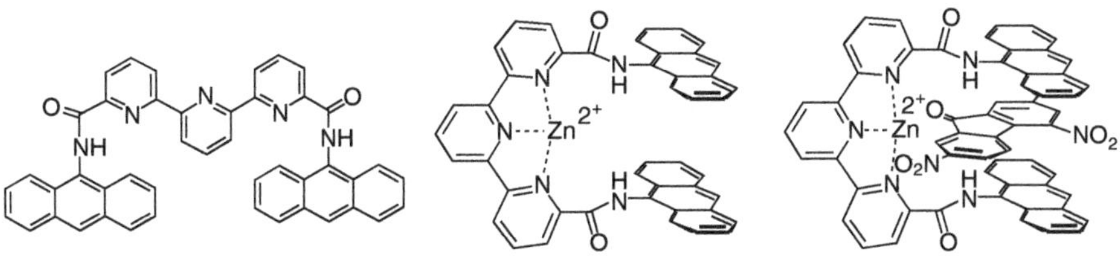

Figure 5.22 The terpyridine-based dynamic host with amide links (left), the Zn^{2+}-activated molecular tweezers (middle), and their double complex with 2,4,7-trinitrofluorenone and Zn^{2+} (right).

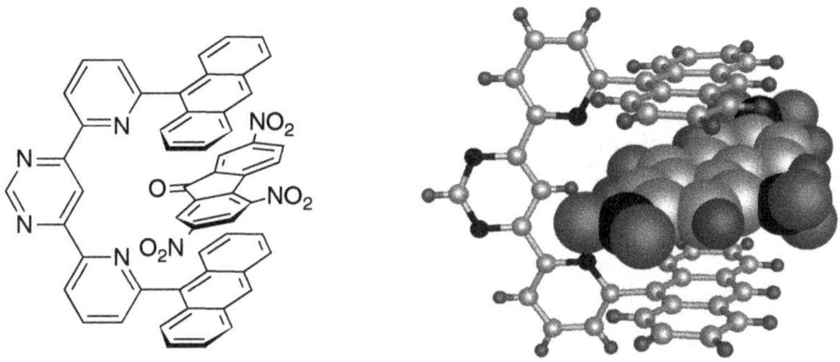

Figure 5.23 The complex of pyrimidine-based dynamic molecular tweezers with 2,4,7-trinitrofluorenone (left) and its X-ray structure (right).

the terpyridine spacer with pyrimidine results in the inversion of structural changes upon complexation. In contrast to the terpyridine systems described earlier, this pyrimidine-bis(pyridine) dynamic host binds a guest molecule only in the absence of the controlling Cu^+ cation. As expected, the electron-rich anthracene pincers provide better affinity toward 2,4,7-trinitrofluorenone ($K_a = 2500\ M^{-1}$) compared to the acridine pincers ($K_a = 50\ M^{-1}$) [29].

The decomplexation of the pyrimidine-bis(pyridine) molecular tweezer complexes can be achieved by the addition of Cu^+ via allosteric mechanism (Scheme 5.2). In this case, coordination of Cu^+ cations with pyridine and pyrimidine units results in a conformational change of the host molecule from U to W shape. The acridine pincers in the W-shaped Cu^+ complex can no longer be engaged in sandwich-type interactions with a guest molecule, thus effectively releasing it into the environment. The NMR titration data indicate that the binding of the two Cu^+ cations is an anticooperative process: the binding of the second metal cation is weaker than the first one. As a result, the full release of the guest molecule can only be achieved by employing a large excess of Cu^+ [29].

Subsequent development of dynamic molecular tweezers led to the synthesis of novel host molecules with electron-deficient NDI pincers (Figure 5.24). The

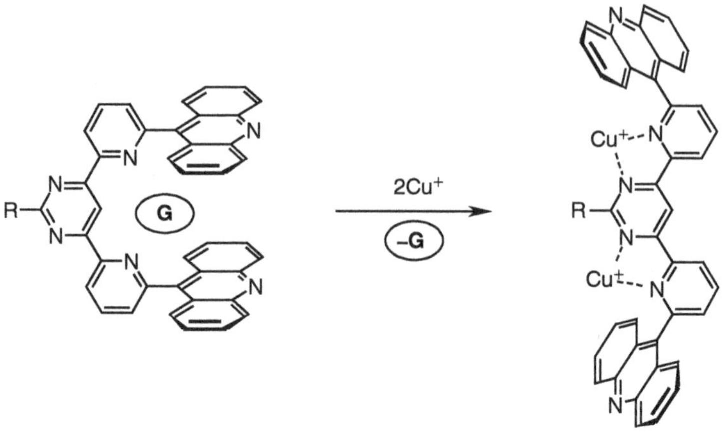

Scheme 5.2 Allosteric release of a guest molecule (G) form the dynamic molecular tweezers upon complexation with Cu^+ cation (R = *n*-Bu).

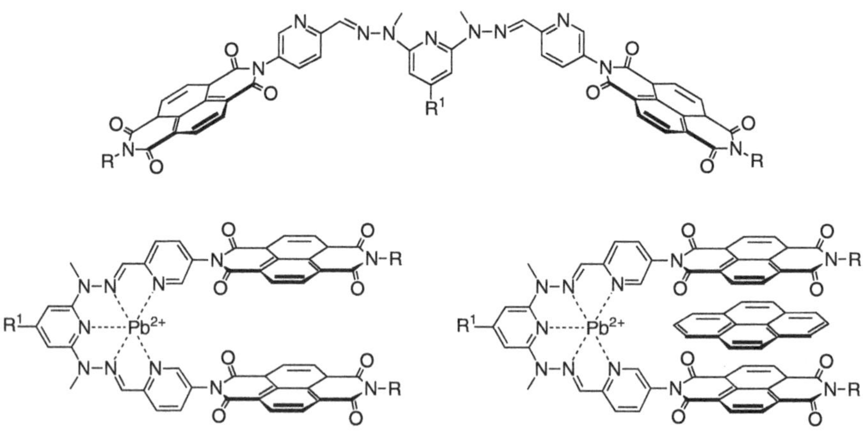

Figure 5.24 The dynamic host with electron-deficient naphthalene diimide pincers (top), the activated molecular tweezer complex with Pb^{2+} (left bottom), and the double complex with pyrene and Pb^{2+} (right bottom); R = *n*-Bu, $R^1 = CH_2OCH(C_2H_5)C_5H_{11}$.

cation-binding spacer in these systems is based on a bis(hydrazone)pyridine framework. The metal-free host in W conformation does not produce complexes with aromatic compounds, whereas the U-shaped molecular tweezer complexes with some divalent metals, such as Zn^{2+} or Pb^{2+}, are suitable for binding with π-electron-rich aromatic compounds. The binding affinity of these molecular tweezer complexes is affected by the size of the metal cation as a control element. In the case of pyrene as a guest molecule, the best results

($K_a = 320\,M^{-1}$) were achieved with large Pb^{2+} cation due to the proper alignment of the NDI pincers [58, 59].

Dynamic molecular tweezers controlled by metal cations are also capable of capturing certain derivatives of biologically important heterocycles, such as 9-ethyladenine (Figure 5.25). This host molecule is designed on the basis of a pyridine–bis(imidazole) spacer and can be activated by the addition of Cu^{2+} salts. In the double tweezer complex, Cu^{2+} cation also interacts with the guest molecule, thus facilitating strong, guest-selective binding ($K_a = 88{,}000\,M^{-1}$ for 9-ethyladenine in acetone). The activated molecular tweezers easily form dimers in the absence of the guest molecules [60].

Also known are dynamic molecular tweezers with metal complex pincer elements [61–63]. A representative example of tweezer host systems employing Pt–salphen pincer units and a terpyridine spacer is shown in Scheme 5.3 [61]. The switching of this dynamic molecular host system from a W-type open conformation to a tweezer U-type close conformation can be accomplished by the addition of Pb^{2+} salts. Reopening of the U form back to the W form without the distortion of Pt–salphen metal complex pincers can be accomplished with tris(2-aminoethyl)amine as a selective competitive ligand [61].

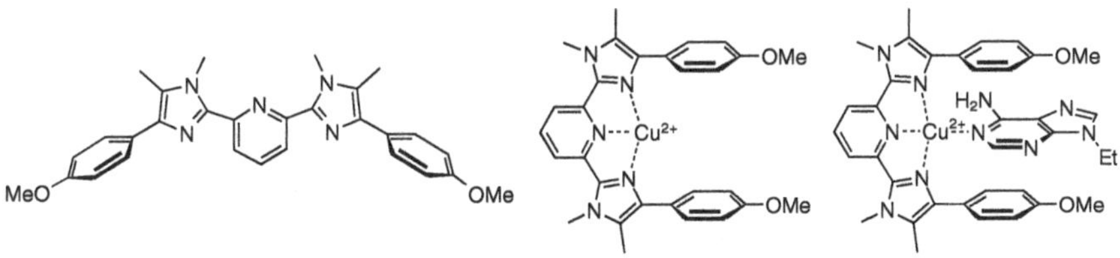

Figure 5.25 The dynamic host with the pyridine–bis(imidazole) spacer (left), the activated molecular tweezer complex with Cu^{2+} (middle), and the double complex with Cu^{2+} and 9-ethyladenine (right).

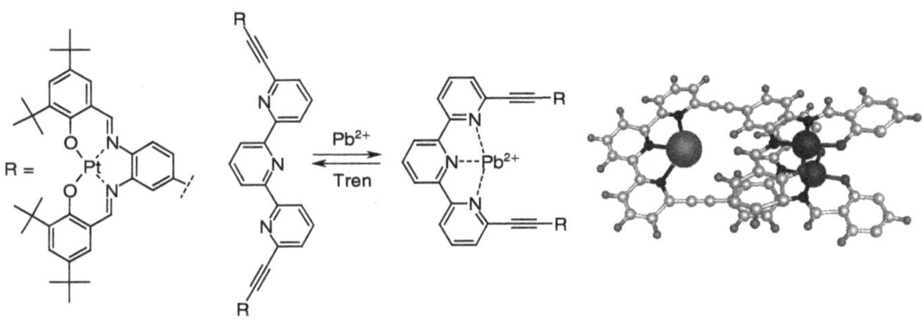

Scheme 5.3 The dynamic host system with the terpyridine spacer and Pt–salphen pincers (left, tren = tris(2-aminoethyl)amine) and the X-ray structure of the U shape Pb^{2+} tweezer complex (right, *t*-Bu groups were removed for clarity and hydrogen atoms were added at idealized positions).

The tricyclic spacer motif frequently employed in dynamic tweezer systems proved to be a useful concept for the construction of other nonswitchable molecular tweezers with metalated pincers [64–66]. A representative example of these host molecules with metal-coordinated terpyridine pincers is shown in Figure 5.26. Pincer elements in these tweezers are attached to the spacer unit via organometallic Pt—C bonds. This tweezer host has reasonable binding affinity toward polyaromatic compounds. The binding constant K_a improves progressively with the size of the guest molecules: naphthalene – $36\,M^{-1}$, anthracene – $910\,M^{-1}$, pyrene – $2451\,M^{-1}$, triphenylene – $3821\,M^{-1}$, perylene – $6528\,M^{-1}$, and coronene – $10{,}240\,M^{-1}$ (NMR titration in $CDCl_3$) [65].

The abovementioned pincer design allows for easy access to rare double-decker tweezer molecules with two cavities for guest accommodation (Figure 5.27). These molecular tweezers exhibit strong tendency toward dimerization in the solid state. As expected, the host molecules produce well-defined 1:2 complexes with complementary organometallic guest molecules. Some of these complexes are very stable with association constants K_a exceeding $10^{13}\,M^{-1}$ [67].

A combination of bowl-shaped corannulene pincers and an organometallic spacer proved to be one of the best design ideas for the construction of molecular tweezers with improved selectivity toward spherical fullerene guest molecules (Figure 5.28) [68]. This Pt-based tweezer molecule has good affinity toward fullerene C_{70} ($K_a = 20{,}700\,M^{-1}$ in toluene-d_8) and relatively high C_{70}/C_{60} selectivity ($K_a = 4600\,M^{-1}$ for C_{60} in toluene-d_8, which corresponds to approximately 4.5:1 C_{70}/C_{60} selectivity).

Further improvements in selectivity and affinity of flexible molecular tweezers can be achieved with the use of metalated porphyrin pincers [69]. The metalated porphyrin rings can be considered as pincers with "teeth" due

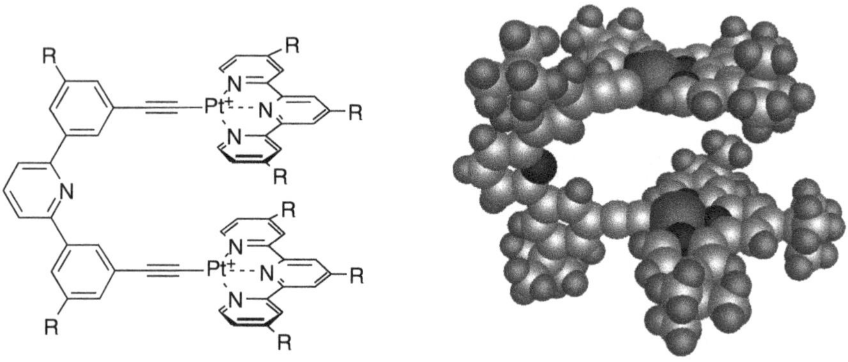

Figure 5.26 The flexible molecular tweezers with Pt–terpyridine pincers (left, R = t-Bu) and their X-ray structure (right) [64].

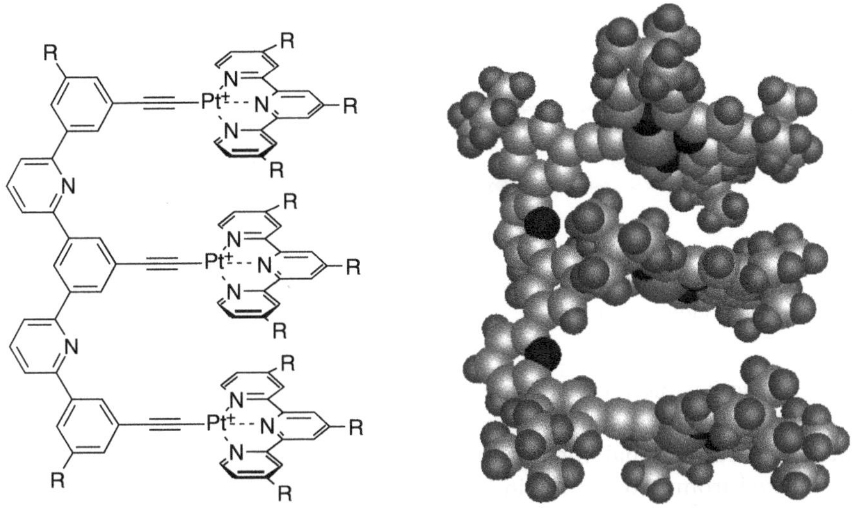

Figure 5.27 The double-decker molecular tweezers with Pt–terpyridine pincers (left, R = *t*-Bu) and their X-ray structure (right, only one monomeric unit of the dimer is shown for clarity).

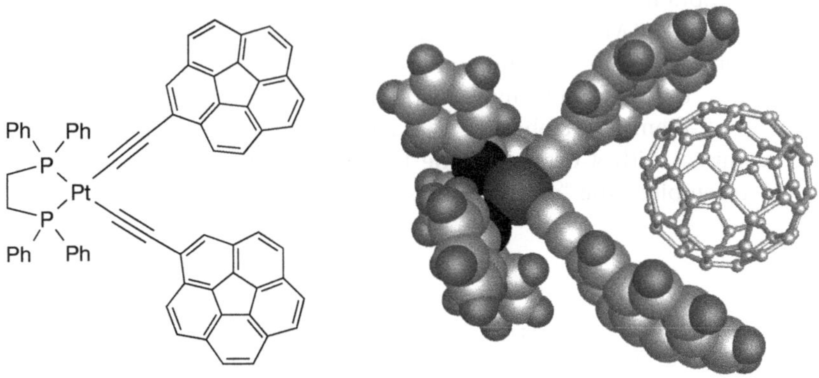

Figure 5.28 The Pt-based tweezer molecule with corannulene pincers (left) and the X-ray structure of its complex with fullerene C_{60} (right).

to the presence of a metal atom in the center of the porphyrin moiety. This type of tweezers is widely used in surgery under the special name "forceps" (Figure 5.29). The enhanced binding properties of porphyrin pincers can be attributed to their ability to engage simultaneously in π–π interactions and in metal-coordination bonding with appropriate guests.

The simplest molecular tweezers with porphyrin pincers can be constructed by simply linking two porphyrin units with a flexible aliphatic

Figure 5.29 Conventional surgical tweezers with teeth, also known as forceps. *Source:* Courtesy of ACE Surgical Supply Co.

chain. These flexible molecular tweezers were successfully employed for the determination of the absolute configuration of certain chiral diamines [70] and other bifunctional molecules [71]. For example, molecular tweezers constructed from two porphyrin units interconnected by a flexible CH_2CH_2 spacer demonstrate high binding affinity ($K_a = 1.25 \times 10^7\,M^{-1}$) toward (1*R*,2*R*)-(−)-1,2-diaminocyclohexane in CH_2Cl_2 (Figure 5.30) [70].

An additional insight into the structural parameters and performance of these porphyrin-based molecular tweezers with the CH_2CH_2 spacer was obtained during the X-ray investigation of their *syn* 1:1 complex ($K_a = 1820\,M^{-1}$) with *o*-phenylenediamine (Figure 5.31) and *anti* 1:2 complex with 4-cyanopyridine (Figure 5.31). Unlike *syn* 1:1 complex with *o*-phenylenediamine, which is structurally similar to the (1*R*,2*R*)-(−)-1,2-diaminocyclohexane tweezer complex described earlier, the *anti* 1:2 complex with 4-cyanopyridine features a wide open conformation no longer resembling the "classic" tweezer motif. A weak *anti* 1:2 complex

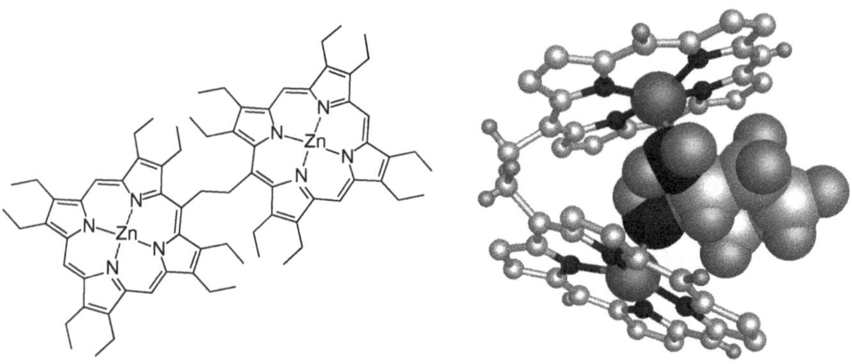

Figure 5.30 Simple flexible porphyrin-based molecular tweezers (left) and the X-ray structure of their complex with (1*R*,2*R*)-(−)-1,2-diaminocyclohexane (right, CH_2CH_3 groups were removed for clarity) [72].

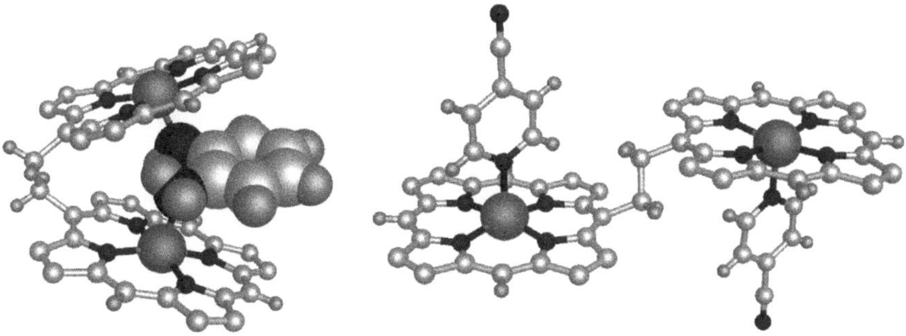

Figure 5.31 The X-ray structures of the *syn* 1:1 complex with *o*-phenylenediamine (left) and the *anti* 1:2 complex with 4-cyanopyridine (right), CH_2CH_3 groups were removed for clarity.

($K_a = 134\,M^{-2}$) was also detected with a large excess of *o*-phenylenediamine [73].

Simple porphyrin-based flexible molecular tweezers discussed earlier can be effectively used only for bifunctional guest molecules, such as diamines or diols. Better binding affinity for other types of guest molecules can be attained with porphyrin-based molecular tweezers having extended spacers. A representative example of such host systems is shown in Figure 5.32. These tweezers form a relatively strong 1:2 complex with pyrene as a fluorescent probe ($1.78 \times 10^5\,M^{-2}$ for 1:2 binding), even though the interplanar distance between the porphyrin pincers (about 6.5 Å) is shorter than the optimal distance [74].

Rigid polycyclic scaffolds equipped with conformationally mobile linking elements can be successfully used for the construction of flexible spacers for porphyrin-based molecular tweezers. An example of such a tweezer system is presented in Figure 5.33. Conformational flexibility of this large molecular tweezer system was achieved via two *p*-phenylene links connecting porphyrin pincers to rigid *exo*-fused polycyclic "arms" of the spacer. A complex of

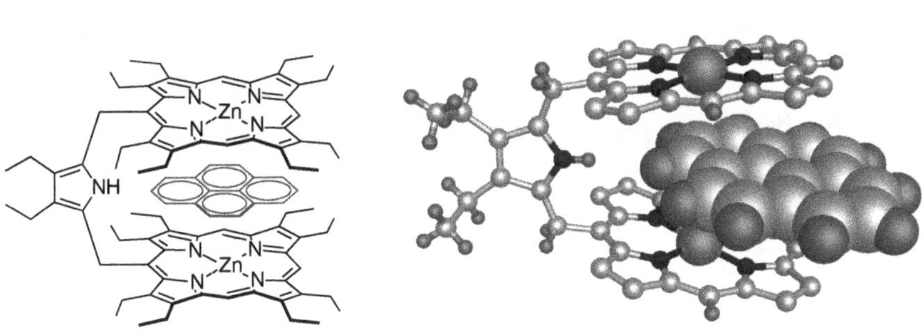

Figure 5.32 The pyrene complex of the flexible porphyrin-based molecular tweezers with the extended spacer unit (left) and its X-ray structure (right, CH_2CH_3 groups in the porphyrin pincers were removed for clarity).

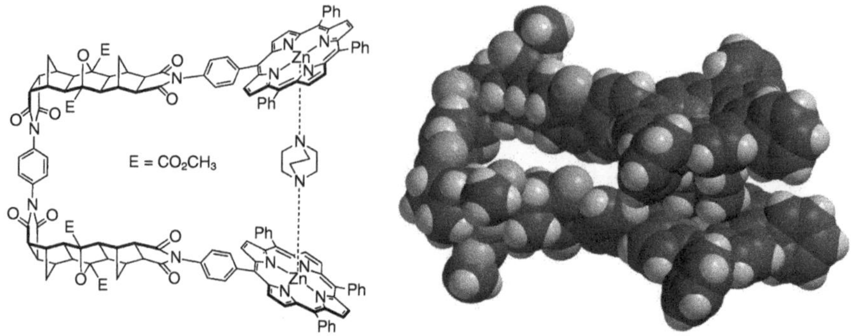

Figure 5.33 The 1,4-diazabicyclo[2.2.2]octane complex of large flexible porphyrin-based molecular tweezers (left) and its energy-minimized 3D structure (right). *Source*: Adapted from Murphy *et al.* 2013 [75]. Reproduced with permission of John Wiley & Sons.

these molecular tweezers with 1,4-diazabicyclo[2.2.2]octane (DABCO) has a relatively high association constant ($K_a = 8.1 \times 10^7\ M^{-1}$) in $CHCl_3$ [75].

A bis-thiophene heterocyclic spacer was successfully employed for the construction of porphyrin-based flexible molecular tweezers (Figure 5.34). This host molecule demonstrates good binding affinity toward both C_{60} and C_{70} fullerenes in solution. The association constant of the fullerene C_{70} complex ($K_a = 9630\ M^{-1}$) was found to be almost twice higher compared to the C_{60} complex ($K_a = 4795\ M^{-1}$) [76]. Substantially better C_{60}/C_{70} selectivity (up to 1:7) was achieved with (thia)calix[4]arene–porphyrin molecular tweezer systems [77].

The incorporation of electroactive bipyridinium units into the spacer unit structure of porphyrin-based molecular hosts allows for the electrochemical control of the complexation process (Figure 5.35). The partial reduction of the bipyridinium fragments results in the formation of strongly associated

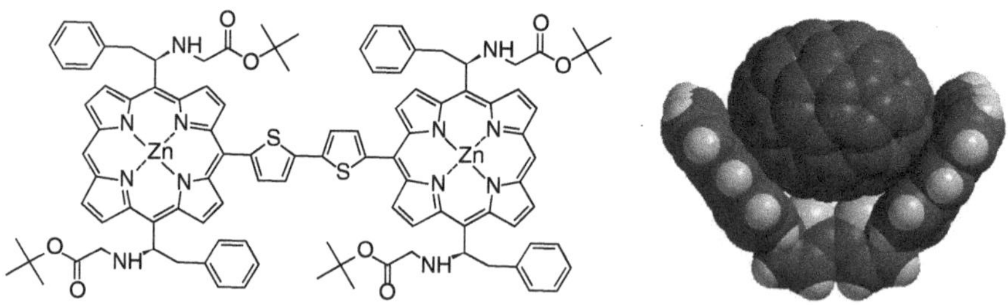

Figure 5.34 The porphyrin-based molecular tweezers with the bis-thiophene spacer (left) and the energy-minimized 3D structure of their complex with fullerene C_{70} (right, $CH(CH_2Ph)NHCH_2CO_2C(Me)_3$ fragments were replaced with hydrogen atoms). *Source*: Adapted from Mukherjee *et al.* 2011 [76], Copyright 2011, with permission from Elsevier.

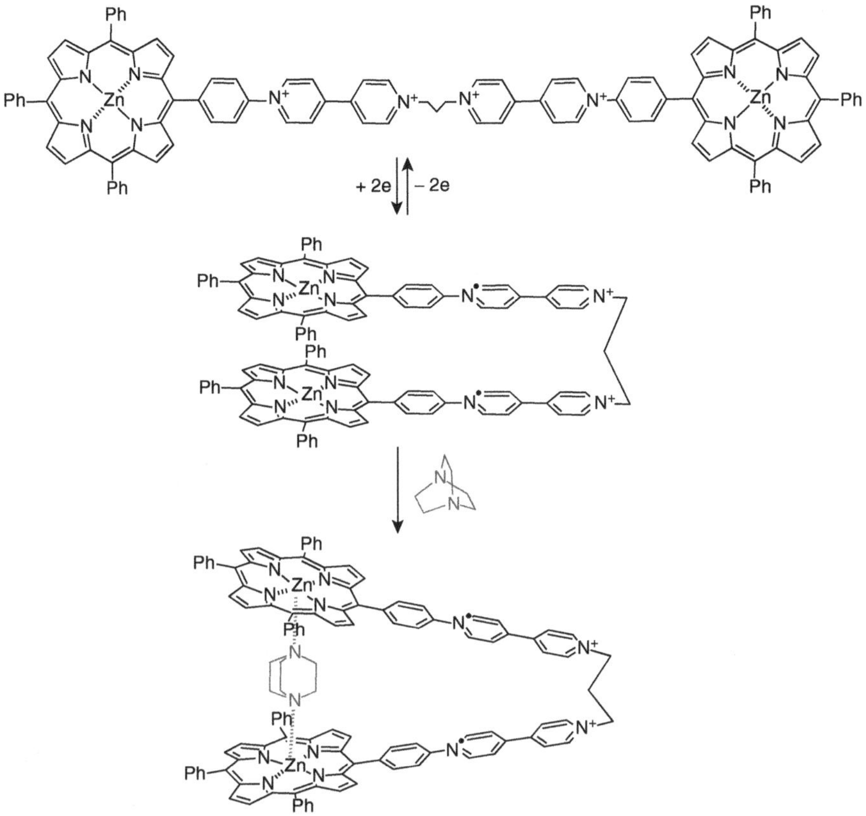

Figure 5.35 The porphyrin-based host with the electroactive bipyridinium spacer (top), the dynamic molecular tweezer cation radical (middle), and its complex with 1,4-diazabicyclo[2.2.2]octane (bottom).

pair of cation radicals forcing the *syn*-conformation of the two porphyrin pincers. This preorganized *syn*-form of the dynamic molecular tweezer host is capable of producing a sandwich complex with DABCO as a bidentate guest ($K_a = 82{,}000\ M^{-1}$) [25].

The presence of heterocyclic fragments in a spacer assembly is the key feature of other programmable, or dynamic porphyrin-based molecular tweezers, which can be turn "on" or "off" by the addition of certain ions. One of the recently developed systems employs a bis-indole spacer unit capable of selective complexation with chloride anion (Scheme 5.4). Without the chloride anion, the host molecule adopts thermodynamically favorable *anti*-conformation (Scheme 5.4). The addition of chloride anion, which can bind to the bis-indole unit via a pair of hydrogen bonds, induces allosteric transformation of the molecule (Scheme 5.4) into preorganized molecular

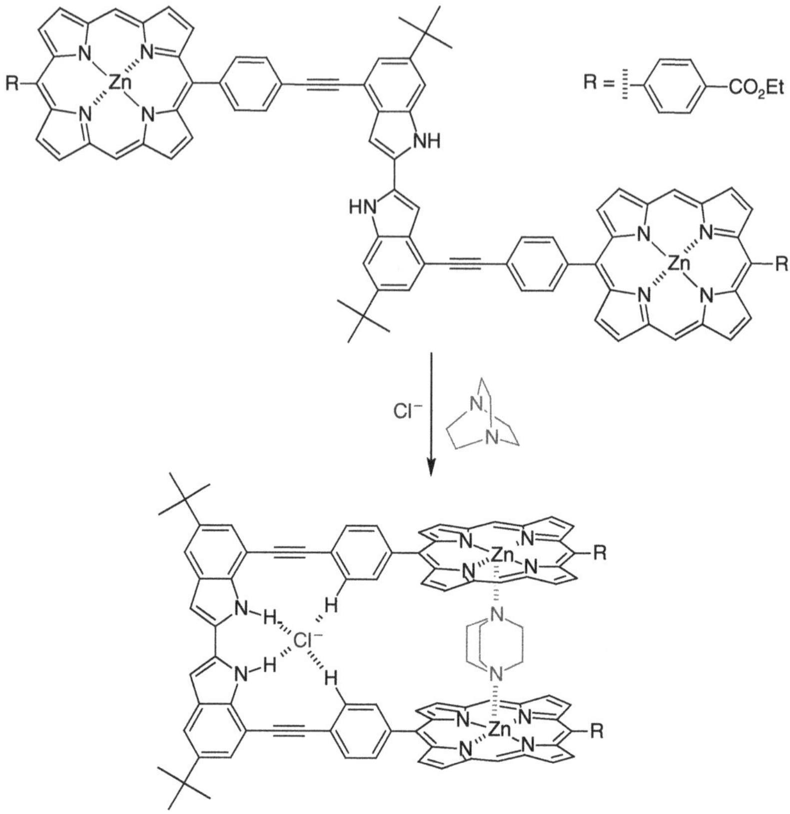

Scheme 5.4 Allosteric activation of the porphyrin-based dynamic molecular tweezers with the bis-indole spacer.

tweezers (*syn*-conformation) having good affinities toward certain bis-amine guests, such as DABCO [78].

Porphyrin-based molecular tweezers with functionalized spacers can provide superior binding sites for guest molecules with similar structural features. In this respect, molecular tweezers with complementary spacer units, designed by analogy to DNA double helix, are of particular interest. A representative example of complementary molecular tweezers with porphyrin pincers is presented in Figure 5.36. This molecular tweezer system with a pyridyl-2,6-dicarboxamide spacer forms a strong sandwich complex with a guest molecule also containing pyridyl-2,6-dicarboxamide moiety ($K_a = 1.3 \times 10^7\ M^{-1}$). Additional stabilization of this complex is achieved via intramolecular hydrogen bonding between the heterocyclic nitrogen atoms and the amide groups in both the host and the guest molecules [79].

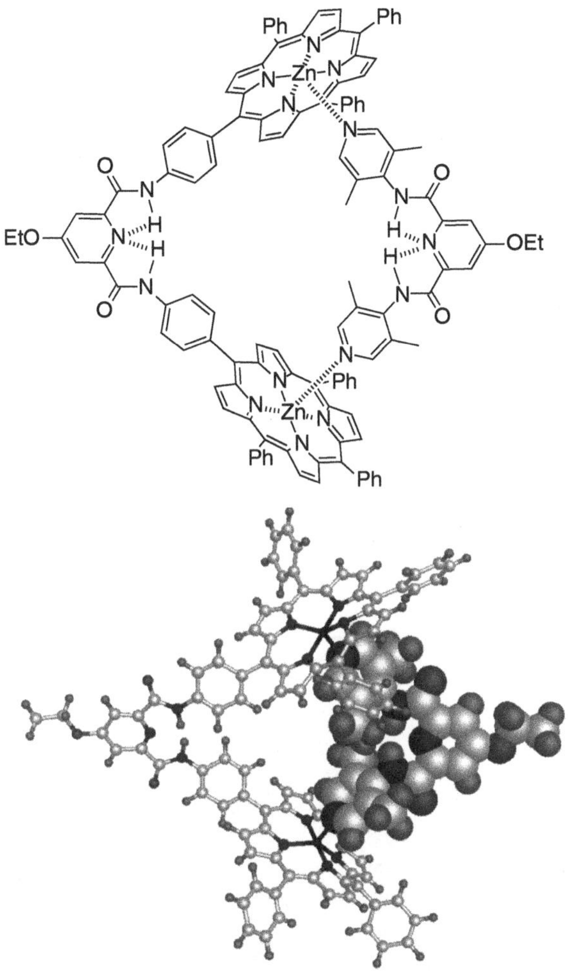

Figure 5.36 An example of the complementary molecular tweezer complex (top) and its energy-minimized 3D structure (bottom).

Molecular tweezers with functionalized spacers can be partially rigidified by a network of intramolecular hydrogen bonds. These semiflexible tweezers can be considered as a transitional type between fully flexible and conformationally restricted systems. Unlike previously reported examples of dynamic tweezers with metal-coordinated spacers, the hydrogen bond network is an internal structural element and does not require additional metal cations to operate. An example of porphyrin-based molecular tweezers rigidified with intramolecular hydrogen bonding and their complex with a dicarboxylate derivative of fullerene C_{60} is presented in Figure 5.37. The complex is stabilized

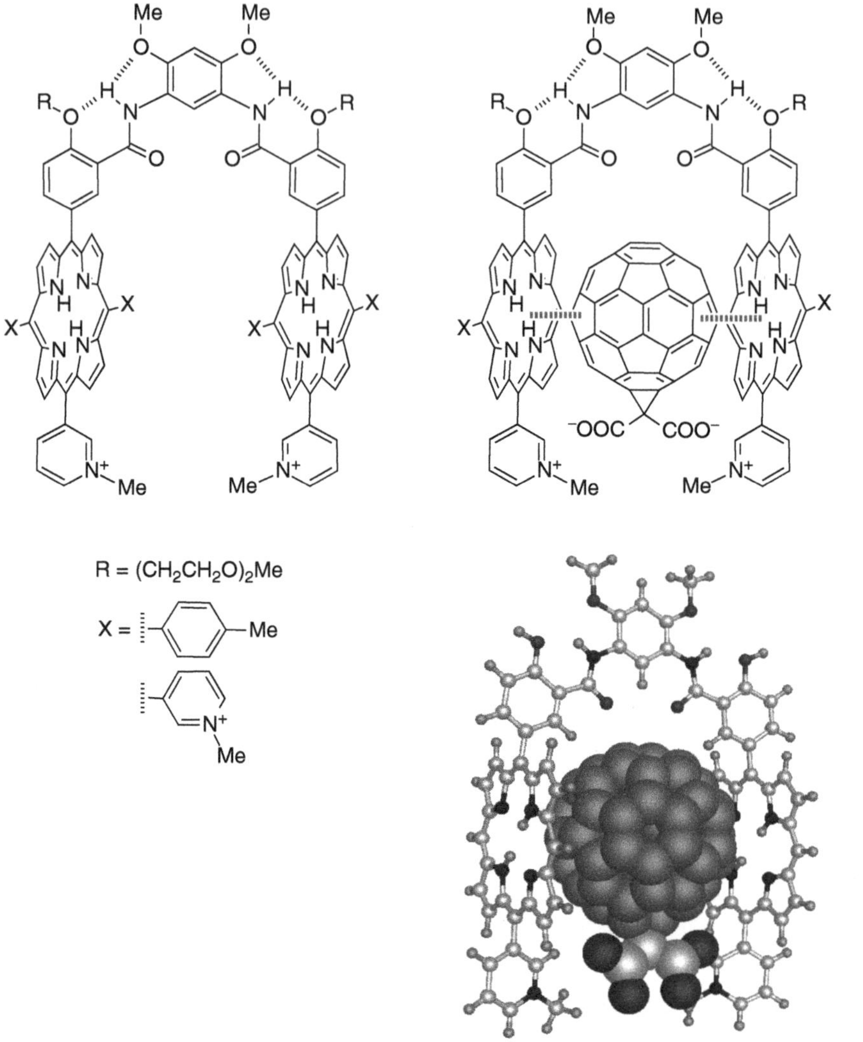

Figure 5.37 The porphyrin-based molecular tweezers rigidified by intramolecular hydrogen bonding (top left), their complex with the fullerene C_{60} dicarboxylate derivative (top right), and the energy-minimized 3D structure of this complex (bottom, R and X substituents were replaced with hydrogen atoms for clarity).

with two favorable cation–anion interactions ($K_a = 1.1 \times 10^5\ M^{-1}$ in water, $X = p$-$PhCH_3$). Even stronger binding affinity is observed in the case of hexacationic molecular tweezers with six 3-(*N*-methylpyridinium) groups ($K_a = 6.9 \times 10^5\ M^{-1}$ in water, $X = m$-$C_5H_4N^+Me$) [80].

Incorporation of metal atoms into the porphyrin fragments improves the binding of fullerenes C_{60} and C_{70} and facilitates effective capture of these molecules without derivatization. An example of metalloporphyrin molecular tweezers rigidified by intramolecular hydrogen bonding is presented in Figure 5.38. The association constants of the molecular tweezers with fullerenes C_{60} and C_{70} were determined by UV–Vis titration ($K_a = 1.0 \times 10^5\ M^{-1}$ for C_{60} and $1.1 \times 10^6\ M^{-1}$ for C_{70} in toluene) [81]. These molecular tweezers also demonstrate high affinity toward some electron-deficient molecular guests containing pentafluorophenyl moieties [82].

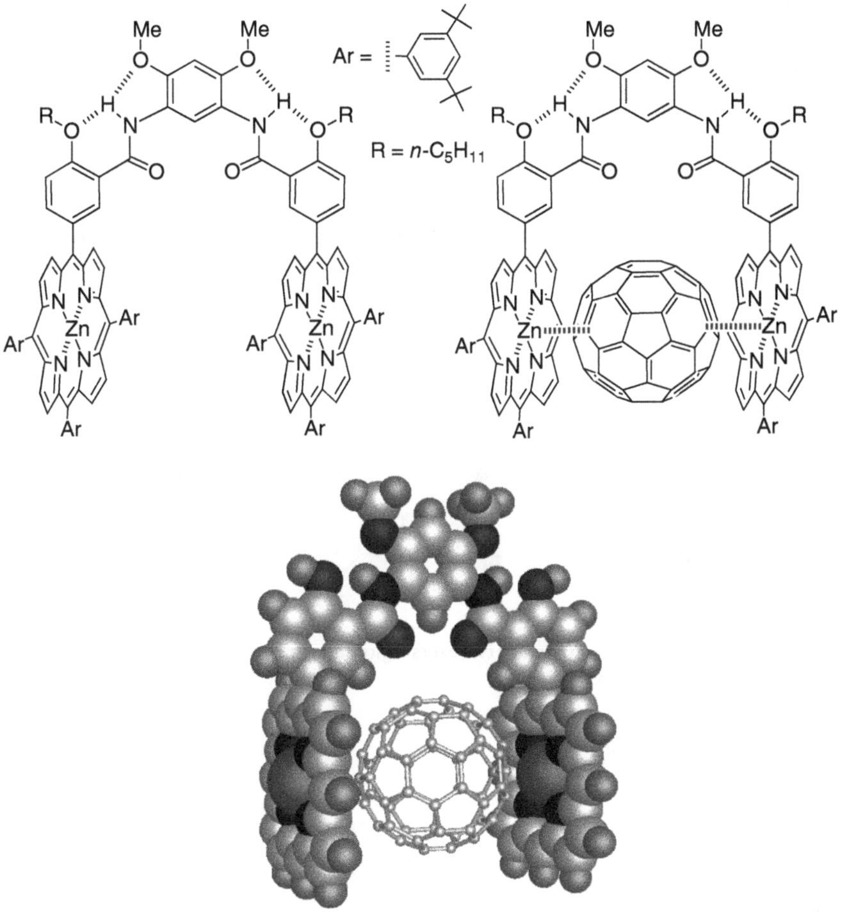

Figure 5.38 The metalloporphyrin molecular tweezers rigidified by intramolecular hydrogen bonding (top left), their complex with the fullerene C_{60} (top right), and the energy-minimized 3D structure of this complex (bottom, R and Ar groups were replaced with hydrogen atoms for clarity).

Similar semiflexible molecular tweezers rigidified by intramolecular hydrogen bonding and equipped with traditional aromatic polycyclic pincers are also known [83, 84]. A representative example of these tweezers is shown in Figure 5.39 [83]. Tweezer-type U conformation is stabilized by the two strong OH· · ·N hydrogen bonds between the phenol units and the neighboring pyrimidine moiety. Unfortunately, the internal cavity of these tweezers is too shallow to accommodate aromatic guest molecules.

Better guest-binding results were achieved with tweezers having extended acetylene connectors between the rigidified spacer unit and the anthracene pincers (Figure 5.40) [84]. These molecular tweezers have larger cavity and form stable sandwich complexes with electron-deficient guests, such as TCNQ, 1,2,4,5-tetracyanobenzene, and 2,4,7-trinitrofluorenone ($K_a = 93\,M^{-1}$, $550\,M^{-1}$, and $2100\,M^{-1}$ in $CDCl_3$, respectively). The addition of tetrabutylammonium fluoride results in the formation of stable OH· · ·F bonds and the disruption of the OH· · ·N intramolecular hydrogen bond network. This, in turn, leads to the release of a guest molecule due to the change of molecular conformation from the tweezer U form to the open W form [84].

Semiflexible molecular tweezers rigidified by intramolecular hydrogen bond network are only operational within the limited pH range. Permanent *syn*-conformation of the pincer elements can only be achieved in

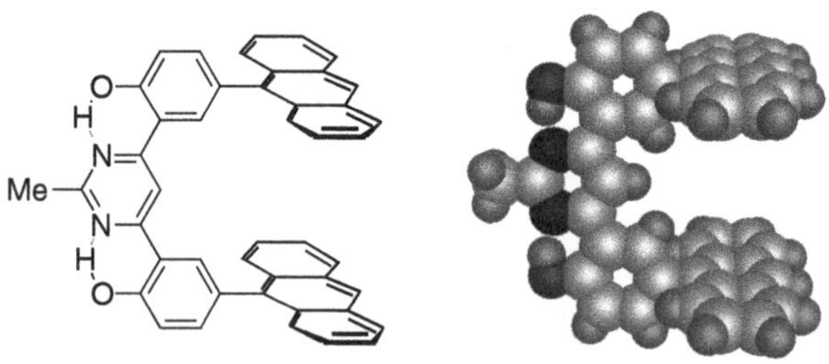

Figure 5.39 The semiflexible molecular tweezers rigidified via hydrogen bonding (left) and their X-ray structure (right) [83].

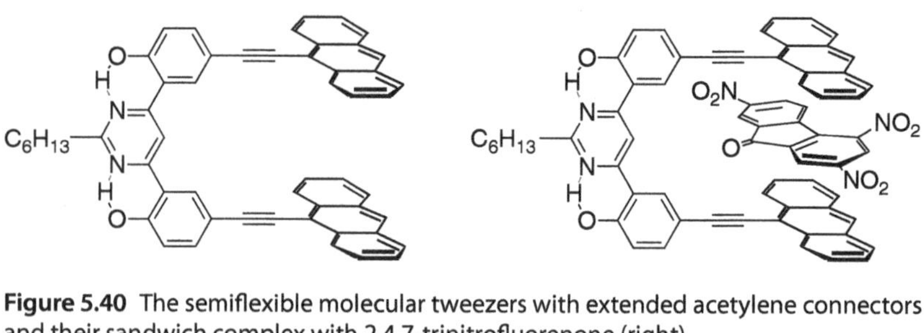

Figure 5.40 The semiflexible molecular tweezers with extended acetylene connectors (left) and their sandwich complex with 2,4,7-trinitrofluorenone (right).

conformationally restricted tweezers, which are discussed in more detail in Section 5.3.

5.3 Conformationally Restricted Molecular Tweezers

It is commonly assumed that the use of rigid spacers and directional links could improve the performance of molecular tweezers by fixing two pincers permanently in *syn* configuration [85]. While this is generally true, fabrication of such tweezers requires precise molecular design since self-organization is limited in conformationally restricted systems. The majority of molecular tweezers of this type are constructed on the basis of fused polycyclic systems as spacer elements.

A representative example of conformationally restricted molecular tweezers with a 5,6,8,9-tetrahydrodibenzo[*c*,*h*]acridine spacer is depicted in Figure 5.41 [86]. Some hosts of this type have high association constants with polynitroaromatic guests, such as 1,3,5-trinitrobenzene, 2,4,7-trinitrofluorenone, and 2,4,5,7-tetranitrofluorenone. Improved binding affinity K_a (up to 20,000 M^{-1}) can be attributed to consummate design of the polyaromatic spacer, which not only enforces a *syn* conformation of the host pincer arms but also provides an optimal 7 Å distance between them [16, 17]. The presence of two single bonds connecting the spacer and the pincers is an important structural feature offering some conformational flexibility, even though the energy barrier for rotation is relatively high, typically in the range of 19–22 kcal/mol [87]. As a result, the exact spatial orientation of pincers can change to better accommodate a guest molecule, thus providing limited self-optimization during the binding process.

Similar to flexible molecular tweezers, conformationally restricted tweezers utilize π-donor–π-acceptor interactions between electron-deficient guests and

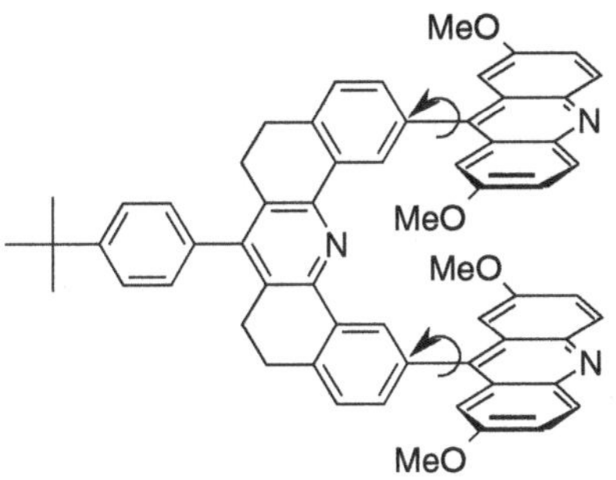

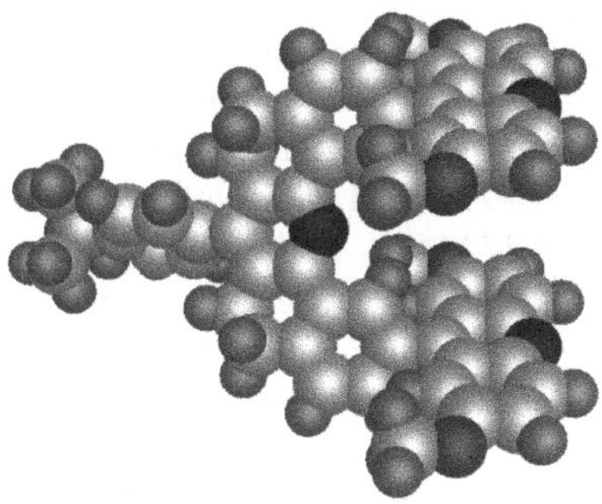

Figure 5.41 A representative example of conformationally restricted molecular tweezers with the 5,6,8,9-tetrahydrodibenzo[*c*,*h*]acridine spacer (left) and their X-ray structure (right, hydrogen atoms were added at idealized positions) [86].

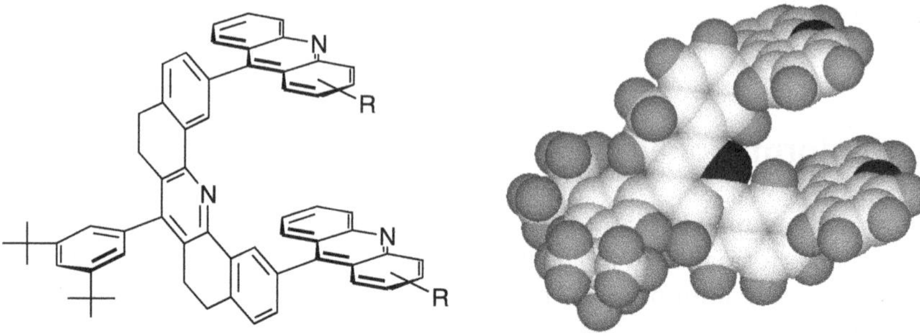

Figure 5.42 The conformationally restricted molecular tweezers with electron-donating groups (left, R = 2-NMe$_2$, 4-OMe, 2-*t*-Bu) and their energy-minimized 3D structure (right, electron-donating groups R were replaced with hydrogen atoms for clarity).

π-electron-rich pincers for their "guest grabbing" action. Not surprisingly, the binding affinity can be improved by adding electron-donating groups to aromatic polycyclic pincers (Figure 5.42). The highest association constant with 2,4,7-trinitrofluorenone guest molecule is attained for R = 2-NMe$_2$ ($K_a = 697\,M^{-1}$ in $CDCl_3$), whereas the association constants with 4-OMe and 2-*t*-Bu donor groups are lower, 320 and 149 M^{-1}, respectively [87].

A much larger set of predominantly electron-rich aromatic guests can be captured using molecular tweezers with electron-deficient aromatic pincers. In this case, pincers, rather than guest molecules, would serve as π-acceptors. Pentafluorophenyl group (C_6F_5) can be considered as an attractive choice for the construction of such tweezer systems due to its excellent electron-withdrawing and π–π-stacking capabilities [15, 88, 89]. A representative example of conformationally restricted molecular tweezers with two electron-deficient pentafluorophenyl pincers is presented in Figure 5.43. The binding affinity of these molecular tweezers toward electron-rich aromatic compounds was evaluated experimentally using ^{19}F NMR titration technique. The best results were achieved for *N,N,N′,N′*-tetramethyl-1,4-phenylenediamine as a guest, $K_a = 0.73\,M^{-1}$. The K_a values for other electron-rich guest molecules, such as 1,3,5-trimethoxybenzene ($K_a = 0.49\,M^{-1}$), 1,3-dimethoxybenzene ($K_a = 0.29\,M^{-1}$), *N,N*-dimethylaniline ($K_a = 0.27\,M^{-1}$), and anisole ($K_a = 0.12\,M^{-1}$), were found to be substantially lower. The association constants (K_a) are correlated with the quadrupole moment of the guest molecules, suggesting considerable quadrupole–quadrupole character of the π–π host–guest interactions. The overall binding data indicate that this molecular tweezer system is a relatively poor host [90].

One of the plausible reasons behind the poor performance of molecular tweezers with pentafluorophenyl pincers is self-dimerization, which is clearly observed in the solid state by X-ray analysis. In crystals, this host molecule

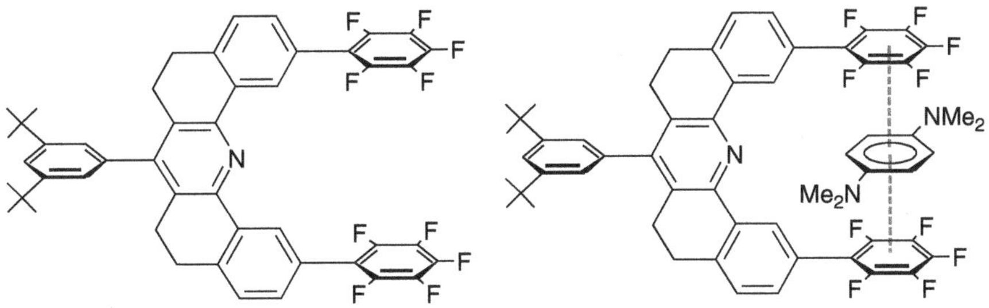

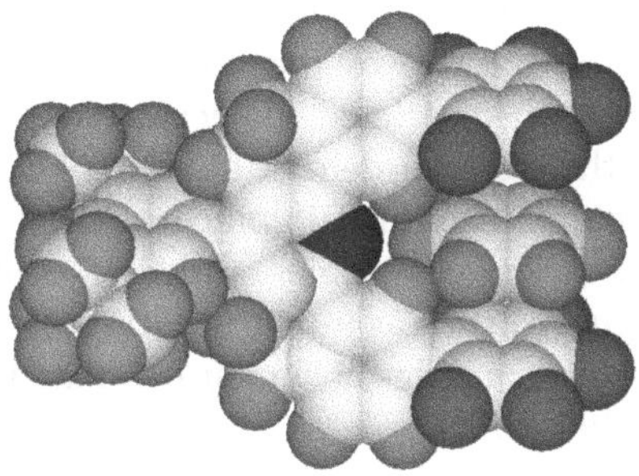

Figure 5.43 The molecular tweezers with pentafluorophenyl pincers (top left), their sandwich complex with *N,N,N′,N′*-tetramethyl-1,4-phenylenediamine (top right), and the energy-minimized 3D structure of this complex (bottom, dimethylamino groups were replaced with hydrogen atoms for clarity).

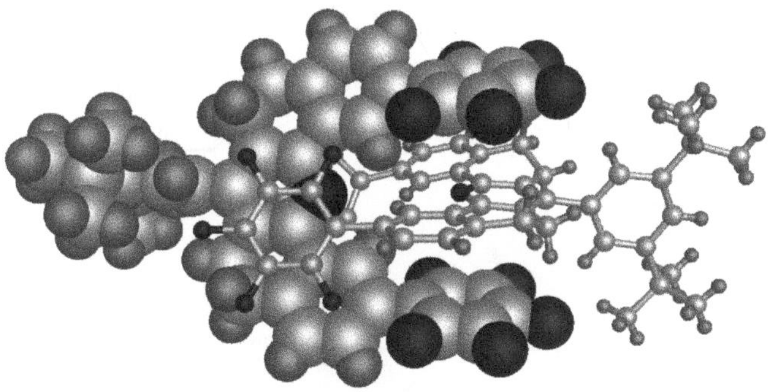

Figure 5.44 The X-ray structure of the dimer of the molecular tweezers with pentafluorophenyl pincers in the solid state [90].

produces well-defined dimers in which the pyridine moiety of the spacer unit is locked between the two C_6F_5 pincers of another molecule (Figure 5.44). This self-dimerization may persist in solutions, thus weakening the complexation with aromatic guest molecules [90]. A subsequent attempt was made to prevent the undesirable dimerization by introducing bulky *o*-substituents into the aryl group attached to the spacer unit [91].

Similar poor results were obtained during an attempt to use molecular tweezers with polyfluorinated heteroaromatic pincers for the capture of electron-rich aromatic guests [92]. Broadening of ^{1}H NMR signals (which could be interpreted as an indication of a complex formation) was only observed with the best electron-rich guest molecule, *N,N,N′,N′*-tetramethyl-1,4-phenylenediamine. No interactions were detected with other electron-rich aromatic compounds, such as 1,4-dimethoxybenzene, 1,3,5-trimethoxybenzene, and *N,N*-dimethylaniline [92].

More encouraging results were obtained with metal-coordinated terpyridine units as pincer elements of conformationally restricted molecular tweezers having a 5,6,8,9-tetrahydrodibenzo[*c,h*]acridine spacer [93–95]. One of the examples of such metalated molecular tweezers is presented in Figure 5.45 [93]. Due to the electron-deficient nature of the positively charged [chloro(2,2′,2″-terpyridine)palladium(II)]$^{+}$ pincers, these molecular tweezers are capable of forming relatively stable sandwich complexes with some electron-rich aromatic guests, such as 9-methylanthracene ($K_a = 650\,M^{-1}$, 1:1 complex in CD_3CN). Other examples of similar metalated tweezer assemblies have also been reported [94, 95].

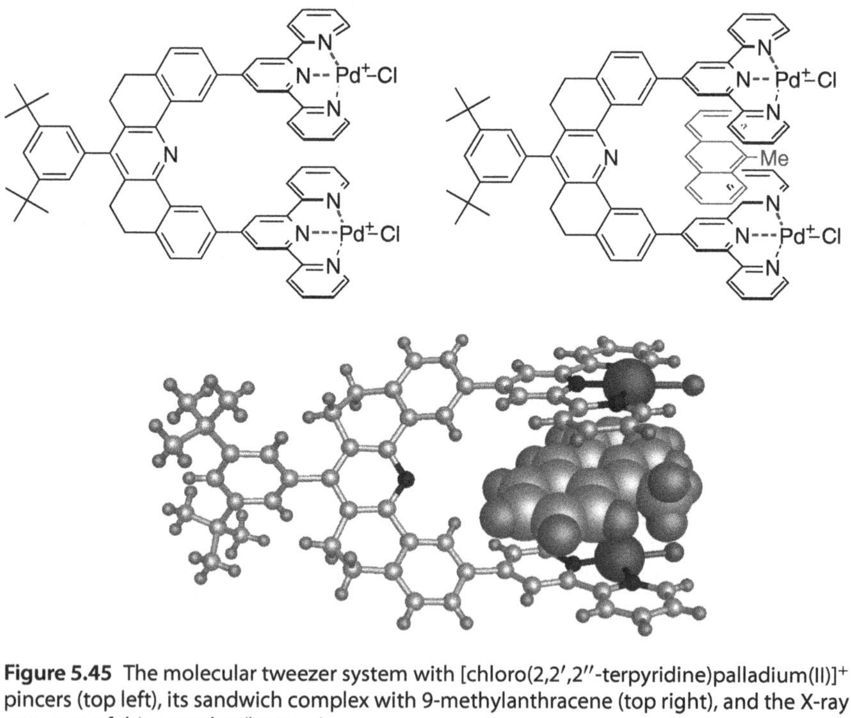

Figure 5.45 The molecular tweezer system with [chloro(2,2′,2″-terpyridine)palladium(II)]$^{+}$ pincers (top left), its sandwich complex with 9-methylanthracene (top right), and the X-ray structure of this complex (bottom).

Substantially better affinity and selectivity can be achieved by adding supplementary binding elements into the cavity of conformationally restricted molecular tweezers [96]. One of the examples of enhanced molecular tweezer systems is presented in Figure 5.46. This tweezer host features a carboxylic group deeply embedded into the host cavity. This modification is especially important for the binding of various nucleotide-based guest molecules, such as adenine derivatives, with high affinity and selectivity (K_a up to 1.2×10^5 M^{-1}) [97, 98]. Computational studies indicate that both hydrogen bonding and π-stacking equally contribute to the observed strong interactions between these molecular tweezers and adenine compounds [19].

The use of porphyrin rings as pincer elements significantly improves the binding affinity of conformationally restricted molecular tweezers toward certain guest molecules. An example of one of these tweezer systems with extended phenylacetylene connectors is presented in Figure 5.47. This host molecule forms strong complexes with both the electron-rich bidentate guest molecules, such as DABCO ($K_a = 1.1 \times 10^8$ M^{-1}), and the electron-deficient guests, such as acridinium cation ($K_a = 1.6 \times 10^7$ M^{-1}) in $CHCl_3$ [99].

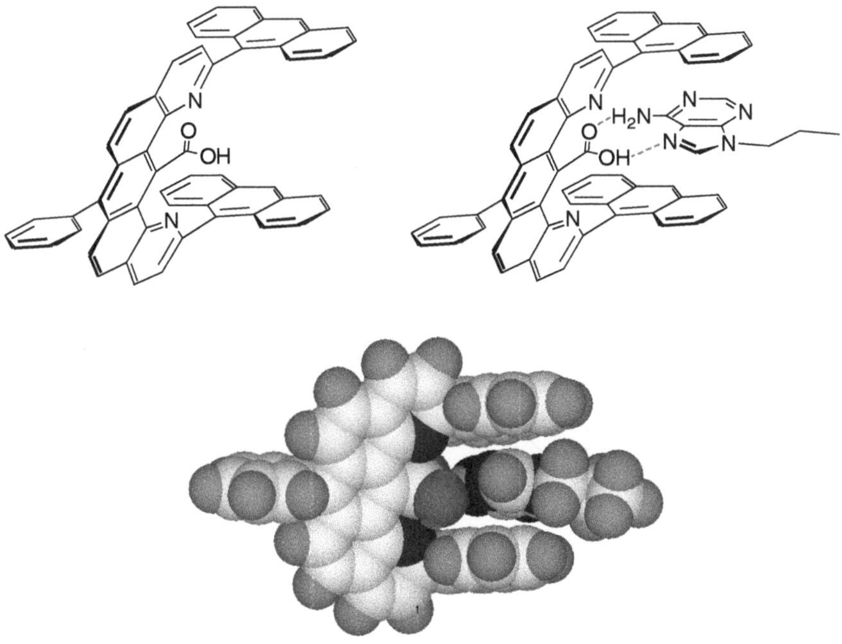

Figure 5.46 A representative example of molecular tweezers with the carboxylic group embedded into the cavity (top left), its complex with 9-propyladenine (top right), and the energy-minimized 3D structure of this complex (bottom).

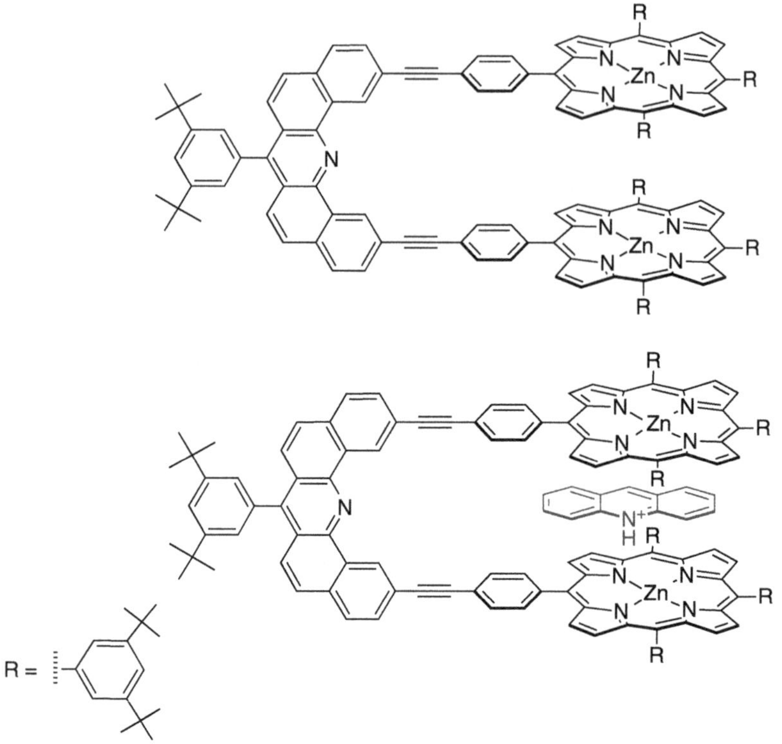

Figure 5.47 The molecular tweezer system with extended phenylacetylene links and porphyrin pincers (top) and its sandwich complex with acridinium cation (bottom).

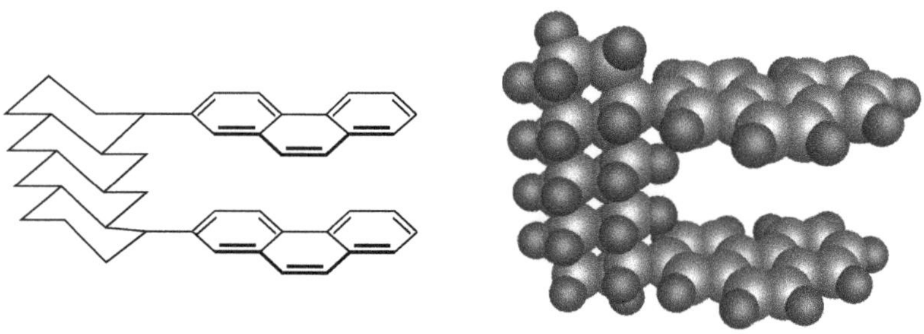

Figure 5.48 The conformationally restricted molecular tweezers with the all-*trans* perhydronaphthacene spacer (left) and their energy-minimized 3D structure (right).

Fused rigid alicyclic systems were also actively explored as spacers for the design of conformationally restricted molecular tweezers. Representative examples include tetra-cyclohexane (all-*trans* perhydronaphthacene) scaffold as a spacer unit (Figure 5.48). However, no measurable binding was

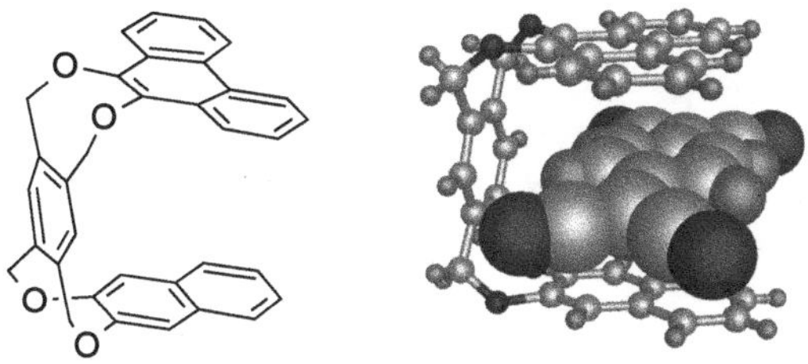

Figure 5.49 The conformationally restricted molecular tweezers with four flexible oxymethylene (OCH_2) connectors (left) and the X-ray structure of their complex with TCNQ (right) [101].

observed in the cases of simple benzene-type pincers. The poor performance of these molecular tweezers can be attributed to a larger-than-optimal interplanar distance between the pincer elements (ca. 7.6 Å), which is further exacerbated by an apparent lack of residual conformational flexibility combined with a shallow internal cavity. Somewhat better results were achieved with tweezer hosts equipped with extended phenanthrene pincers. These molecular tweezers show a moderate affinity toward electron-deficient 1-cyano-3,5-dinitrobenzene ($K_a = 34\,M^{-1}$ in $CDCl_3$) [100].

Further restriction of conformational mobility can be achieved by connecting a pincer unit with two flexible links to a rigid spacer. This design hinders the rotation of the pincer elements while retaining substantial level of conformational flexibility via "wing-flapping" motion. An example of molecular tweezers with two oxymethylene (OCH_2) links is presented in Figure 5.49 [101–103]. This host molecule forms relatively strong sandwich complexes with TCNQ ($K_a = 130\,M^{-1}$) and 1,2,4,5-tetracyanobenzene ($K_a = 1000\,M^{-1}$). The single-crystal X-ray analysis of a TCNQ complex reveals a relatively short (about 6.5 Å) interplanar distance between the polyaromatic pincer units [101].

Ethylene links (CH=CH) can also be used for the design of conformationally restricted molecular tweezers (Figure 5.50). These molecules exhibit *syn–anti* isomerism in solutions. Both the isomers can be easily detected by NMR means due to a relatively high interconversion energy barrier (in 16–21 kcal/mol range). The sandwich complexes of the tweezer-type *syn*-isomers with 1,2,4,5-tetracyanobenzene ($K_a = 1100\,M^{-1}$) display rather unique thermochromic properties [104].

Similar molecular tweezers with bowl-shaped corannulene pincers are of particular interest for capturing spherical aromatic guest molecules, including fullerenes [14]. An example of molecular tweezers with corannulene pincer units, the so-called buckycatcher, is shown in Figure 5.51 [105]. A reasonably good affinity toward fullerene C_{60} was achieved by the combination of complementary geometric features of corannulene pincer elements and limited

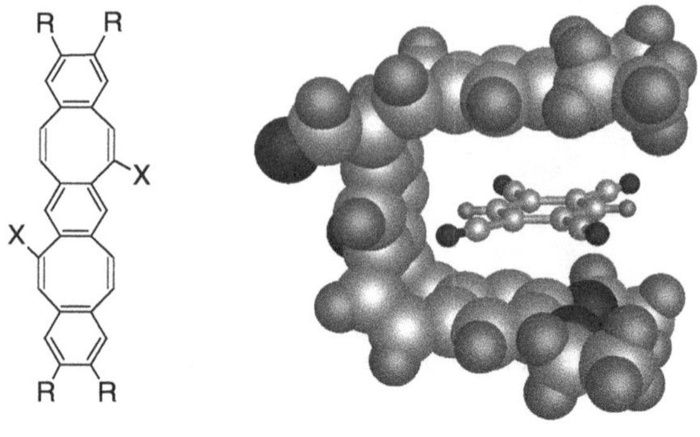

Figure 5.50 Conformationally restricted molecular tweezers with four ethylene links (left, R = OMe, OEt; X = CN, CHO, CH_2OMe) and the X-ray structure of the sandwich complex with 1,2,4,5-tetracyanobenzene (right, R = OEt, X = CHO).

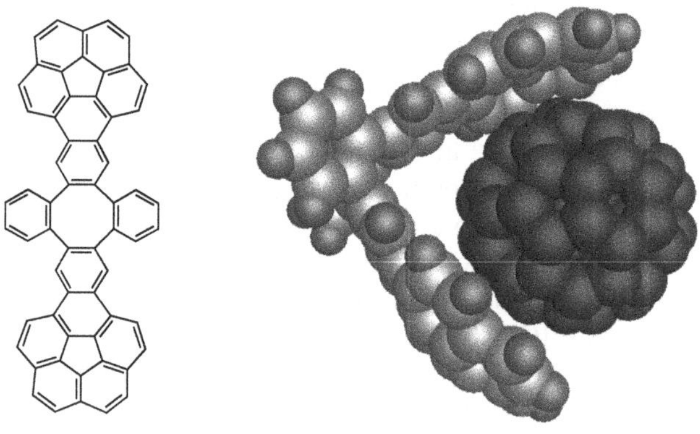

Figure 5.51 The buckycatcher molecular tweezers with corannulene pincers (left) and the X-ray structure of their complex with fullerene C_{60} (right, C_{60} guest molecules are orientationally disordered in the crystal) [105].

flexibility of the spacer scaffold allowing optimal self-organization of the complex ($K_a = 4300\,M^{-1}$ at 288 °K in toluene-d_8, NMR titration). A slightly better binding affinity was observed for fullerene C_{70} ($K_a = 5200\,M^{-1}$ at 288 °K in toluene-d_8, NMR titration) [106].

The buckycatcher tweezer host can also accommodate electron-deficient guest molecules, such as nitrobenzene and 1,3,5-trinitrobenzene (Figure 5.52). The host molecule in the 1,3,5-trinitrobenzene complex has much smaller cavity compared to the fullerene C_{60} complex, further underscoring the importance of restricted conformational flexibility for proper self-organization of the host–guest assembly. It is also interesting to note that the nitrobenzene complex not only contains two guest molecules but also has an inverted conformation of one of the bowl-shaped corannulene pincers [107].

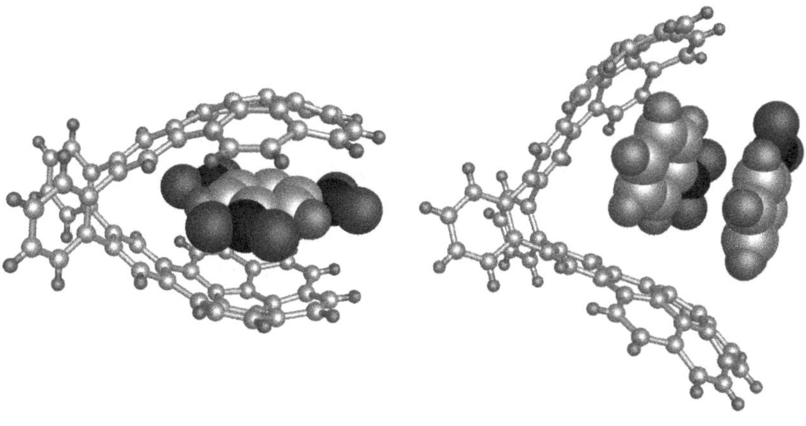

Figure 5.52 The X-ray structures of the buckycatcher molecular tweezer complexes with 1,3,5-trinitrobenzene (left) and with two nitrobenzene guest molecules (right).

A different design idea was successfully used for the construction of conformationally restricted molecular tweezer systems on the basis of oxa-calix[4]arenes [108, 109]. Partial conformational mobility of these systems can be attributed to four oxygen links between the aromatic subunits. A representative example of one of these tweezer systems is shown in Figure 5.53. The X-ray analysis reveals that the distance between two naphthalene pincer units defining the molecular tweezer cavity is about 7.0 Å, with almost coplanar *syn*-orientation. This tweezer host molecule fails to form observable complexes with toluene or benzonitrile. However, more electron-deficient guest molecules containing hydrogen bond donor groups, such as *o*-salicylic acid, can produce well-defined sandwich complexes. The association constants (K_a), measured in CH_2Cl_2–CD_2Cl_2 mixtures by ^1H NMR spectroscopy, were found to be about 300 M^{-1}. The authors suggested that both π–π interactions and hydrogen bonding might play a role in guest binding [108]. A detailed X-ray

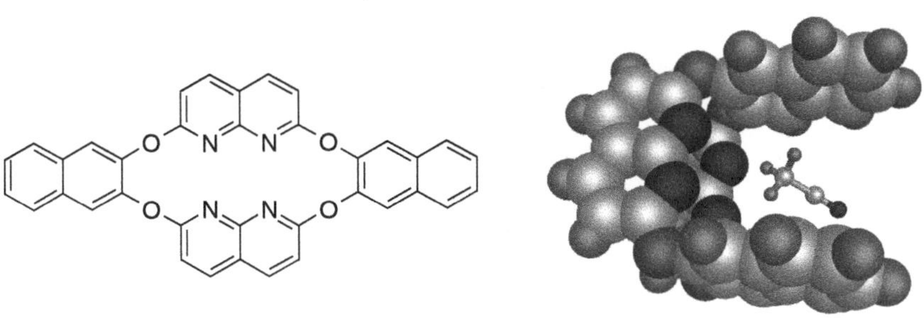

Figure 5.53 An example of the conformationally restricted molecular tweezers with four oxygen links (left) and the X-ray structure of their inclusion complex with CH_3CN (right) [108].

analysis of the oxacalix[4]arene tweezer sandwich complexes reveals structural features indicative of preferential planar guest molecule binding [109].

Glycoluril is a popular scaffold, which is frequently employed as a spacer for the construction of conformationally restricted molecular tweezer systems [4, 110, 111]. This functionalized spacer has two carbonyl groups and four nitrogen atoms and can provide additional binding sites for hydrogen bonds. Although glycoluril-based molecular tweezers are frequently referred to as clips or clefts due to a substantial angle between the pincer elements, the generic *molecular tweezers* term is used throughout the text for consistency.

Earlier examples of glycoluril-based tweezer molecules are presented in Figure 5.54 [112–114]. Although these tweezer hosts have a relatively shallow cavity, some of them demonstrate good binding affinity toward aromatic guests with two hydroxy groups, such as 2,7-dihydroxynaphthalene ($K_a = 7100\ M^{-1}$ in $CDCl_3$, R = OMe, Figure 5.54) [114]. The recent X-ray reinvestigation of the tetramethoxy derivative revealed a relatively large dihedral angle of about 40° between the benzene pincer units with a distance of about 6.7 Å between the centers of the benzene rings [115].

Further insight into the nature of the host–guest interactions in diphenylglycoluril-based molecular tweezer complexes can be attained from the X-ray analysis of a sandwich complex with nitrobenzene shown in Figure 5.55 [9]. The dihedral angle between the naphthalene pincers is about 27°, with relatively tight contacts (ca. 3.3 Å) between the naphthalene pincer elements and the nitrobenzene guest molecule.

The affinity of the diphenylglycoluril-based molecular tweezers toward various dihydroxybenzene guests has been studied in detail [116, 117]. These studies confirm that the host-binding capacity toward these guest molecules is determined by the combination of hydrogen bonding, π–π-stacking interactions, and a "cavity effect". The computational modeling indicates that these molecular tweezers have moderate conformational mobility, which can

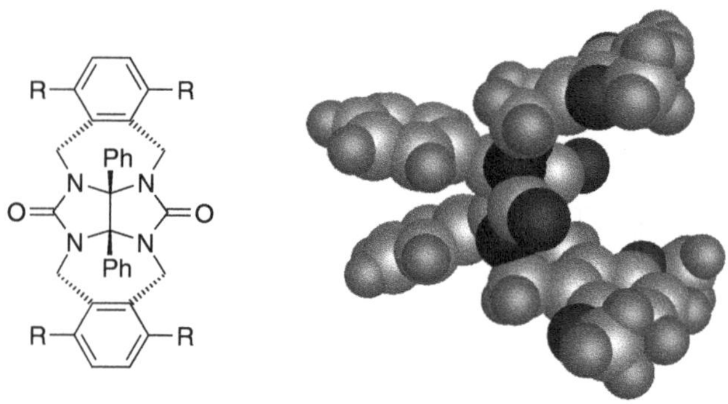

Figure 5.54 Early examples of the diphenylglycoluril-based molecular tweezers (left, R = H, Me, OH, OMe) and the X-ray structure of the tetramethoxy derivative (right, R = OMe) [115].

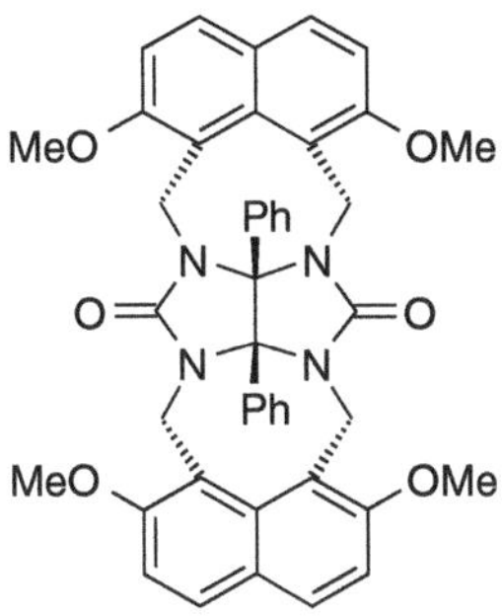

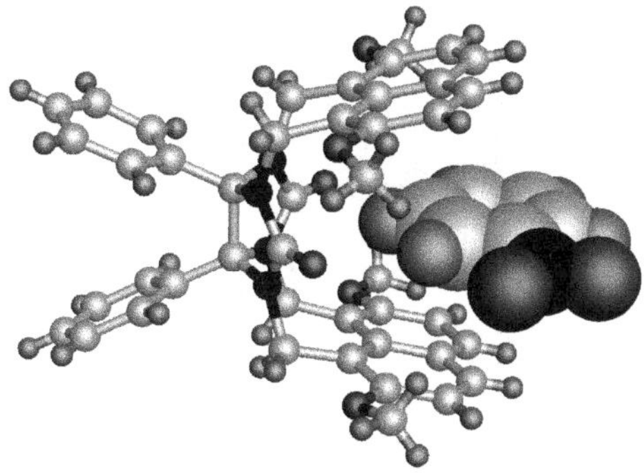

Figure 5.55 The diphenylglycoluril-based molecular tweezers with naphthalene pincers (left) and the X-ray structure of their complex with nitrobenzene (right).

facilitate binding via an "induced-fit" mechanism. It was also noted that the introduction of the π-donor substituents on the benzene pincers resulted in the improvement of the binding affinity, despite the presence of similar substituents in the target guest molecules. Further attempts to improve the binding affinities by employing larger polyaromatic pincers were not successful. Some examples of molecular tweezers with extended pincers are presented in Figure 5.56 [118–120].

The incorporation of pyrazine moieties into the pincer elements permits complexation-induced allosteric modulation of the cavity geometry in the presence of certain inorganic cations, such as Ag^+ or NH_4^+. A representative example of these tunable diphenylglycoluril-based molecular tweezers is shown in Figure 5.57 [28]. The reported dihedral angle between the pincer units in the NH_4^+ PF_6^- complex is relatively small, about 33°.

An additional functionalization of the glycoluril spacer can be accomplished by the incorporation of 2,2′-bipyridine moiety as a potent secondary binding site for metal cations (Figure 5.58). These molecular tweezers have moderate binding affinities toward dihydroxybenzenes. The best results were achieved with 5-cyanoresorcinol ($K_a = 240\,M^{-1}$ in $CDCl_3$). Subsequent complexation of this tweezer host molecule with $[Ru(2,2'\text{-bipyridine})_2]Cl_2$ provides an easy access to water-soluble molecular tweezers. The NMR dilution titration

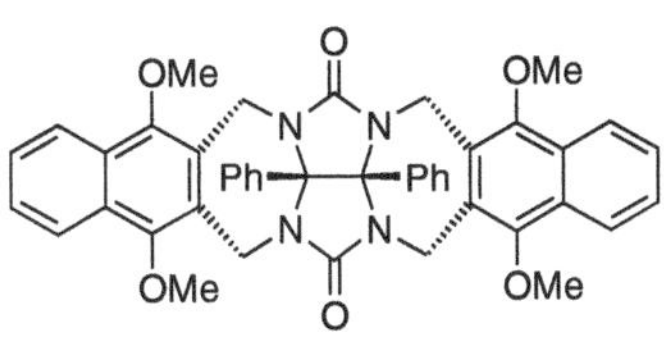

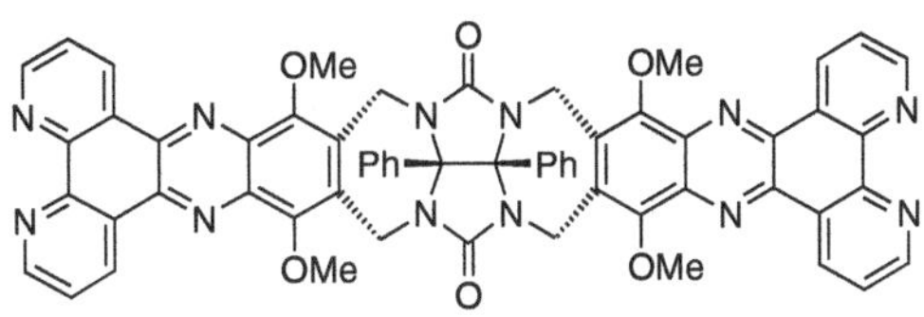

Figure 5.56 The diphenylglycoluril-based molecular tweezers with extended polyaromatic pincers.

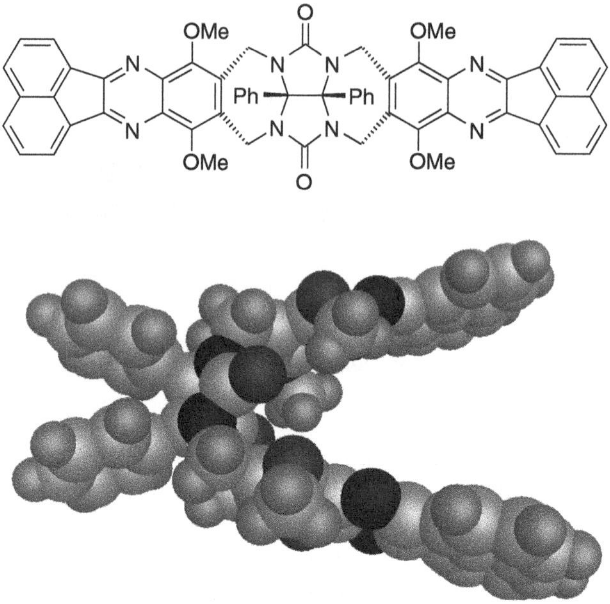

Figure 5.57 The tunable diphenylglycoluril-based molecular tweezers (top) and the X-ray structure of their NH_4^+ PF_6^- complex (bottom, only the host molecule is shown for clarity).

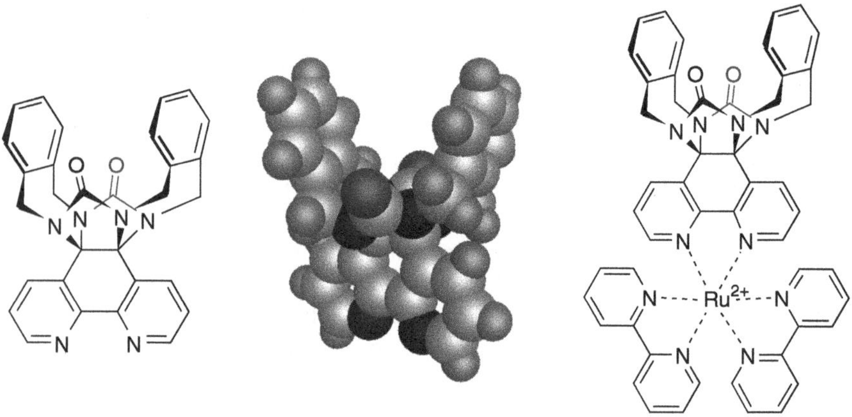

Figure 5.58 The bipyridine-glycoluril molecular tweezers (left), their X-ray structure (middle), and the complex with $[Ru(2,2'\text{-bipyridine})_2]Cl_2$ (right).

studies indicate strong self-association of the complex in solutions, with the self-association constant of about 50 M^{-1} in D_2O [121].

Further improvements in the performance of glycoluril-based molecular tweezers, including redox control, can be attained with the proper choice of pincer elements. Several tweezer molecules with tetrathiafulvalene pincers have been reported recently [122–126]. These tweezer hosts bind with electron-deficient aromatic compounds, such as *N,N′*-dimethyl-4,4′-bipyridinium salts [123], 1,3-dinitrobenzene [124], or 7,7,8,8-tetracyano-2,3,5,6-tetrafluoroquinodimethane [125], and can be oxidized electrochemically in a reversible manner. A representative example of the glycoluril-tetrathiafulvalene molecular tweezers is shown in Figure 5.59 [124]. Tetrathiafulvalene pincers have also been successfully employed for the construction of molecular tweezers with a calix[4]arene spacer backbone [127–129].

The introduction of porphyrin rings alters the flexibility and the binding properties of the diphenylglycoluril molecular tweezers. These porphyrin tweezers exhibit an extended set of interactions with guest molecules, such as hydrogen bonding, π–π, and dipole–dipole attractions, and, in the case of a Zn(II)-loaded porphyrin, complexation to the metal center [130, 131]. One of the diphenylglycoluril molecular tweezers with the porphyrin pincers is presented in Figure 5.60. This tweezer host produces strong sandwich 1:1 complexes with various diamines, particularly 4,4′-bipyridine [130]. Also known is a reverse design of porphyrin tweezers, which entails the use of

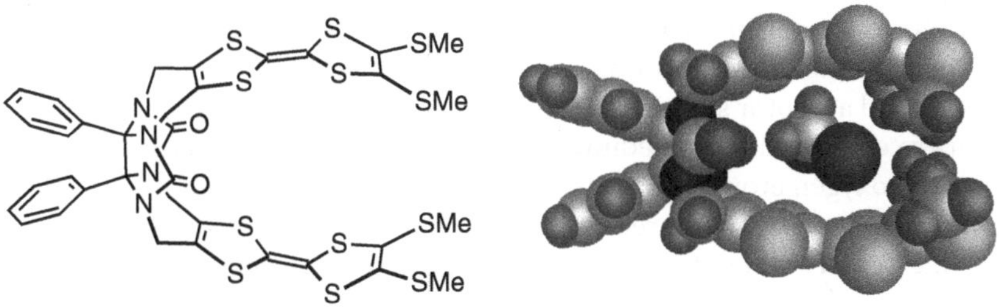

Figure 5.59 The diphenylglycoluril-based molecular tweezers with tetrathiafulvalene pincers (left) and the X-ray structure of their inclusion complex with CH_2Cl_2 (right) [124].

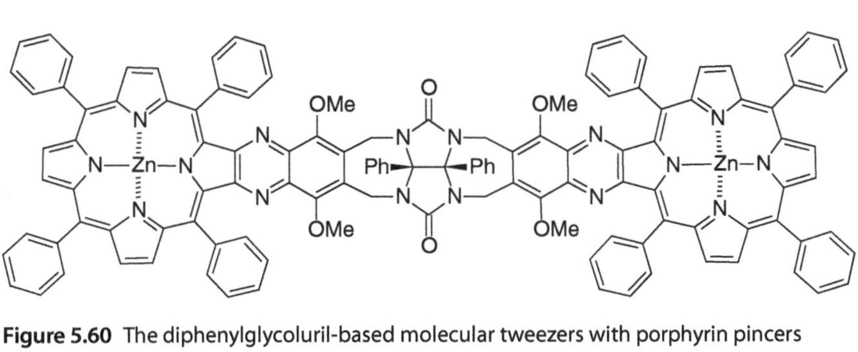

Figure 5.60 The diphenylglycoluril-based molecular tweezers with porphyrin pincers [130, 131].

glycoluril moieties as pincer elements and the porphyrin macrocycle as a spacer [132].

One of the major disadvantages of the diphenylglycoluril molecular tweezers discussed earlier is their relatively low solubility in organic solvents. The solubility can be improved by replacing the phenyl groups attached to the spacer unit with ethoxycarbonyl moieties [133, 134]. Although the majority of these ethoxycarbonyl derivatives tend to self-dimerize in the solid state, this dimerization is less pronounced in solutions. A representative example of bis(ethoxycarbonyl)glycoluril molecules with tweezer functions is shown in Figure 5.61. The cavity shape of this host molecule is almost indistinguishable from its diphenylglycoluril analog. The dihedral angle between the benzene pincers is approximately 42°, and the distance between the pincer centroids is about 6.7 Å [133].

A somewhat different synthetic approach was developed for the construction of aminal-type connectors between the bis(ethoxycarbonyl)glycoluril spacer and the aromatic pincers [135, 136]. These tweezer molecules can be easily prepared via a one-pot reaction of bis(ethoxycarbonyl)glycoluril, formaldehyde, and aromatic amines (Scheme 5.5).

Very good binding affinities toward some resorcinol derivatives are observed with propanediurea-based molecular tweezers (Figure 5.62, $K_a = 2.4 \times 10^6\ M^{-1}$ for 5-cyanoresorcinol in $CDCl_3$ at room temperature). This enhanced affinity can be attributed to the optimal spatial orientation of the oxygen atoms as hydrogen bond acceptor sites in the host molecule, even though the differences between glycoluril and propanediurea spacers are relatively small (the distance between the oxygen atoms is only 0.3 Å shorter in the propanediurea spacer). The affinity of the propanediurea tweezer receptors toward resorcinols can be further fine-tuned by the appropriate substitution of the propanediurea spacer

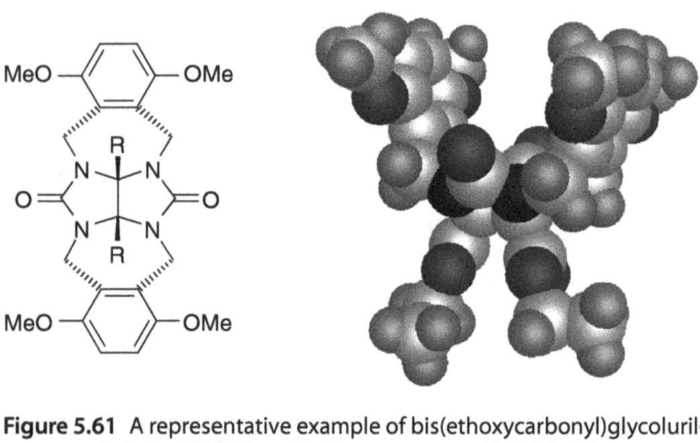

Figure 5.61 A representative example of bis(ethoxycarbonyl)glycoluril-based molecular tweezers (left, R = COOEt) and its X-ray structure (right) [133].

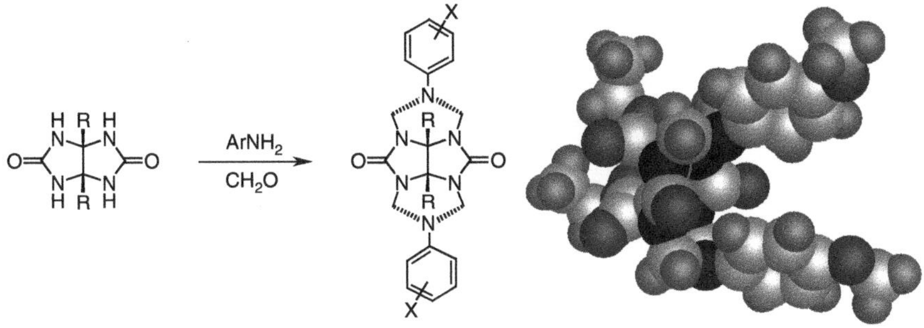

Scheme 5.5 One-pot synthesis of glycoluril molecular tweezers with aminal-type links (left, R = CO_2Et, X = Me, *i*-Pr, OMe, Cl, Br, I) and the X-ray structure of one of these compounds (right, R = CO_2Et, X = 4-OMe) [135].

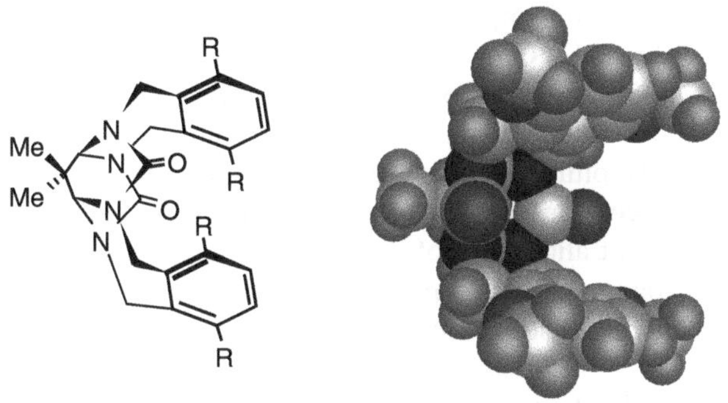

Figure 5.62 A representative example of the propanediurea-based molecular tweezers (left, R = OMe) and its X-ray structure (right).

unit. These observations highlight the importance of host–guest complementarity for the formation of molecular tweezer complexes [137].

Bis(ethoxycarbonyl)glycoluril tweezers with heterocyclic indazol-3-one pincers display strong fluorescence and can be used as selective chemosensors for Fe^{3+} in the presence of other metal cations (K^+, Mg^{2+}, Hg^{2+}, Cd^{2+}, Zn^{2+}, Co^{2+}, Ni^{2+}, Cu^{2+}, Pb^{2+}, Cr^{3+}, and Ce^{3+}). The two isomers of these molecular tweezers have very different spatial characteristics of their internal cavities. In the isomer with higher fluorescence emission and *anti* orientation of indazol-3-one carbonyl groups, the dihedral angle between the benzene pincer units is 79°, and the distance between their centroids is 9.1 Å. In the isomer with lower fluorescence emission and *syn* orientation of the carbonyl groups (Figure 5.63), the distance between the pincer centroids is 7.6 Å, and the dihedral angle is 49° [138].

Oligomeric glycoluril molecular tweezers featuring several glycoluril units provide additional binding advantages owing to the presence of multiple

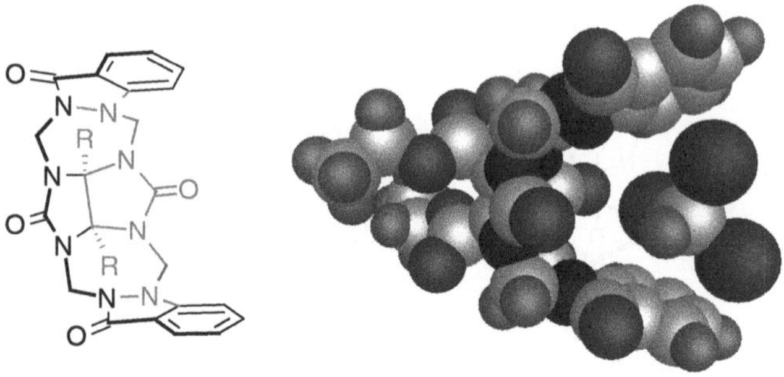

Figure 5.63 The fluorescent bis(ethoxycarbonyl)glycoluril tweezers with *syn* indazol-3-one pincers (left, R = COOEt) and the X-ray structure of their inclusion complex with $CHCl_3$ (right).

hydrogen bonding acceptor sites and improved "wrap-around" geometry of the host internal cavities [139–141]. One example of oligomeric glycoluril molecular tweezers with two bis(ethoxycarbonyl)glycoluril units is presented in Figure 5.64 [139]. Unlike previously discussed mono-glycoluril hosts, the tweezer molecules with two glycoluril units exhibit selective affinity toward hydroquinone due to the larger cavity size (7.3–7.6 Å distance between the pincer elements) and different alignment of the four oxygen atoms in the bis-glycoluril spacer as hydrogen bond acceptors. Self-dimerization in the solid state is also noted as a characteristic feature of these tweezer compounds [139, 140].

Oligomeric glycoluril molecular tweezers with three [142] and four [143] glycoluril units are also known. An example of tweezer molecules with three glycoluril units is shown in Figure 5.65 [142]. This tweezer host forms a well-defined 1:1 complex with *N,N′*-dimethyl-4,4′-bipyridinium salts ($K_a = 7.5 \times 10^4\ M^{-1}$) [142].

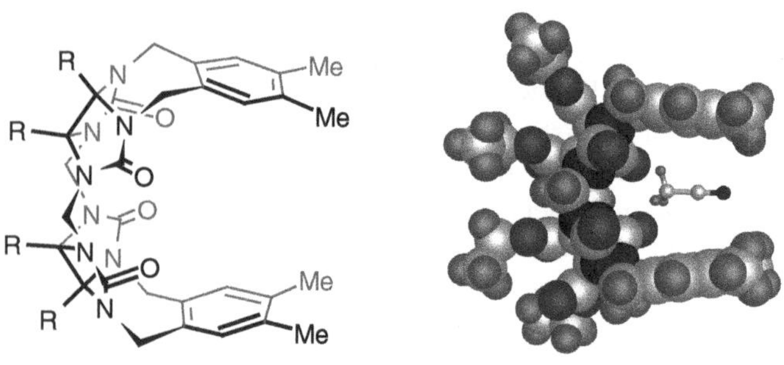

Figure 5.64 Oligomeric glycoluril molecular tweezers with two glycoluril units (left, R = COOEt) and the X-ray structure of their inclusion complex with CH_3CN (right) [139].

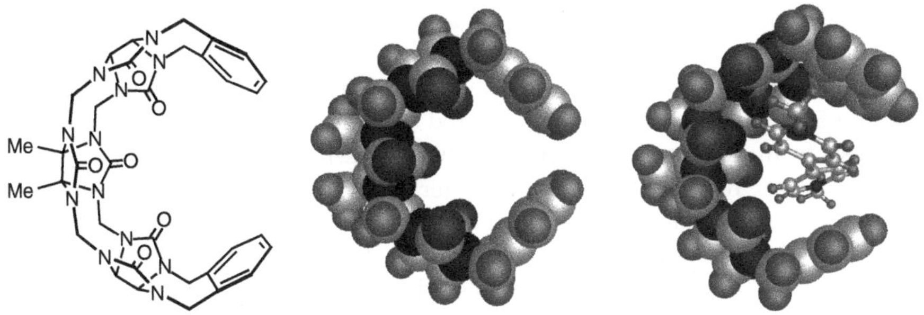

Figure 5.65 The oligomeric glycoluril molecular tweezers with three glycoluril units (left), their X-ray structure (middle), and the X-ray structure of the complex with the *N,N′*-dimethyl-4,4′-bipyridinium dication (right) [142].

5.4 Rigid Molecular Tweezers

Total conformational control is only feasible in rigid molecular tweezers, sometimes called clips due to the specific angular 3D shape of their concave cavities. In strict terms, totally rigid host molecules capable of producing sandwich complexes with guest molecules cannot be asserted as molecular tweezers because they lack the "squeeze grabbing" function of their conventional mechanical counterparts. Nevertheless, for the purpose of this chapter, we will classify them as molecular tweezers since they still possess a limited conformational mobility due to the elastic nature of covalent bonds and therefore can utilize a limited induced-fit tweezer mode of action.

The conformational control of molecular tweezers can be achieved with the use of specially designed reinforced 3D linking elements between the spacer and the pincer units. In the preponderance of these cases, rigidified bicyclic moieties, such as bicyclo[2.2.1]heptane [1, 3, 144], were employed. Some popular conformationally constrained bicyclic moieties frequently used as 3D linkers for the construction of the rigid molecular tweezers are presented in Figure 5.66.

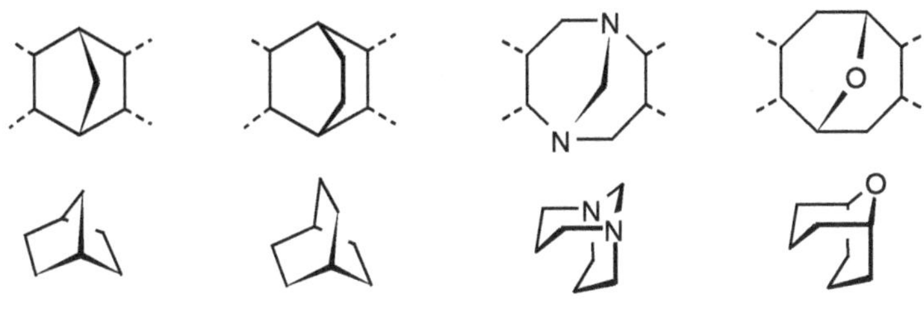

Figure 5.66 Some examples of conformationally constrained bicyclic linking units frequently used for the construction of rigid molecular tweezers.

Typical rigid molecular tweezers usually contain several of these 3D links to achieve proper coplanar orientation of the pincers. One of the simplest rigid molecular tweezers with two bicyclo[2.2.1]heptane links is presented in Scheme 5.6. The compound was prepared via a simple one-step Diels–Alder synthesis from easily accessible intermediates [145].

This tweezer host molecule is characterized by a relatively large dihedral angle between the naphthalene pincers (above 50°) with a distance between the pincer tips of about 11.4 Å (Figure 5.67). The geometry of these molecular tweezers undergoes considerable changes upon complexation with electron-deficient aromatic guests. For example, the dihedral angle between the naphthalene pincers in the *N*-ethyl-4-ethoxycarbonylpyridinium complex ($K_a = 137\,M^{-1}$) is substantially smaller (less than 30°), with a tip-to-tip pincer distance of approximately 8.2 Å. This is closer to the optimal values for π–π charge–transfer aromatic stacks (Figure 5.67). Examination of the X-ray data indicates that the induced-fit geometry changes are possible due to substantial bond bending in the spacer units. Given the strain energy increase in the

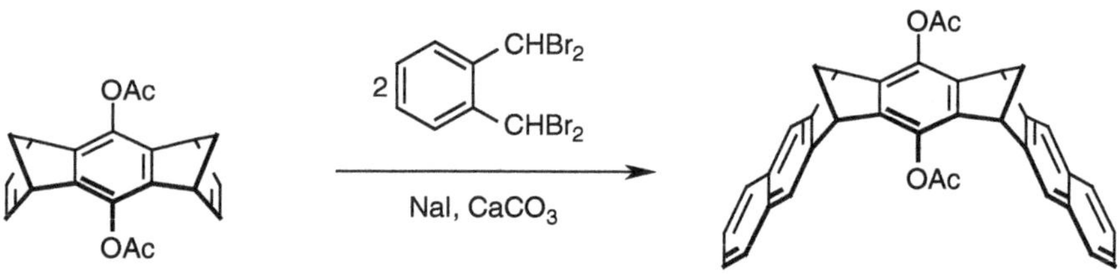

Scheme 5.6 Synthesis of the simple rigid molecular tweezers with two bicyclo[2.2.1]heptane links and naphthalene pincers.

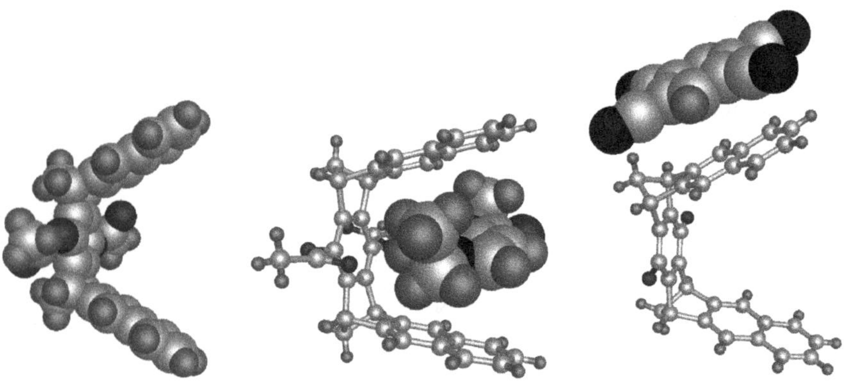

Figure 5.67 The X-ray structures of the rigid molecular tweezers with two bicyclo[2.2.1]heptane links (left), their sandwich complex with *N*-ethyl-4-ethoxycarbonylpyridinium (middle), and the outside-cavity complex with 1,2,4,5-tetracyanobenzene (right, disordered CH_3CO fragments were removed for clarity).

tweezer system required for optimal host–guest sandwich complexation, it is not surprising that less electron-deficient aromatic guests bind to naphthalene pincers outside the tweezer cavity (Figure 5.67). The geometry of the molecular tweezers in this non-sandwich complex is about the average between the geometries of the free host molecule and the sandwich complex [145].

The induced-fit effect is even more pronounced in molecular tweezers containing longer anthracene pincer units (Figure 5.68) [146]. In the empty host molecule, the distance between the pincer tips is about 14.5 Å. Complexation with 1,2,4,5-tetracyanobenzene reduces this distance to 6.5 Å. This conformational change results in a substantial strain energy increase, which weakens the guest binding ($K_a = 140\,M^{-1}$ vs $690\,M^{-1}$ for the similar diacetoxy derivative with naphthalene pincers in $CDCl_3$) [146]. The complex is also notable for its long-wavelength fluorescence at 668 nm [147]. Similar molecular tweezers with extended benzo[*k*]fluoranthene pincers are also known [148].

Substantially better alignment of the pincer elements can be achieved in rigid molecular tweezers with three bicyclo[2.2.1]heptane links. The representative examples of these tweezer molecules are shown in Figure 5.69. The X-ray studies of the tweezer system with two benzene pincers (Figure 5.69) reveal a relatively small dihedral angle between the pincer units and the optimal tip-to-tip distance of approximately 7.2 Å. The binding affinities of the molecular tweezers with longer naphthalene pincers (Figure 5.69) toward the electron-deficient aromatic guests were explored in detail. The best 1:1 association constant ($K_a = 1.43 \times 10^7\,M^{-1}$ in $CHCl_3$ at 25 °C) is observed for the 1,2,4,5-tetracyanobenzene complex [149].

Incorporation of four bicyclo[2.2.1]heptane links is also possible, but these molecular tweezers are structurally more related to partially opened molecular belt compounds (see Chapter 4) than to molecular tweezers. The representative

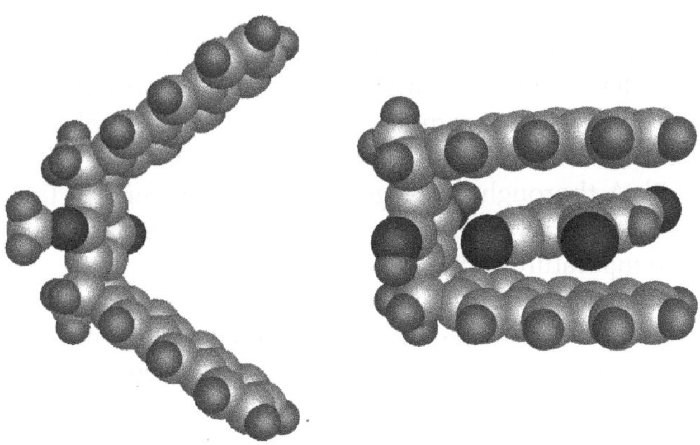

Figure 5.68 The X-ray structures of the rigid molecular tweezers with anthracene pincers (left) and their sandwich complex with 1,2,4,5-tetracyanobenzene (right) [146].

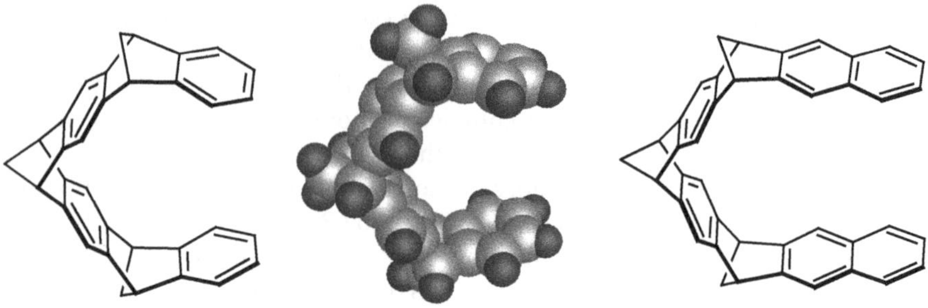

Figure 5.69 The rigid molecular tweezers with three bicyclo[2.2.1]heptane links and benzene pincers (left), their X-ray structure (middle), and the analog tweezer host molecule with larger naphthalene pincers (right).

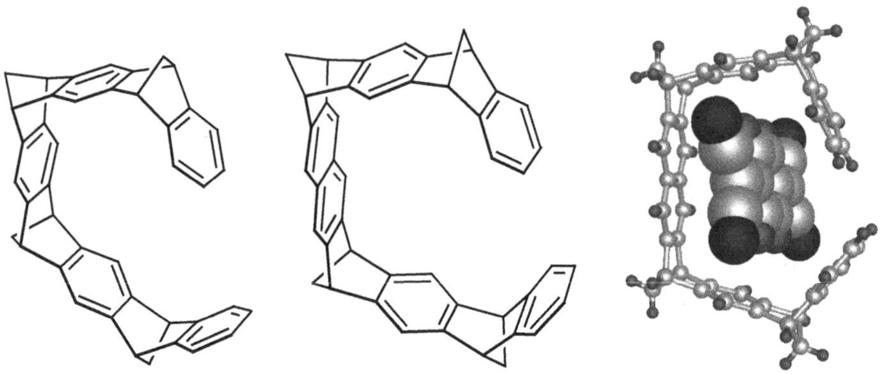

Figure 5.70 The molecular tweezers with four bicyclo[2.2.1]heptane links and benzene (left) or naphthalene (middle) spacers and the X-ray structure of the CNQ complex of the naphthalene-based host molecule (right).

examples of these molecules are shown in Figure 5.70. The belt-like character of the molecular hosts is evident from the examination of the X-ray structure of the TCNQ complex (Figure 5.70). In this complex, the aromatic guest molecule is oriented almost perpendicular to one of the pincers and is coplanar with the aromatic spacer unit [150]. A thorough investigation of the dynamics of this complex by ^{1}H NMR in solutions reveals rapid rotation of the guest molecule inside the cavity at room temperatures [151].

Substantial enhancement in binding selectivity as well as water solubility can be achieved by the introduction of polar and ionic functional groups into the spacer units of the rigid molecular tweezers (Figure 5.71) [152]. Due to their unique binding mode, these functionalized tweezer molecules demonstrate good affinity toward certain heterocyclic compounds of biomedical importance, such as nicotinamide adenine dinucleotide (NAD^{+}) [153, 154] and 3′-azido-2′,3′-dideoxythymidine (AZT) [155]. ^{1}H NMR and computational

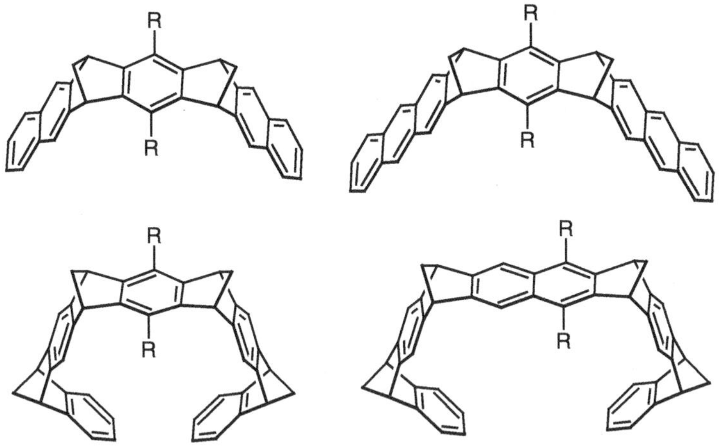

Figure 5.71 Functionalized molecular tweezers with bicyclo[2.2.1]heptane links ($R = OP(CH_3)O_2^-\ Li^+$ [156, 157], $OP(Ph)O_2^-\ n\text{-}Bu_4N^+$ [154], $OPO_2^{2-}\ 2Li^+$ [152], $OSO_3^-\ Na^+$ [152], $OCH_2CO_2^-\ Na^+$ [152]).

studies revealed a significant role of nonclassical hydrophobic effect affecting the host–guest tweezer binding and dimerization in water [156]. More details regarding biomedical applications of these functionalized tweezer compounds are presented in Section 5.5 at the end of this chapter.

Also known are tweezer compounds incorporating several bicyclo[2.2.1]heptane and 7-oxabicyclo[2.2.1]heptane moieties. A recent example of rigid molecular tweezers with two bicyclo[2.2.1]heptane and two 7-oxabicyclo[2.2.1]heptane units is shown in Figure 5.72. These molecular tweezers exhibit a strong tendency to dimerize in the solid state [158].

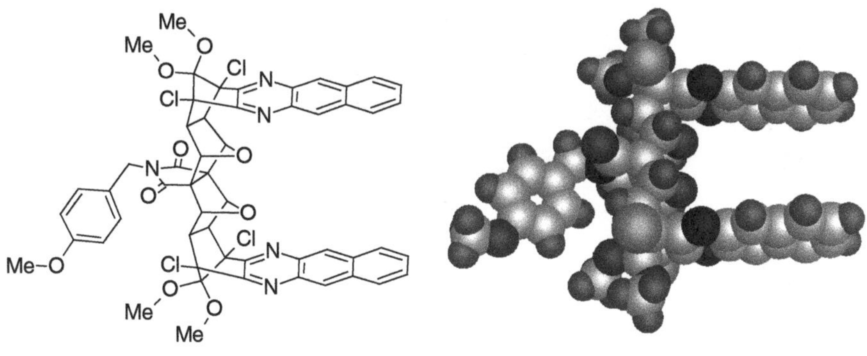

Figure 5.72 The molecular tweezers with two bicyclo[2.2.1]heptane and two 7-oxabicyclo[2.2.1]heptane links (left) and their X-ray structure (right).

A combination of rigidified bicyclo[2.2.1]heptane links and corannulene bowl-shaped pincers was successfully used for the construction of tweezer molecular hosts tailored to capture ball-shaped fullerene molecules [159, 160]. One example of these tweezer hosts is presented in Figure 5.73 [160]. The host has good binding affinity toward spherical fullerene C_{60} and even better affinity toward ellipsoidal fullerene C_{70} ($K_a = 2.0 \times 10^5\,M^{-1}$ in C_6D_5Cl at room temperature). These high affinities can be attributed not only to the proper preorganization of the corannulene pincers but also to favorable π–π interactions between the guest fullerene molecules and the aromatic spacer of the tweezer host [160].

The set of popular rigidified bicyclic links frequently used for the construction of rigid molecular tweezers also includes 9-oxabicyclo[3.3.1]nonane moieties [2]. They can be easily produced via the acid-catalyzed intramolecular heterocyclization of properly functionalized dimethyl acetals (Scheme 5.7) [161, 162]. The chiral molecular tweezers with naphthalene pincers form well-defined sandwich complexes with maleic anhydride and 4-chloronitrobenzene [163].

The X-ray studies of a 1,3,5-trinitrobenzene complex of rigid molecular tweezers with 9-oxabicyclo[3.3.1]nonane links and dibenzofuran pincers reveal almost parallel orientation of the pincer planes (Figure 5.74). The interplanar distance between the pincer elements is close to optimal for sandwich charge-transfer complexes (about 6.6 Å) [162]. This complex is further stabilized by hydrogen–π interactions between a partially positively charged hydrogen of the 1,3,5-trinitrobenzene guest molecule and the benzene

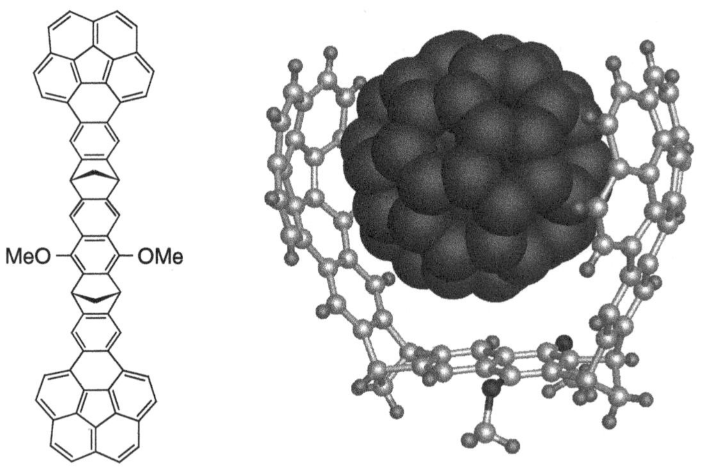

Figure 5.73 The molecular tweezers with two rigidified bicyclo[2.2.1]heptane links and corannulene bowl-shaped pincers (left) and the X-ray structure of their complex with fullerene C_{60} (right).

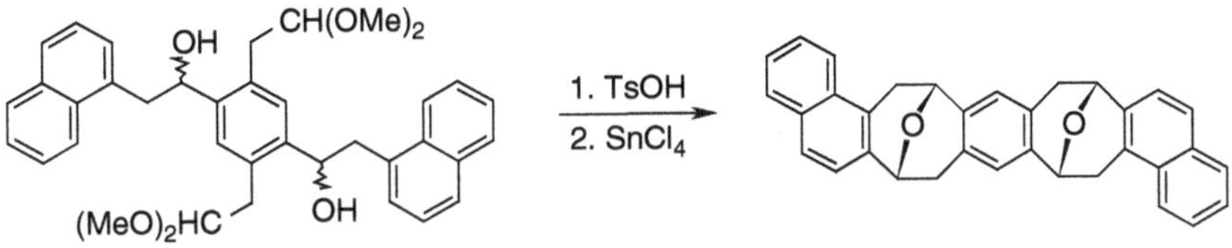

Scheme 5.7 Synthesis of the rigid molecular tweezers with two 9-oxabicyclo[3.3.1]nonane links and naphthalene pincers [163].

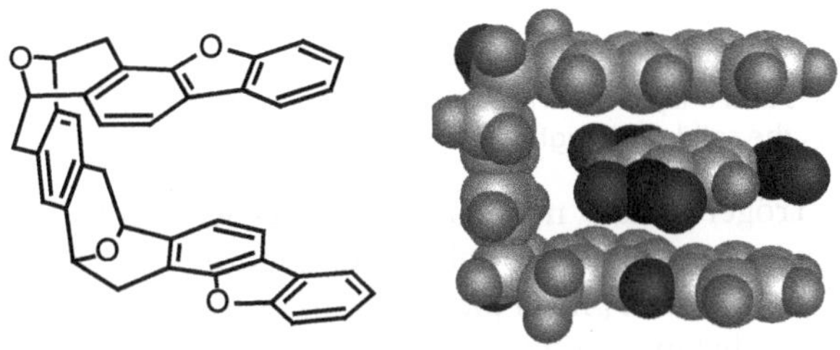

Figure 5.74 The rigid molecular tweezers with 9-oxabicyclo[3.3.1]nonane links and dibenzofuran pincers (left) and the X-ray structure of their complex with 1,3,5-trinitrobenzene (right) [162].

ring of the spacer unit. The association constant K_a of the complex is about 2000 M^{-1} in chloroform [2].

Another popular rigidified bicyclic linking moiety frequently used for the construction of inherently chiral rigid molecular tweezers is 1,5-diazabicyclo[3.3.1]nonane [164]. Similar to the molecular tweezers with 9-oxabicyclo[3.3.1]nonane links discussed earlier, these tweezers can be easily produced via the acid-catalyzed heterocyclization reaction of properly functionalized aromatic diamines with formaldehyde sources, such as hexamethylenetetramine or paraformaldehyde (Scheme 5.8) [165–167]. The reaction usually produces a mixture of *syn* and *anti* isomers, of which only *syn* isomers have geometric features of molecular tweezers. One of these so-called bis-Tröger's base molecular tweezers with naphthalene pincers is known to form sandwich complexes with nitrobenzene [168].

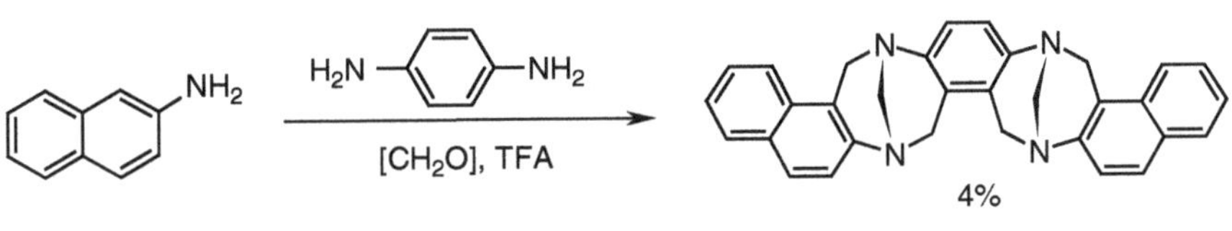

Scheme 5.8 One-step synthesis of the rigid molecular tweezers with two 1,5-diazabicyclo[3.3.1]nonane links and naphthalene pincers (TFA = trifluoroacetic acid) [167].

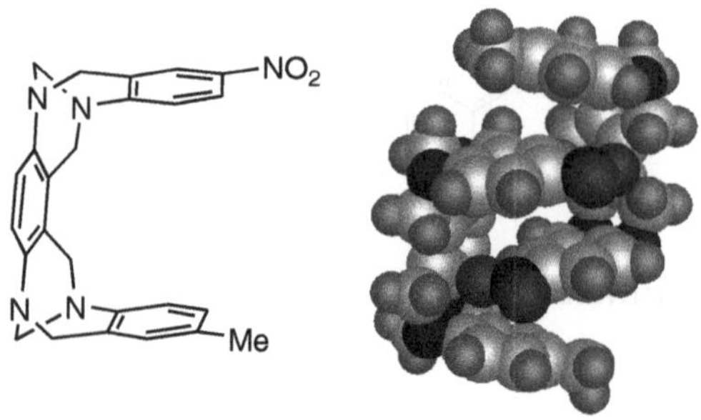

Figure 5.75 The "bis-Tröger's base" molecular tweezers with two 1,5-diazabicyclo[3.3.1]nonane links and unsymmetrical donor–acceptor pincers (left) and the X-ray structure of the dimer in the solid state (right).

Dimerization of the "bis-Tröger's base" molecular tweezers with different donor–acceptor pincers is quite notable in the solid state (Figure 5.75) [169]. The dihedral angle between the aromatic pincer planes of this tweezer molecule is relatively small (23.1°). The distance between the aromatic pincer tips is about 7.2 Å. In the dimer, the nitrobenzene pincer of one tweezer molecule is located in the cavity of another tweezer molecule. A similar self-dimerization was also observed for molecular tweezers with two nitrobenzene pincers [165].

Expanded "tris-Tröger's base" molecular tweezers contain three 1,5-diazabicy clo[3.3.1]nonane bridges [170, 171]. The multistep synthesis of "tris-Tröger's bases" usually produces a complex mixture of *anti–anti, anti–syn,* and *syn–syn* isomers. Fortunately, the yield of the desired tweezer *syn–syn* isomers can be improved by subsequent acid-catalyzed isomerization in the trifluoroacetic–hydrochloric acid mixture [172]. The X-ray structure of one of these "tris-Tröger's base" molecular tweezers with naphthalene pincers is presented in Figure 5.76. According to the X-ray data, the dihedral angle between the naphthalene pincer planes is approximately 40°, with the shortest tip-to-tip pincer distance of about 6.2 Å. This tweezer host molecule forms a relatively stable complex with 1,2,4,5-tetracyanobenzene ($K_a = 615\ M^{-1}$) [172].

Complex mixed-type molecular tweezers with 1,5-diazabicyclo[3.3.1]nonane and bicyclo[2.2.2]octane links were actually among the first rigid molecular tweezers ever prepared [173]. The recent representative examples of the molecular tweezers with one 1,5-diazabicyclo[3.3.1]nonane and two bicyclo[2.2.2]octane bridges are shown in Figure 5.77 [174]. The distance between the benzene pincer units is relatively large, above 9 Å. This tweezer molecule forms self-complementary dimers in the solid state [174].

All-carbon bicyclo[3.3.1]nonane links are rarely used for the construction of rigid molecular tweezers, presumably due to the synthetic challenges associated with the preparation of these compounds. Inherent chirality is one of the distinct features of these rigidified bicyclic linking elements [36, 175]. Some of these tweezer molecules can be considered as topological analogs of helicene

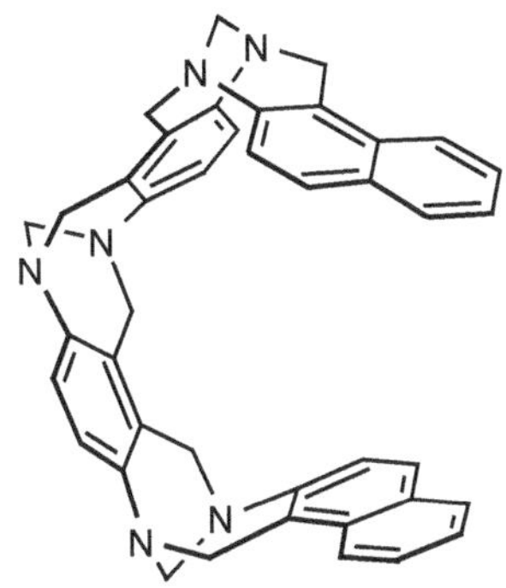

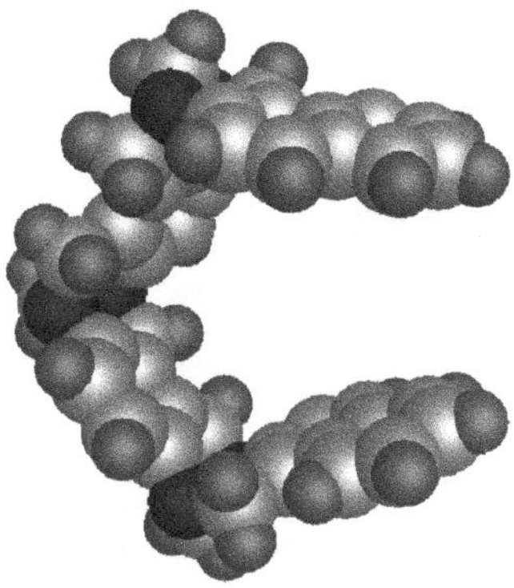

Figure 5.76 The "tris-Tröger's base" molecular tweezers with three 1,5-diazabicyclo[3.3.1]nonane links and naphthalene pincers (left) and their X-ray structure (right) [172].

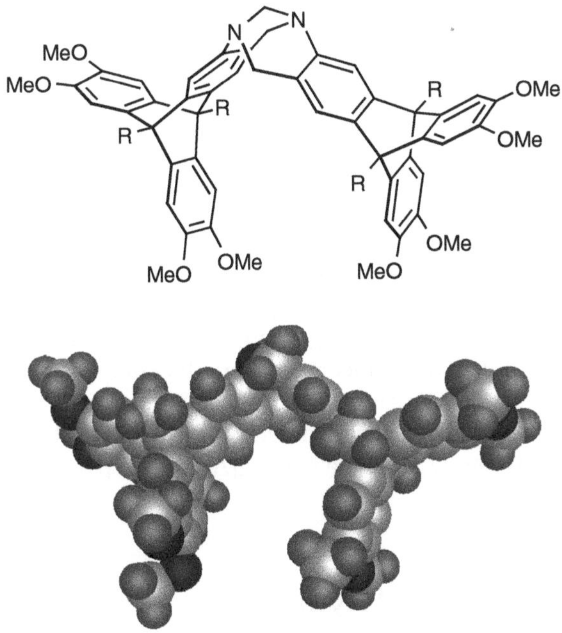

Figure 5.77 The mixed-type molecular tweezers with one 1,5-diazabicyclo[3.3.1]nonane and two bicyclo[2.2.2]octane links (top, $R = CH_3$, $CH_2CH_2CH_3$) and the X-ray structure (bottom, $R = CH_3$) [174].

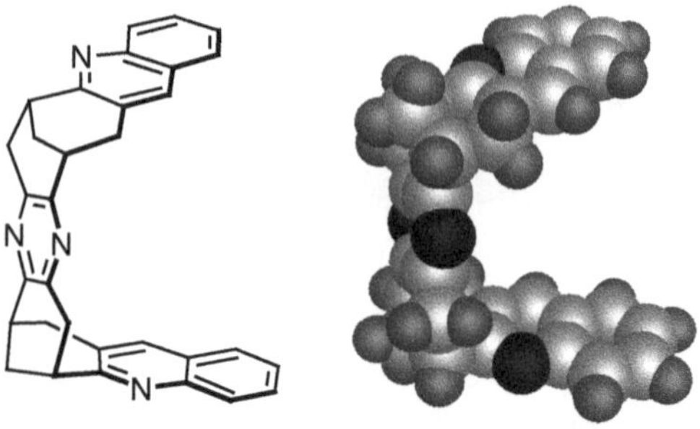

Figure 5.78 The rigid heterocyclic molecular tweezers with two all-carbon bicyclo[3.3.1]nonane links and quinoline pincers (left) and their X-ray structure (right) [36].

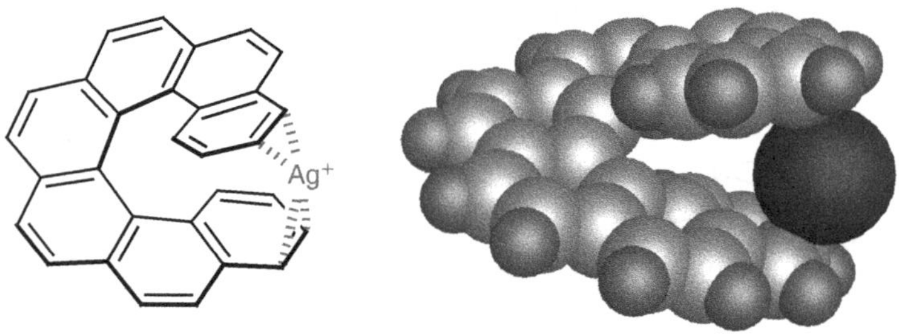

Figure 5.79 The [7]helicene tweezer complex with silver cation (left) and its X-ray structure (right).

[175]. An example of the chiral heterocyclic molecular tweezers incorporating two bicyclo[3.3.1]nonane fragments is presented in Figure 5.78 [36].

Chiral [7]helicene itself represents an interesting example of rigid molecular tweezers with identical spacer, linker, and pincer elements. Its complexation with a proper metal cation (such as a silver cation) results in intercalation of the cation between the spiral edges of the [7]helicene molecule (Figure 5.79) [176]. The association constant K_a between [7]helicene and $Ag[Al\{OC(CF_3)_3\}_4]$ was found to be about 269 M^{-1} in CH_2Cl_2.

5.5 Claw Grabbers, Grippers, and Flytraps

Molecular claw grabbers, grippers, and flytraps are far less established technomimetic concepts compared to other types of molecular tweezers. The name "flytrap" actually has biological, not technological, origin. It refers to the specific mode of action of an unusual plant, *Dionaea muscipula* Ellis (*Venus flytrap*), which can catch insects with a jaw-like deliberate closing of its trap [177]. There are also complex molecular systems of biological origin, which are collectively called Venus Flytrap Module (VFTM). These systems are characterized by the presence of two lobes with a deep cleft between them and are capable of catching certain guest molecules and ions in the cavity [178]. VFTM is a common domain of some G-protein-coupled receptors (GPCRs) [178], calcium-sensing receptor (CaR) [179], glutamate/glycine-gated ion channels, pheromone receptors, α-amino-3-hydroxy-5-methyl-4-isoxazolepropionic acid (AMPA) channels [180], and others.

Although many flexible molecular tweezers are capable of catching guest molecules using very similar modes of action (see Section 5.2, flexible molecular tweezers), the major distinction of molecular grippers, claw grabbers, and flytraps presented in this section is the ability to grab guest molecules using multiple pincers. By definition, these technomimetic molecules contain at least three pincers attached to a spacer unit. Conventional mechanical prototypes include fruit picker tools, robotic arms, and crane grabbers (Figure 5.80).

The simplest molecular claw grabber systems can be built on the basis of symmetrical 1,3,5-triethyl-2,4,6-trisubstituted benzene spacers in which all three functionalized substituents adopt *syn*-conformation due to "steric gearing" [181]. A representative example of simple claw grabber hosts equipped with three guanidiniocarbonylpyrrole pincers is shown in Figure 5.81 [182]. UV–Vis and fluorescence titration studies confirm that this molecular claw grabber is capable of capturing citrate and other tricarboxylate guest molecules with association constants K_a in excess of $10^5\ M^{-1}$ [182, 183]. The binding of tricarboxylates with the host molecule was investigated in detail using a combination of NMR (NOESY) experiments and molecular modeling calculations. Ion pairing and hydrogen bonding between the carboxylate fragments and the guanidiniocarbonylpyrrole pincers were found to be the major binding forces, further amplified by the nonpolar microenvironment of the cavity. The presence of an aromatic fragment in the trimesic acid trianion as a guest molecule improves the strength of the complex via additional π–π aromatic interactions ($K_a = 3.4 \times 10^5\ M^{-1}$ in water at pH 6.3, according to the UV–Vis titration data) [182].

Figure 5.80 Examples of conventional mechanical devices with several pincers. *Credits*: The American Society of Mechanical Engineers, Imbondeiro Waste Management.

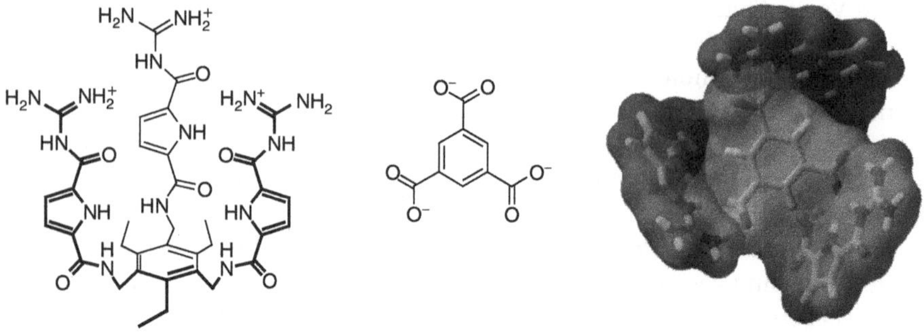

Figure 5.81 The molecular claw grabber host with three guanidiniocarbonylpyrrole pincers (left), the trimesic acid trianion guest (middle), and their complex (right). *Source*: Schmuck and Schwegmann 2005 [182]. Reproduced with permission of American Chemical Society.

Also known are similar tripodal molecular host systems with cationic pincers that are specifically designed for inorganic anions [184]. One of them, with urea-type pincer elements ($R = CONHC_6H_4CH_3$) [185], is presented in Figure 5.82. Another claw grabber host system equipped with electroactive ferrocene ($R = CH_2C_5H_4FeC_5H_5$) pincers demonstrates good binding selectivity toward chloride anion [186].

A similar idea was used for the construction of molecular grippers with three electroactive pyrrolotetrathiafulvalene pincers (Figure 5.83). This host molecule exhibits good binding affinity to certain electron-deficient aromatic compounds, such as TCNQ and *N*-alkylpyridinium salts [187].

Tripodal claw grabbers with imidazole pincers can be constructed on the basis of hexahomotrioxacalix[3]arene (Figure 5.84). This molecular host in *cone* conformation demonstrates good ionophoric affinity toward Cu^{2+} ($K_a = 4.72 \times 10^4\,M^{-1}$), $CH_3(CH_2)_3NH_3^+$ ($K_a = 1980\,M^{-1}$), and $Cr_2O_7^{2-}$ (at low pH) [188].

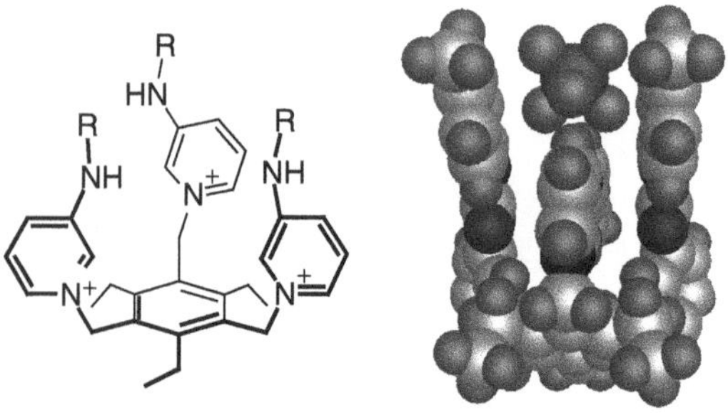

Figure 5.82 The molecular claw grabber hosts with three cationic pincer elements (left, $R = CONHC_6H_4CH_3$, $CH_2C_5H_4FeC_5H_5$) and the X-ray structure of a $3 \times PF_6^-$ salt (right, $R = CONHC_6H_4CH_3$, two PF_6^- anions outside the cavity were removed for clarity) [185].

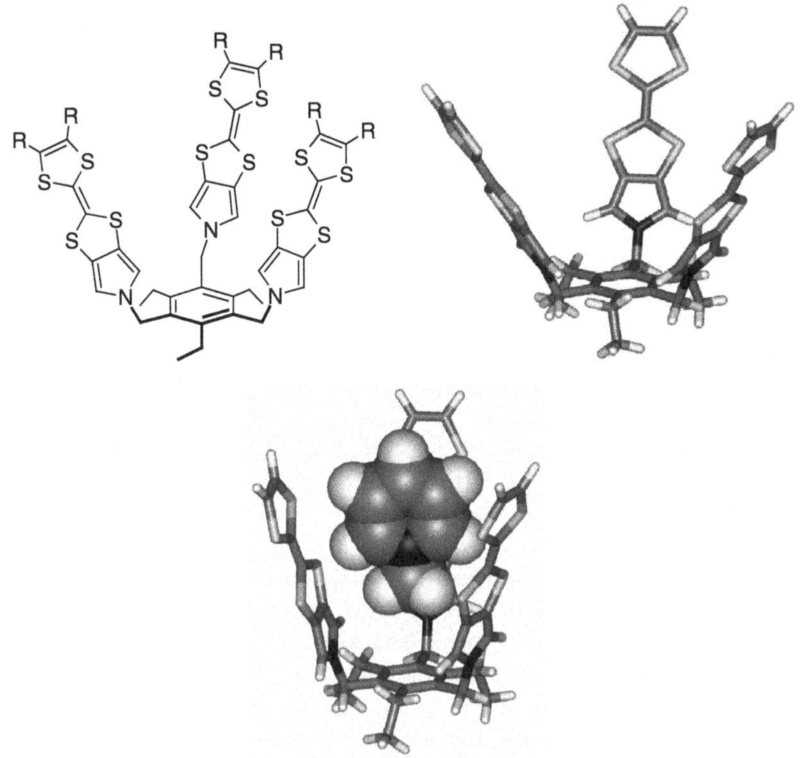

Figure 5.83 The molecular gripper hosts with three pyrrolotetrathiafulvalene pincers (top left, R = H, $SCH_2CH_2CH_3$), the energy-minimized structures of the free host (top right, R = H), and its complex with *N*-methylpyridinium cation (bottom, R = H). *Source*: Reprinted from Watat *et al.* 2014 [187]; Copyright 2014, with permission from Elsevier.

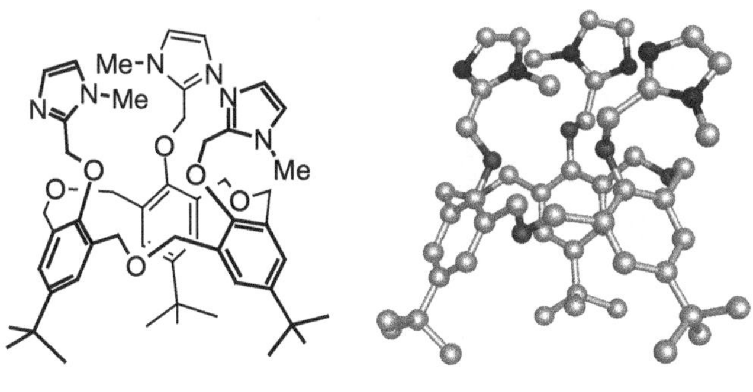

Figure 5.84 The tripodal claw grabber with three imidazole pincer elements (left) and its X-ray structure (right, hydrogen atoms were removed for clarity).

A cyclotriveratrylene spacer was successfully used for the construction of a tridentate molecular flytrap with three bowl-shaped corannulene pincers (Figure 5.85). It was expected that three corannulene units would provide better binding for spherical fullerene guests compared to molecular tweezers with only two corannulene units. However, this particular tridentate system did not deliver better results ($K_a = 1500\,M^{-1}$ for fullerene C_{60} and $1180\,M^{-1}$ for fullerene C_{70}) than the buckycatcher bidentate system discussed earlier (Section 5.3) [189].

Also known are molecular flytraps with four pincers. Recent examples include a set of Ag^+-specific flytrap host molecules, "argentivorous molecules". They consist of a macrocycle with several aromatic side arms as leaves attached to the ring with flexible links. These molecular devices work similar to the Venus flytrap plant by grabbing and holding Ag^+ "insect" species from the solution. The process of complexation can be easily monitored by NMR and UV–Vis spectroscopy in solution. The X-ray study of one of the complexes in solid state shows that a metal cation in the center of this molecular flytrap is completely covered by four aromatic "petals" (Figure 5.86). The 1:1 complex is quite strong ($K_a = 6.3 \times 10^6\,M^{-1}$) due to a combination of crown-ether-type interactions between Ag^+ and four nitrogen atoms of the macrocyclic polyamine supplemented by four $Ag^+–\pi$ interactions with the aromatic pincers. This flytrap host forms even stronger complexes with Hg^{2+}, Cu^{2+}, and Zn^{2+} ions, but conformational changes indicative of the Venus flytrap mechanism of action are observed only for Ag^+ [190].

Calix[*n*]arene spacers in *cone* conformations can provide for good spatial alignment of four and more pincer elements. Full functionality of these gripper hosts can be achieved via electrochemical [191] or allosteric [192] conformation control. A representative example of molecular grippers utilizing the calix[4]pyrrole platform is shown in Figure 5.87. In this case, the

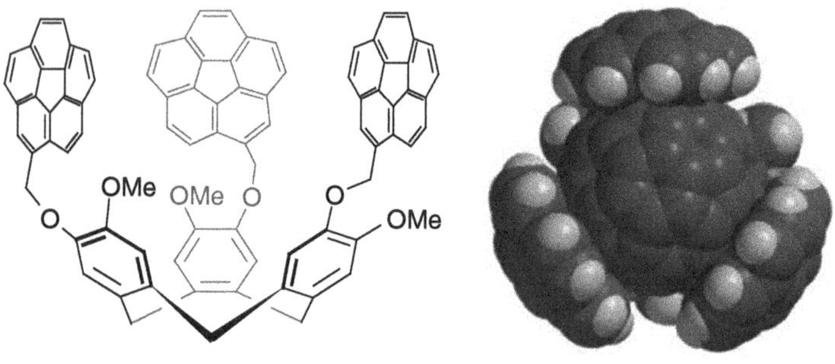

Figure 5.85 The tridentate molecular flytrap with the cyclotriveratrylene spacer and the three corannulene pincers (left) and the energy-minimized structure of its complex with fullerene C_{60} (right). *Source*: Reprinted from Yanney and Sygula 2013 [189]; Copyright 2013, with permission from Elsevier.

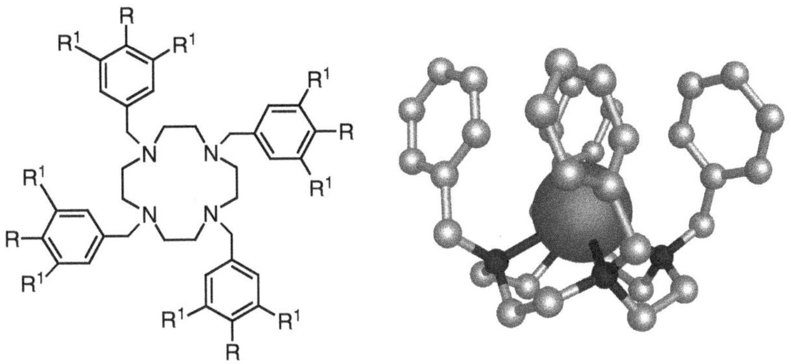

Figure 5.86 The flytrap host with four aromatic pincers (left, R = H, F, Me, OMe, NO_2, OPh; R^1 = H, F) and the X-ray structure of its complex with Ag^+ (right, R = R^1 = H; hydrogen atoms were removed for clarity).

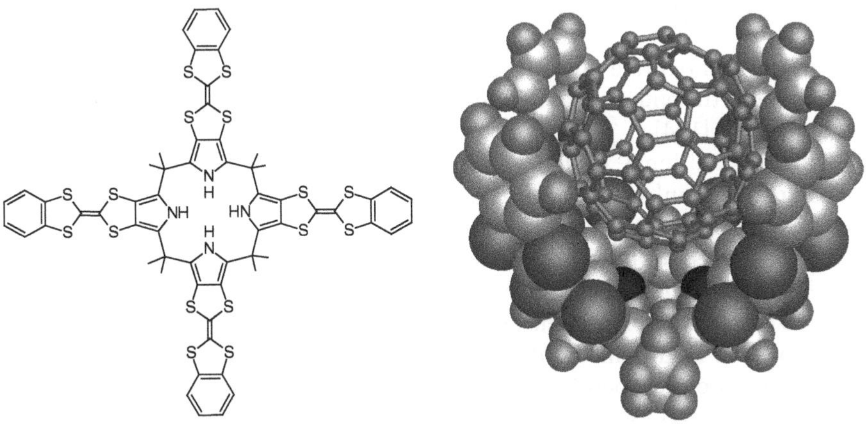

Figure 5.87 The molecular gripper host with four benzotetrathiafulvalene pincers (left) and the X-ray structure of its complex with fullerene C_{70} (right) [193].

conformation of pincer elements can be controlled allosterically via external coordination with chloride or fluoride anions. Allosteric modulation of this gripper host system allows to achieve very strong binding with fullerene C_{60} (K_a up to $8.1 \times 10^6\ M^{-1}$) in the presence of fluoride anion and strong binding with fullerene C_{70} (K_a up to $3.3 \times 10^6\ M^{-1}$) in the presence of chloride anion [193].

Even more sophisticated external conformational control can be implemented for complexes with electroactive guest systems, such as endohedral fullerene Li@C_{60} (Figure 5.88). This complex can be best described in terms of an anion-radical–cation-radical pair. The addition of a tetraethylammonium

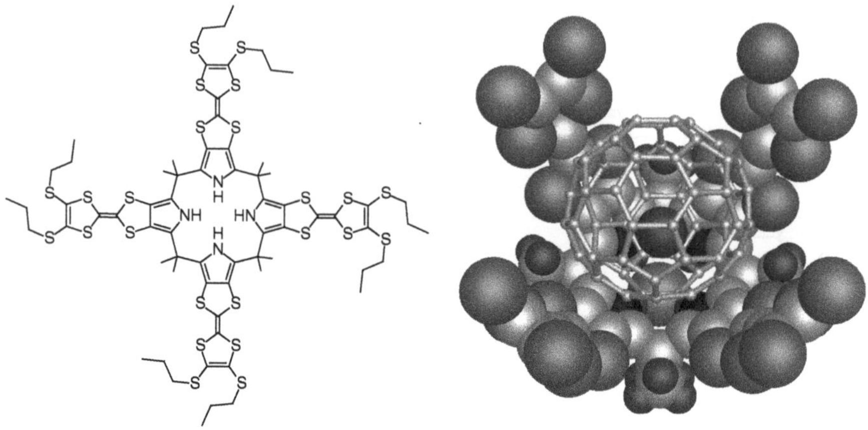

Figure 5.88 The molecular gripper host with four tetrathiafulvalene pincers (left) and the X-ray structure of its radical pair complex with endohedral fullerene Li@C_{60} and chloride anion (right, $CH_2CH_2CH_3$ groups were removed for clarity).

cation as a competitive guest leads to the restoration of the initial oxidation states of both the endohedral fullerene Li@C_{60} guest and the tetrathiafulvalene host molecule [194].

A more traditional design of the redox-controlled molecular grippers utilizing a resorcin[4]arene spacer has been reported recently. In the reduced form, this host adopts a vase conformation due to the network of hydrogen bonding between the pincer elements. The reduced host makes complexes with a variety of organic guest molecules of matching sizes, such as cyclooctane ($K_a = 83\,M^{-1}$) and 1,4-cyclohexanedione ($K_a = 1080\,M^{-1}$). Oxidation of the host molecule results in the disruption of the hydrogen bond network with the formation of an open kite conformation and release of a guest molecule (Scheme 5.9) [191]. Similar design ideas are utilized for the construction of tetraferrocenyl-resorcin[4]arene, in which ferrocene units serve as electrochemical controllers for the selective binding and release of alkylammonium cations [195].

5.6 Applications

The ability to selectively capture certain guest molecules from the environment provides attractive opportunities for the utilization of molecular tweezers as valuable components of chemical sensors (detectors and molecular probes), as well as bioactive compounds for medicinal applications. Further improvements in selectivity and sensitivity of these molecular devices can be achieved by incorporating additional functional groups with the desired optical or

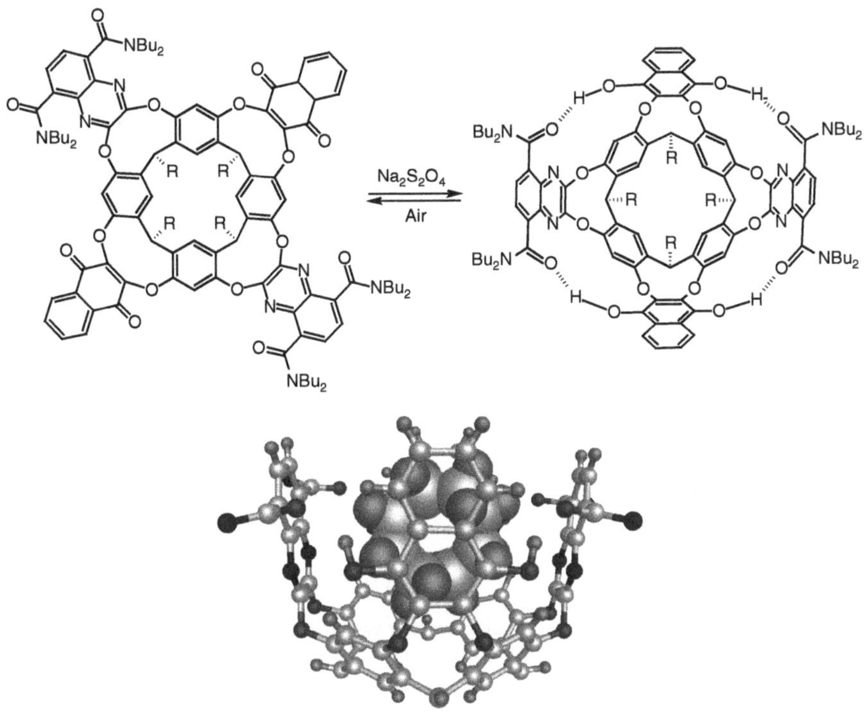

Scheme 5.9 Redox control of the molecular gripper with the resorcin[4]arene spacer (top, $R = CH_2(CH_2)_4CH_3$) and the X-ray structure of the reduced complex with cyclooctane (bottom, alkyl groups were removed for clarity) [191].

guest recognition properties. While the preponderance of current applications utilizes molecular tweezers in solutions, solid-supported molecular tweezers already demonstrated their utility as high-performance liquid chromatography (HPLC) stationary phases [196] and optical sensors [197].

The performance of solid-supported molecular tweezer systems as HPLC stationary phases was evaluated in detail for conformationally rigid tweezer molecules presented in Figure 5.89. The best separation results were achieved for simple dicyano- and dinitroaromatic compounds (Figure 5.89). The observed selectivity can be attributed to a combination of electrostatic and steric complementarity factors [198].

The abovementioned selective binding of electron-deficient aromatic compounds is one of the most distinguishing characteristics of molecular tweezers. Not surprisingly, substantial effort was directed to the development of analytical methods for the detection and quantitative analysis of industrially important nitroaromatic explosives using tweezer molecules [199]. A representative example of molecular tweezer systems with polyaromatic fluorescent

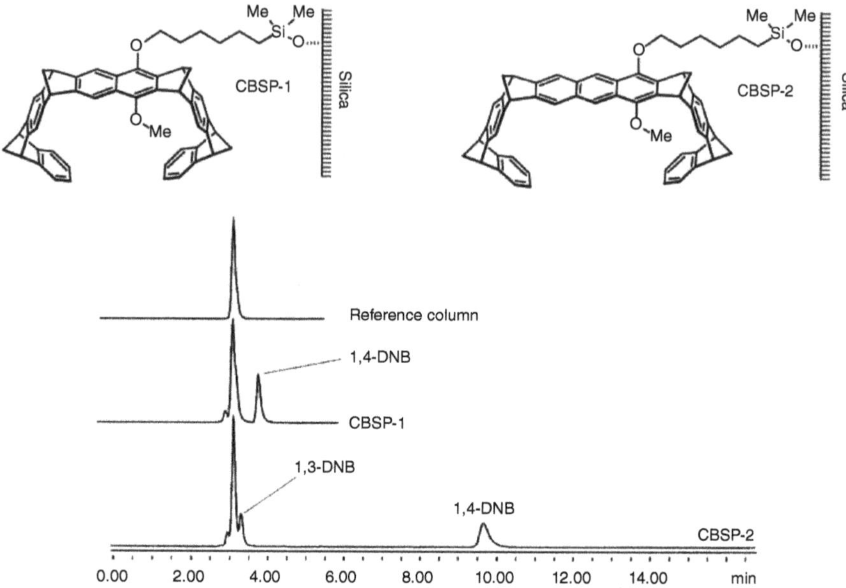

Figure 5.89 The solid-supported molecular tweezer systems as HPLC stationary phases (top) and the HPLC chromatograms of dinitroaromatic compound mixtures (bottom, 1,3-DNB = 1,3-dinitrobenzene; 1,4-DNB = 1,4-dinitrobenzene). *Source*: Kamieth *et al.* 1999 [198]. Reproduced with permission of John Wiley & Sons.

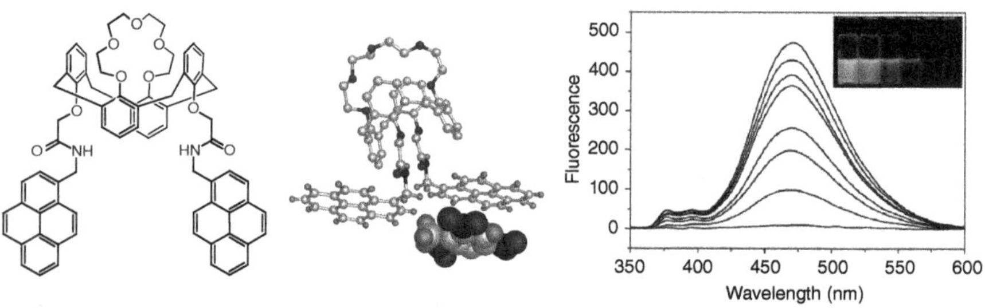

Figure 5.90 The fluorescent molecular tweezers for the detection of nitroaromatic compounds (left), the X-ray structure of their complex with 2,4,6-trinitrotoluene (middle, some hydrogen atoms were omitted for clarity), and fluorescence quenching upon addition of 2,4,6-trinitrotoluene (right, *Source*: Lee *et al.* 2010 [200]. Reproduced with permission of John Wiley & Sons).

pincers and calix[4]arene spacers is presented in Figure 5.90 [200]. Similar tweezer molecules with flexible spacers [201], anthracene pincers [202], and polymer-supported systems [203] are also known. The fluorescence of these compounds is quenched in the presence of nitroaromatic guest molecules due to the formation of charge–transfer complexes.

The flytrap molecular tweezer design, which entails the use of multiple polyaromatic pincers, is also a popular approach for the construction of chemical sensors [204, 205]. A representative example of these tripodal fluorescent sensors is shown in Figure 5.91 [204]. The presence of several fluorophoric moieties can improve sensitivity of the sensors and lower the minimum detection level of nitroaromatic compounds. The best reported tweezer systems can detect industrially important polynitroaromatic compounds, such as picric acid, in nanomolar concentrations [205].

A different approach was employed for the design of tetrathiafulvalene calix[4]pyrroles as colorimetric chemosensors for nitroaromatic explosives [206]. These tweezer molecules produce deep-colored charge–transfer sandwich complexes in the presence of polynitroaromatic compounds [207]. One of the examples of the molecular detection systems is presented in Figure 5.92 [208].

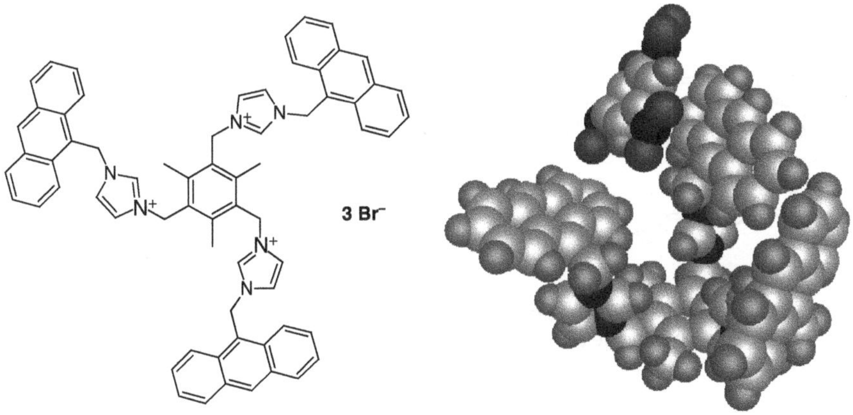

Figure 5.91 The fluorescent flytrap molecular tweezers with anthracene pincers for the detection of nitroaromatic compounds (left) and the X-ray structure of their complex with picric acid (right, bromide counteranions were removed for clarity) [204].

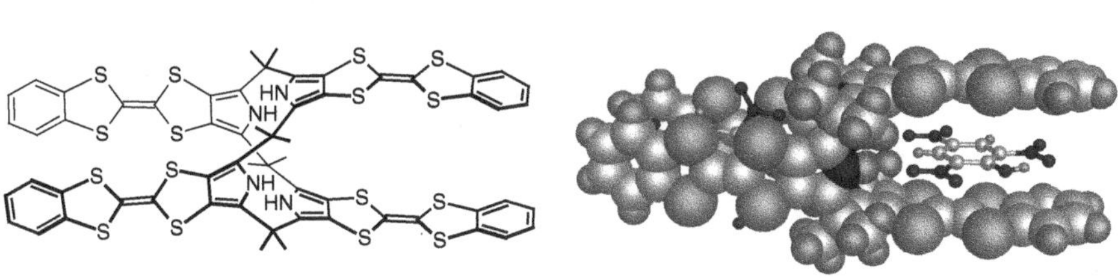

Figure 5.92 The 1,3-alternate tetrathiafulvalene calix[4]pyrrole molecular tweezers as chemosensors for colorimetric detection of nitroaromatic explosives (left) and the X-ray structure of the sandwich complex with two molecules of picric acid (right) [208].

Adequate results were also achieved with similar molecular tweezers having only two tetrathiafulvalene pincers. Addition of 1,3,5-trinitrobenzene results in deep green coloration of the solution ($\lambda_{max} = 649$ nm) due to the formation of a 1:1 sandwich complex (Figure 5.93) [209].

Molecular tweezers for fluorescence analysis include allosteric "twist-on" systems, which adopt tweezer-like *syn*-conformation only in the presence of certain anionic guests, such as chloride anion. In the absence of an appropriate guest, this sensor host molecule exists in the most thermodynamically stable *anti*-conformation with weak fluorescence. Upon complexation with the guest anion, the system "twists on" into a tweezer-like *syn*-conformation, yielding enhanced fluorescence due to mutual proximity of the two aromatic pincers (Figure 5.94). As expected, the complex strength falls in the F > Cl > Br progression due to the weakening of the anion hydrogen bonding ($K_a = 10{,}233 > 361 > 36\ M^{-1}$ in $CHCl_3$/DMSO, respectively) [210].

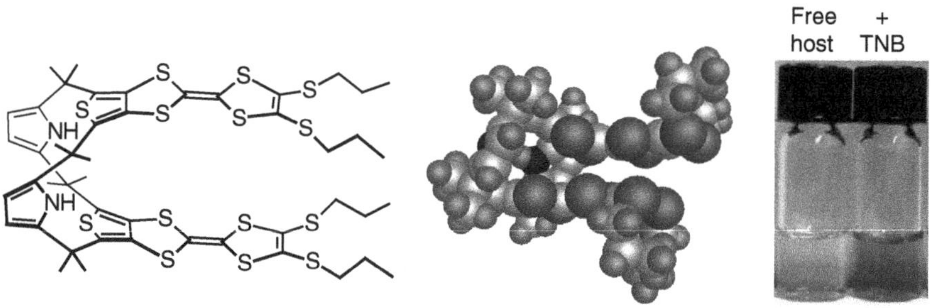

Figure 5.93 The tetrathiafulvalene molecular tweezers (left), their X-ray structure (middle), and the solution color change after addition of 1,3,5-trinitrobenzene (TNB, right). *Source*: Adapted with permission of Springer from Kim *et al.* 2009 [209]. Copyright 2009 Springer.

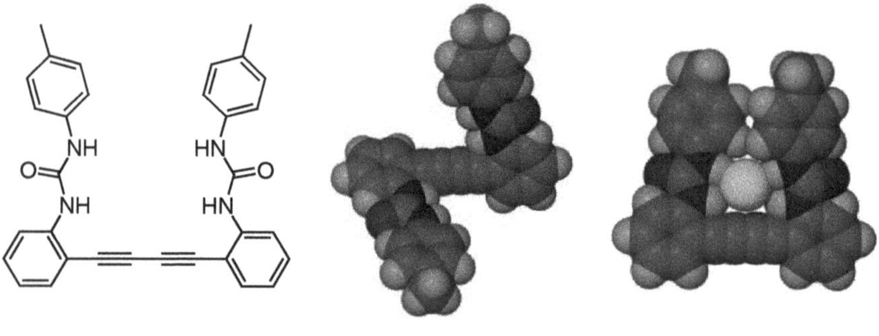

Figure 5.94 The twist-on molecular tweezer sensor system (left), its energy-minimized *anti*-conformation (middle), and the *syn*-complex with chloride anion (right). *Source*: Adapted from Swinburne *et al.* 2010 [210]. Reproduced with permission of John Wiley & Sons.

Deoxycholic acid-based molecular tweezers were successfully employed as a sensor element in the development of carbonate-selective electrodes for the determination of oceanic carbon dioxide. Two electron-deficient trifluoroacetylated pincers are attached in *syn*-configuration to a rigid steroid backbone (Figure 5.95). They serve as a preorganized cavity for the electron-rich carbonate anion. The system has good carbonate detection limit (5.8×10^{-7} M) and selectivity in the presence of other common anions in the ocean water [211, 212].

Tripodal flytrap molecules were successfully employed as turn-on fluorescence chemosensors for the detection of phosphate anions in aqueous solutions (Figure 5.96) [213]. These molecules exhibit up to 200-fold increase in fluorescence intensity in the presence of $H_2PO_4^-$, $HP_2O_7^{3-}$, and AcO^- anions. The binding affinities decrease in the following general order: $H_2PO_4^- > HP_2O_7^{3-} > AcO^- \gg Cl^- > Br^-$ with association constants K_a in the range of 10–100 M^{-1} for Cl^- and up to 5.0×10^6 M^{-1} for $H_2PO_4^-$, $HP_2O_7^{3-}$, AcO^-, and F^-. Further improvements in anion detection limits can be achieved through the incorporation of sensor molecules into hydrophilic polyurethane matrices as array-type assemblies.

Molecular tweezers with NDI pincers are capable of detecting minor quantities of fluoride in the presence of other common anions. This anion selectivity can be attributed to the formation of an orange-colored $NDI^{\bullet-}$ radical anion and then, subsequently, a pink-colored NDI^{2-} dianion due to the unprecedented anion–π interactions and charge/electron transfer between the host molecule and the fluoride anion (Figure 5.97) [214].

Molecular tweezers with nitrogen-containing links were successfully employed as metal-sensitive chemosensors. These chemosensors are activated

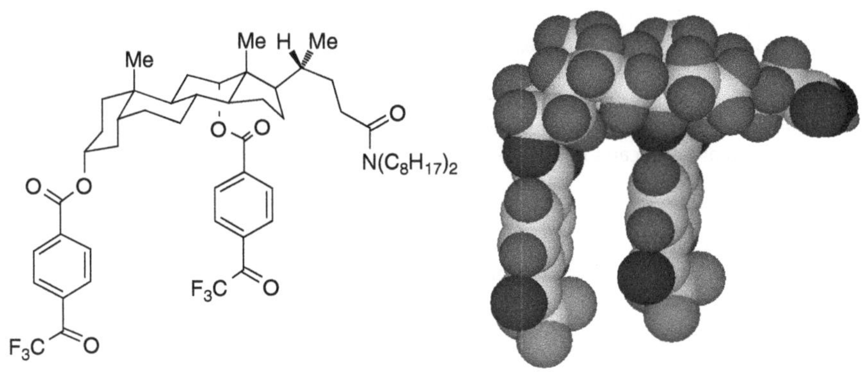

Figure 5.95 The deoxycholic acid-based molecular tweezers under development for carbonate-selective electrodes (left) and their energy-minimized 3D structure (right, C_8H_{17} fragments were removed for clarity).

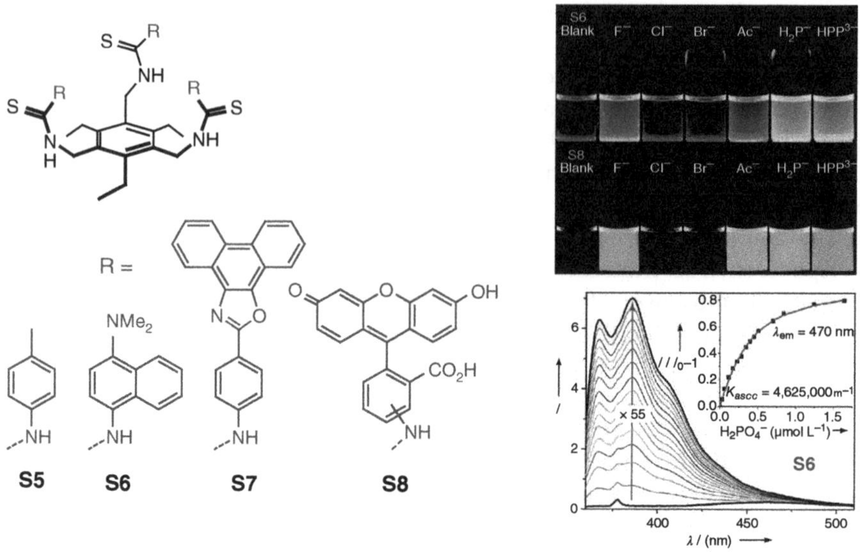

Figure 5.96 The fluorescence turn-on tripodal flytrap chemosensor systems (left) and the fluorescence response of S6 and S8 molecules in the presence of anions (right, $H_2P^- = H_2PO_4^-$, $HPP^{3-} = HP_2O_7^{3-}$). *Source*: Adapted from Zyryanov *et al.* 2007 [213]. Reproduced with permission of John Wiley & Sons.

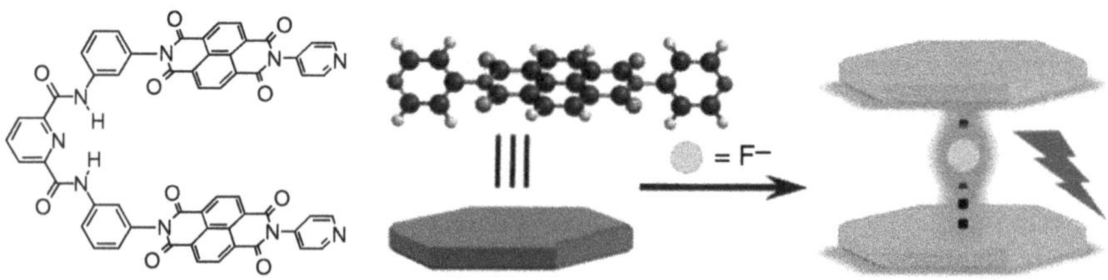

Figure 5.97 The sensor system for fluoride ion based on molecular tweezers with naphthalene diimide pincers. *Source*: Adapted from Guha and Saha 2010 [214]. Reproduced with permission of American Chemical Society.

by metal cations due to a complex formation involving the nitrogen atoms of the linking fragments. This complexation alters the relative positions of the fluorophoric pincers, which results in notable fluorescence changes [215, 216]. One of these tweezer chemosensors with pyrene pincers and 1,2,3-triazole links is presented in Scheme 5.10 [215]. Binding of silver cations leads to the separation of the pyrene fluorophoric pincers. This separation is associated with an increase in the pyrene monomer emission at 374 and 394 nm and a decrease of the excimer emission at 484 nm [215].

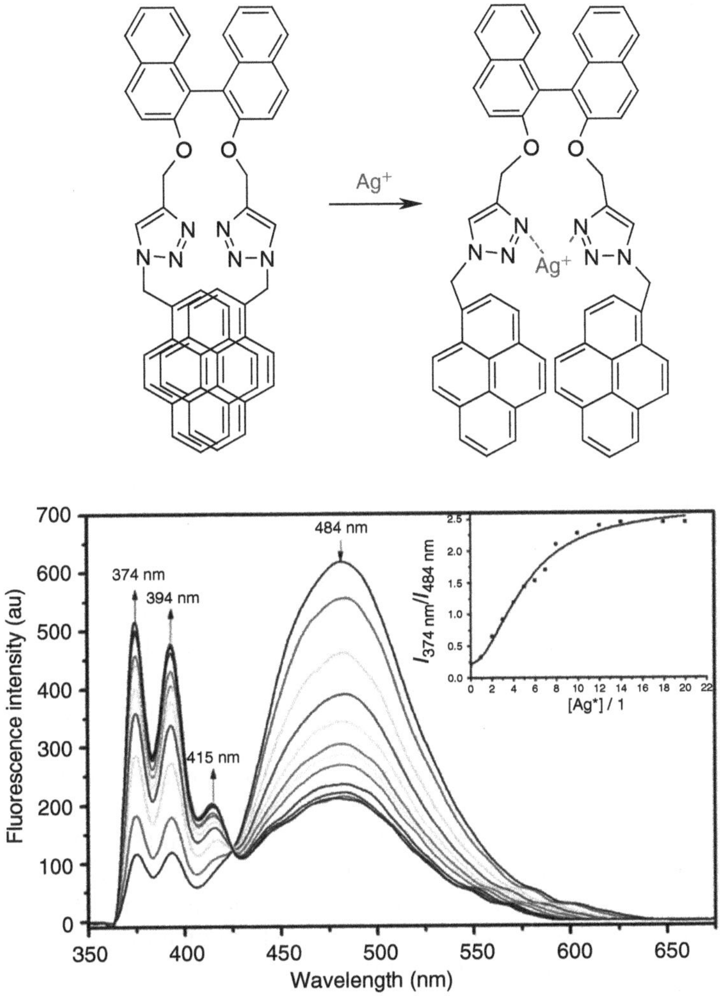

Scheme 5.10 The molecular tweezers with 1,2,3-triazole links and pyrene pincers as a selective chemosensor for silver cations (top) and the changes in fluorescence upon the addition of increasing amounts of silver perchlorate (bottom, *Source*: Liu *et al.* 2011 [215]. Reproduced with permission of Elsevier).

Flexible molecular tweezers with fluorescent pyrene pincers and arylboric receptor elements were proposed as analytical probes for carbohydrates. One of these molecular tweezers (Figure 5.98) shows good selectivity toward some simple sugar molecules. For example, this host shows good binding affinity toward D-glucose ($K_a = 2660\,M^{-1}$) and D-fructose ($K_a = 2850\,M^{-1}$) compared to D-galactose ($K_a = 440\,M^{-1}$) and D-mannose ($K_a = 200\,M^{-1}$) in a

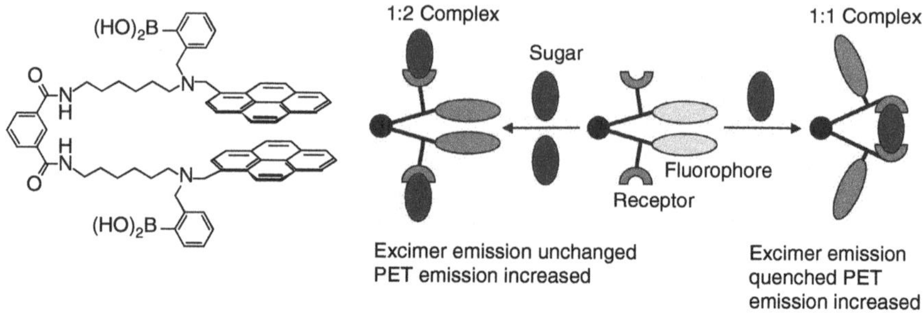

Figure 5.98 The flexible molecular tweezers with pyrene pincers and arylboric receptor elements as an analytical fluorescent probe for carbohydrates (left) and the proposed mode of action (right, *Source*: Adapted from Phillips *et al.* 2009 [34]. Reproduced with permission of Royal Society of Chemistry).

water–methanol buffer. The capture of a sugar molecule results in the increase of photoinduced electron transfer (PET) emission at 377 nm. The excimer emission at 470 nm is quenched in 1:1 sandwich complexes (e.g., D-glucose), but remains unchanged in 1:2 complexes (e.g., D-fructose) [34].

Good sensitivity and selectivity toward *α* and *β* sugar molecules at micromolar concentrations were achieved with tripodal claw grabber chemosensors (Figure 5.99) [217]. Generally, these chemosensors bind much stronger with β-monosaccharides than with *α*-monosaccharides due to a better structural complementarity between the host and the guest molecules.

Bisporphyrin molecular tweezers [69] were extensively studied for the determination of the absolute configuration of chiral compounds using circular dichroism (CD) [218–220], exciton-coupled circular dichroism (ECCD) [71], and magnetic circular dichroism (MCD) [72] methods. One of the recent examples of bisporphyrin molecular tweezer systems, which demonstrate substantial improvement in the CD signal amplitude of chiral diamine sandwich complexes, is presented in Figure 5.100 [219].

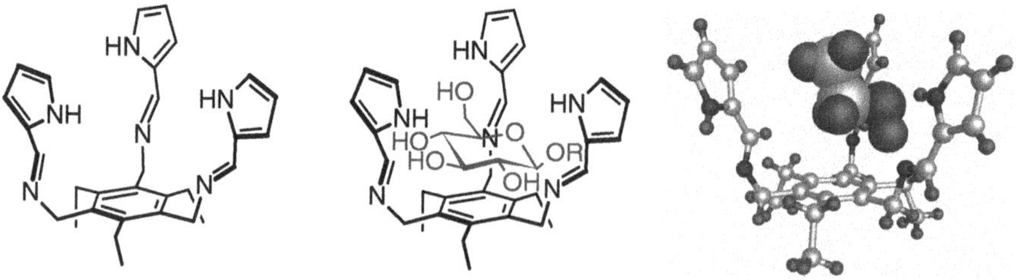

Figure 5.99 The tripodal claw grabber chemosensor for sugar molecules (left), its complex with a model β-monosaccharide (middle, $R = n\text{-}C_8H_{17}$), and the X-ray structure of its inclusion complex with ethanol (right).

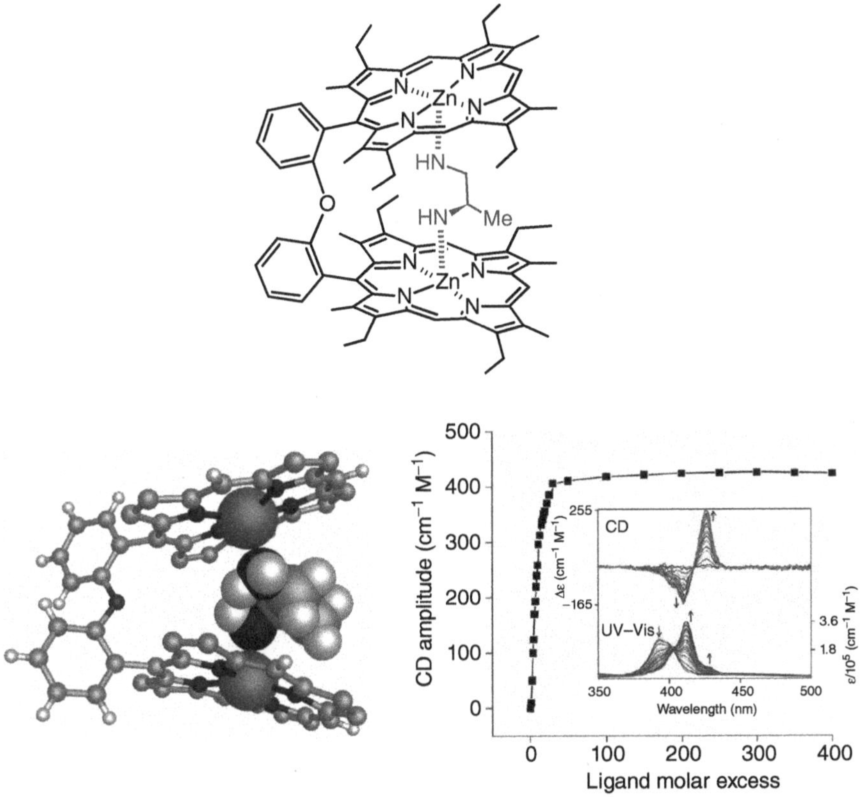

Figure 5.100 The bisporphyrin molecular tweezer complex with (*R*)-1,2-diaminopropane (top), its X-ray structure (bottom left, CH_3 and CH_2CH_3 groups were removed for clarity), and the CD spectra enhancement of (2*S*)-3-phenyl-1,2-propanediamine (bottom right). *Source*: Brahma *et al.* 2014 [219]. Reproduced with permission of American Chemical Society.

Bisporphyrin-ferrocene molecular tweezers presented in Figure 5.101 were successfully employed for CD spectrum enhancement. The experiments show that the CD spectrum amplitude of selected chiral diamines can be improved on an order of magnitude due to good host–guest matching. However, full CD spectrum enhancement can be achieved only with a large excess of the tweezer host (20:1) [220].

Several chiral molecular tweezers were specifically developed for NMR enantiodiscrimination of stereoisomers. One of them is presented in Figure 5.102. This molecular detection system has four chiral centers (two in each α,α'-bis(trifluoromethyl)-9,10-anthracenedimethanol pincer units). Chiral resolution is further facilitated by the presence of two acidic hydroxy groups capable of forming strong hydrogen bonds with an appropriate substrate due to the proximity of electron-withdrawing α-trifluoromethyl groups. The

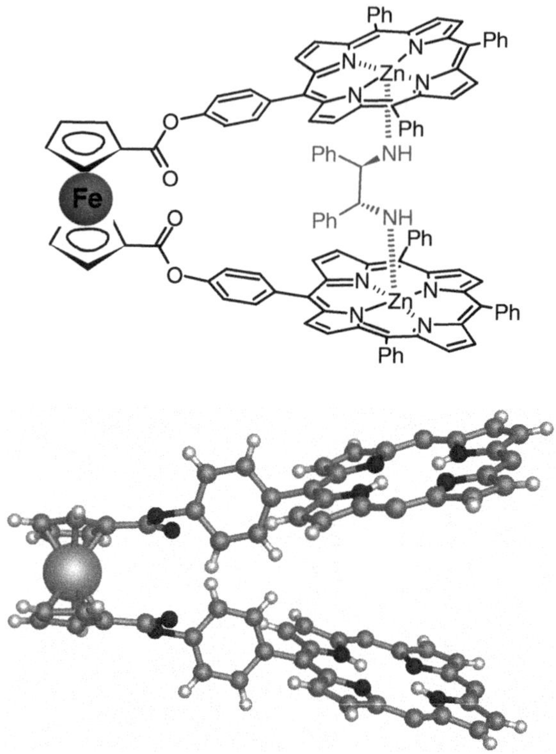

Figure 5.101 The bisporphyrin-ferrocene molecular tweezer complex with (1*R*,2*R*)-1,2-diphenylethylenediamine (top) and the X-ray structure of the Zn-free host molecule (bottom, Ph groups were removed for clarity).

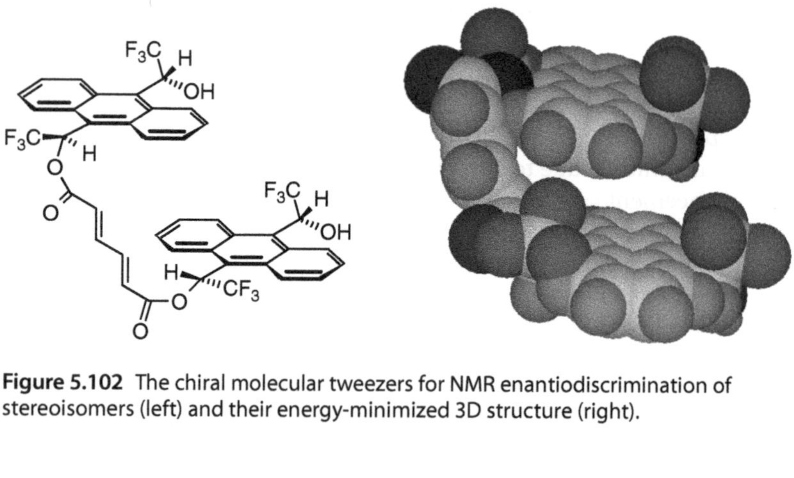

Figure 5.102 The chiral molecular tweezers for NMR enantiodiscrimination of stereoisomers (left) and their energy-minimized 3D structure (right).

spacer unit is constructed from commercially available *trans,trans*-muconic acid, which provides the optimal ~7 Å separation between the pincer elements [221, 222].

An interesting example of sequence-specific binding between molecular tweezers with pyrene pincers and pyromellitimide units of oligomeric molecules is shown in Figure 5.103. In essence, oligomer sequences can be "read" by a mechanism that entails multiple binding of the tweezer host molecules to the diimide residues of the short polymer chains. This sequence-specific binding is a useful feature for the construction of memory devices and other technomimetic applications [223].

Flexible molecular tweezers with long aliphatic spacers were successfully used for the separation of metallic and semiconducting single-walled carbon nanotubes (Figure 5.104). The flexible spacers allow for effective "wrap-around" noncovalent binding between the molecular tweezers and the semiconducting nanotubes. This tweezer complexation facilitates the separation of fractions after centrifugation in a tetrahydrofuran–toluene (80:20) mixed solvent [224]. Similar wrap-around separation methodology was implemented for the optical enrichment of single-walled carbon nanotubes using chiral bisporphyrin

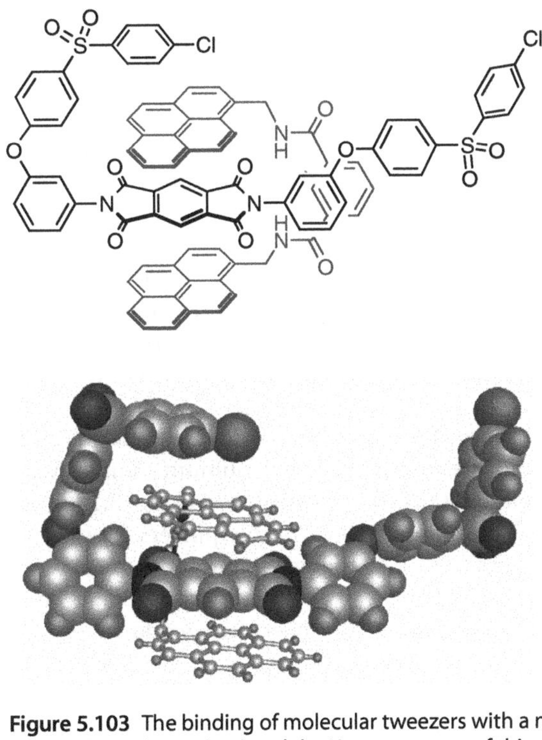

Figure 5.103 The binding of molecular tweezers with a model oligomeric molecule at pyromellitimide site (top) and the X-ray structure of this complex (bottom).

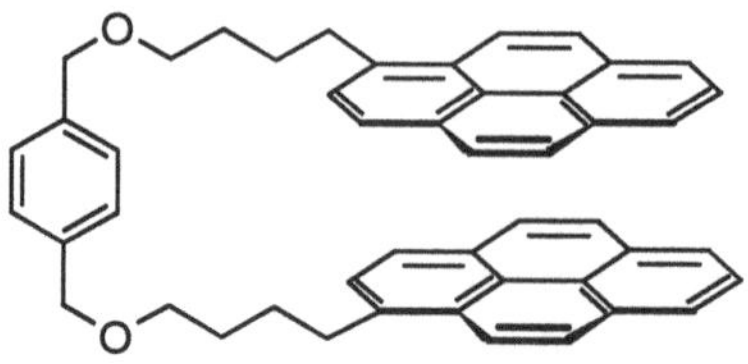

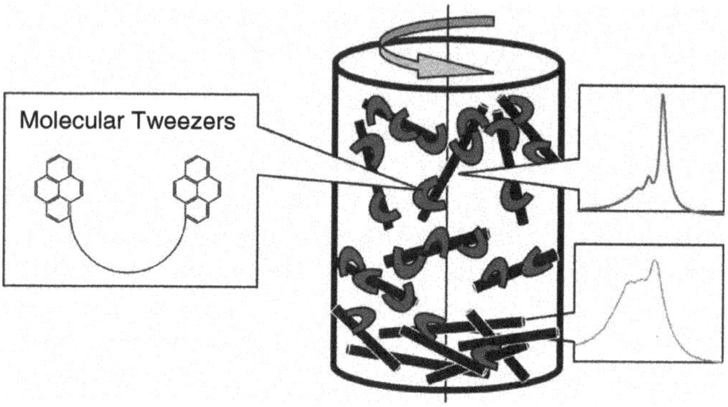

Figure 5.104 The separation of single-walled carbon nanotubes using flexible molecular tweezers. *Source*: Adapted from Yang *et al*. 2012 [224]. Reproduced with permission of Royal Society of Chemistry.

molecular tweezers [225]. In addition, some oligomeric glycoluril molecular tweezers can be used for the solubilization of carbon nanotubes in water at very low concentrations [226].

Several molecular tweezers were successfully adapted for medicinal applications. Early examples include cyclic peptide antibiotics echinomycin and quinomycin [227]. Echinomycin and quinomycin antibiotics have distinct tweezer-like structures and act as DNA bis-intercalators. The structures of a quinomycin analog, antibiotic UK-63052, and its oligonucleotide complex are presented in Figure 5.105 [228].

Similar modes of action are observed for other bis-intercalators, such as anticancer drug ditercalinium. The X-ray structure of a ditercalinium complex with a double-stranded DNA fragment $[d(CGCG)]_2$ is presented in Figure 5.106. The separation distance between the two 7*H*-pyridocarbazole pincers is approximately 10.4 Å, which is larger than the 6–7 Å distance found in many "classic" molecular tweezers. This difference is likely a result of the specific double-helix DNA structure that requires a larger pincer-to-pincer distance for adequate binding. The crystal structure of the ditercalinium complex with the DNA fragment clearly shows two pairs of DNA bases within the cavity of the molecular tweezers (Figure 5.106). The DNA fragment in the complex with ditercalinium is unwound by 36° [229].

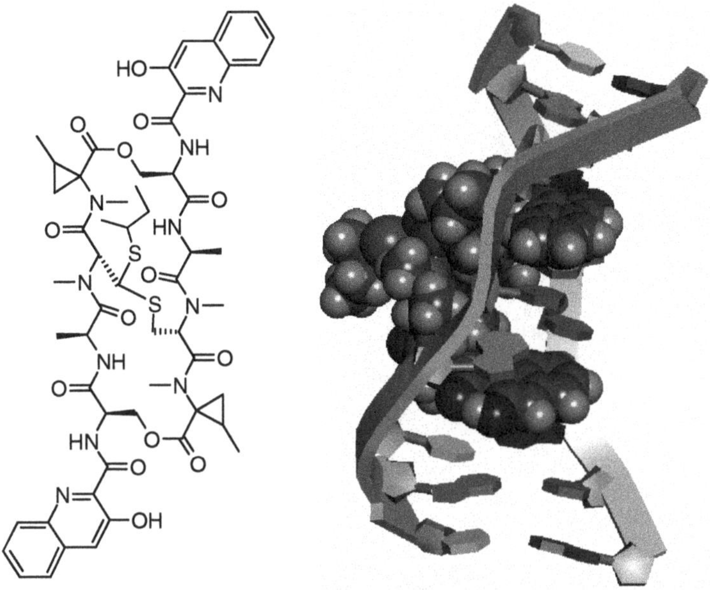

Figure 5.105 The quinomycin analog, antibiotic UK-63052 (left) and the X-ray structure of its oligonucleotide complex (right).

Molecular tweezers with rigid [*n*]polynorbornane spacers can also act as effective DNA bis-intercalators. The structure of the optimized compound ($K_{app} = 32{,}800\,M^{-1}$ for *Escherichia coli* DNA) is presented in Figure 5.107 [230]. The separation distance between the two acridine pincer moieties is approximately 10.3 Å, similar to the ditercalinium–DNA complex discussed earlier.

One of the most promising molecular tweezer drug candidates is a relatively simple rigid molecular clip, which has selective affinity toward the natural amino acid lysine and lysine residues in proteins (Figure 5.108). This compound interferes with abnormal protein folding and aggregation implicated in more than 30 human diseases, including Alzheimer's and Parkinson's diseases [231]. In subsequent studies, this new amyloid assembly inhibitor demonstrates superior binding properties compared to other known inhibitors, such as *scyllo*-inositol and epigallocatechin gallate [232]. Similar to amyloid assembly inhibition, the compound also acts as an inhibitor of α-synuclein neurotoxicity *in vitro* and *in vivo* [233]. In addition, this molecular clip is capable of modulating 14-3-3 protein–protein interactions via binding to a single surface-exposed lysine in proximity to the central channel (Figure 5.108) [234]. Various derivatives of this compound, including polyglycerol-tagged molecular tweezers, were also evaluated [152, 157, 235, 236].

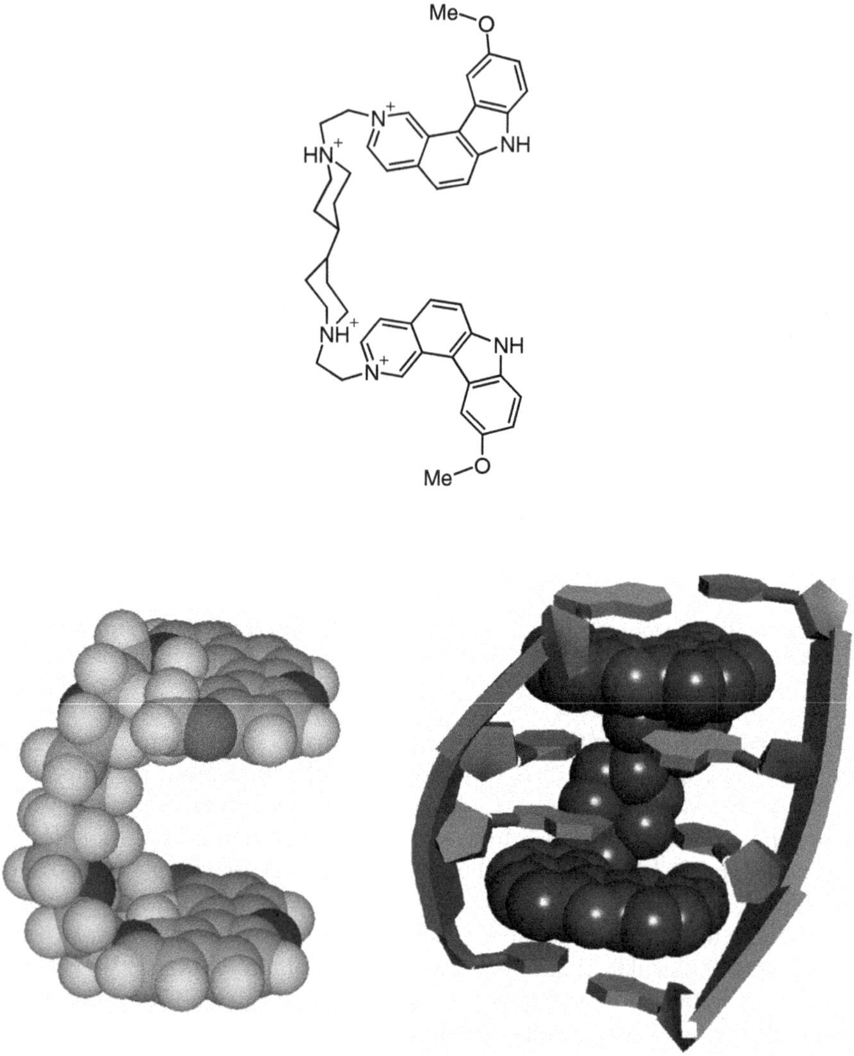

Figure 5.106 The anticancer drug ditercalinium (top), its X-ray structure (bottom left), and the X-ray structure of the complex with the double-stranded DNA fragment $[d(CGCG)]_2$ (bottom right, hydrogen atoms were removed for clarity).

Recent progress in applied biomedical research was achieved with the use of oligomeric glycoluril tweezers. A list of the most promising applications includes solubilization of poorly soluble drugs [237], drug delivery systems [238], and activity reversal of drugs commonly used during and after surgery [239]. The structures of two bioactive tweezer molecules, calabadion 1 and

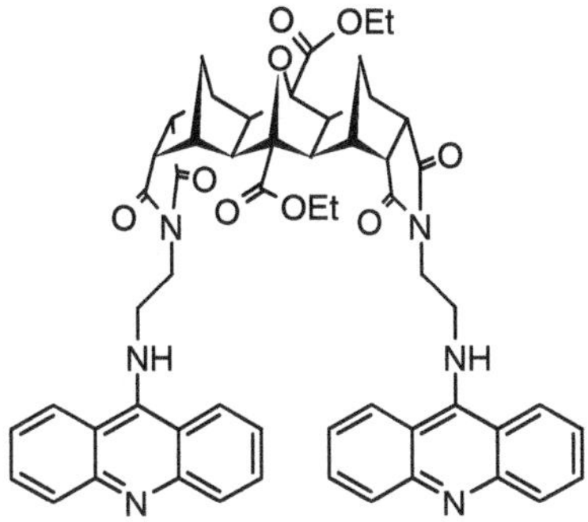

Figure 5.107 The molecular tweezers with the [3]polynorbornane spacer that can act as a DNA bis-intercalator.

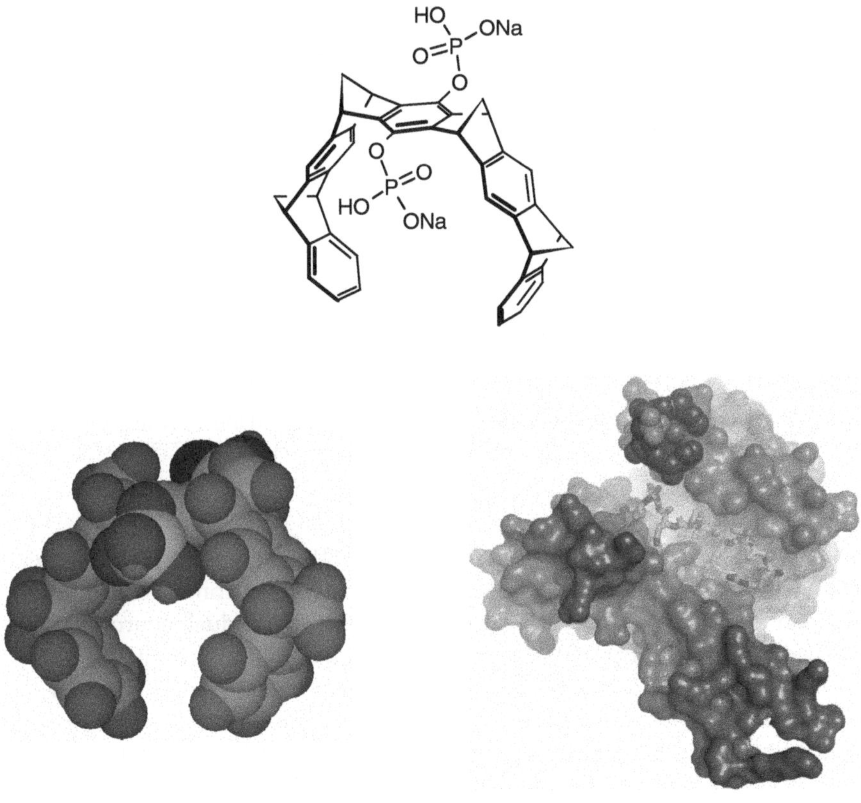

Figure 5.108 The simple molecular tweezers with selective affinity toward the natural amino acid lysine and lysine residues in proteins (top, bottom left) and their complex with a 14-3-3 adapter protein–partner protein assembly (bottom right, *Source*: Bier *et al.* 2013 [234]. Reproduced with permission of Nature Publishing Group).

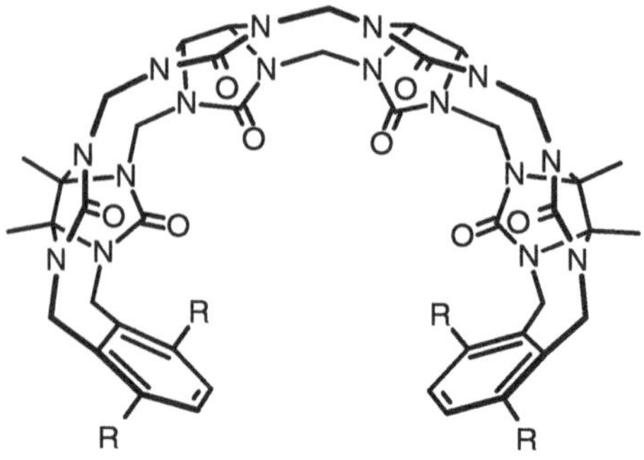

Calabadion 1

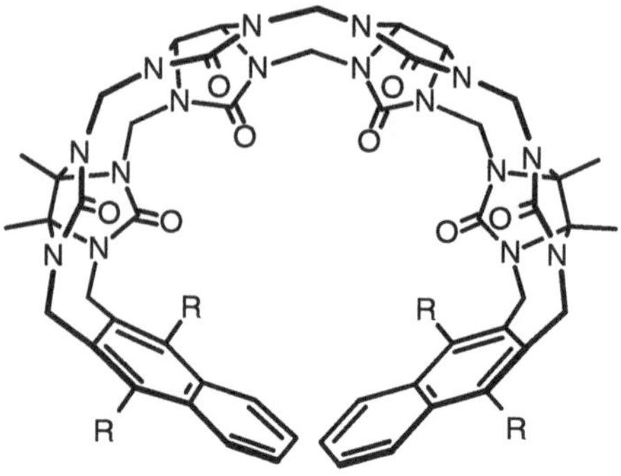

Calabadion 2

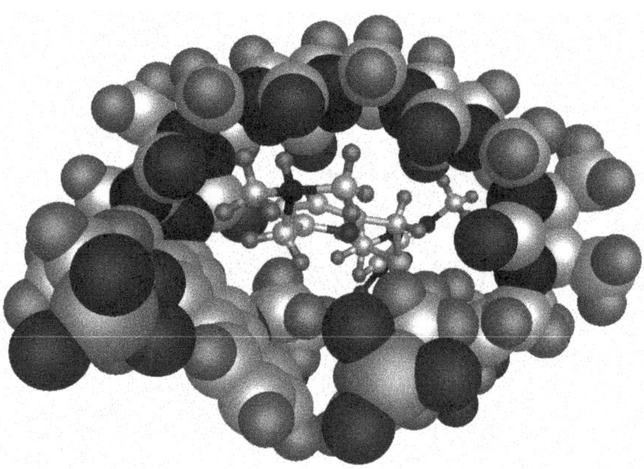

Figure 5.109 Calabadion 1 (top left), calabadion 2 (top right), and the X-ray structure of the calabadion 2 complex with ranitidine (bottom, R = $OCH_2CH_2CH_2SO_3Na$) [239].

calabadion 2, are depicted in Figure 5.109 [240]. These compounds bind selectively with cationic steroidal and benzylisoquinoline neuromuscular blockers and can be used for the reversal of postoperative residual neuromuscular blockade [241, 242] from cisatracurium ($K_a = 4.8 \times 10^6\ M^{-1}$ in water at pH 7.4 for calabadion 2) and rocuronium ($K_a = 3.4 \times 10^9\ M^{-1}$ in water at pH 7.4 for calabadion 2) [240]. The major advantages of shape-flexible oligomeric glycoluril tweezers compared to rigid cyclic cucurbit[*n*]urils are better solubility and improved binding of large drug molecules due to the induced-fit effect [243].

Molecular tweezers with selective affinity toward cholesterol were integrated into a macroporous polymer network employing "pseudo" molecular imprinting protocol (Figure 5.110). The best results for cholesterol rebinding

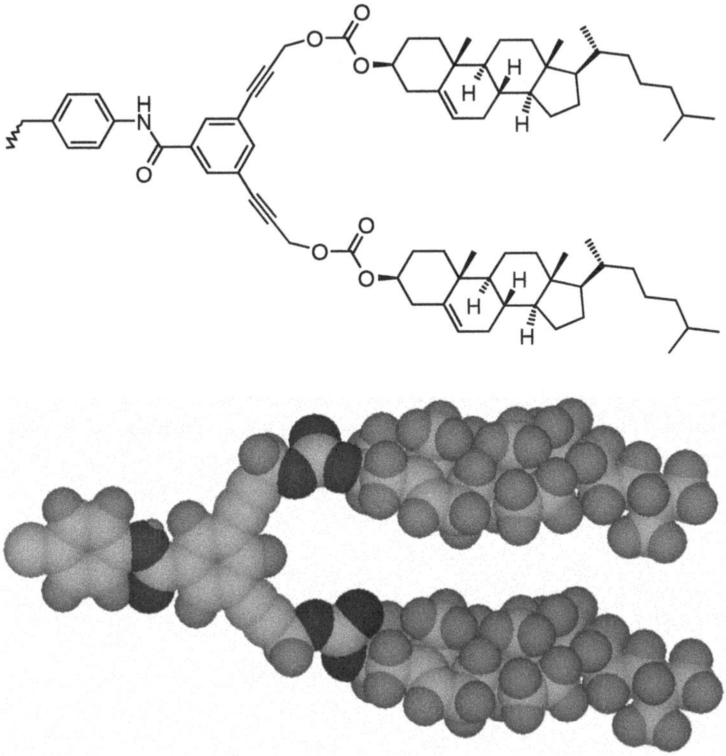

Figure 5.110 A molecular tweezer unit incorporated into a macroporous polymer network with selective affinity toward cholesterol (top) and its energy-minimized 3D structure (bottom).

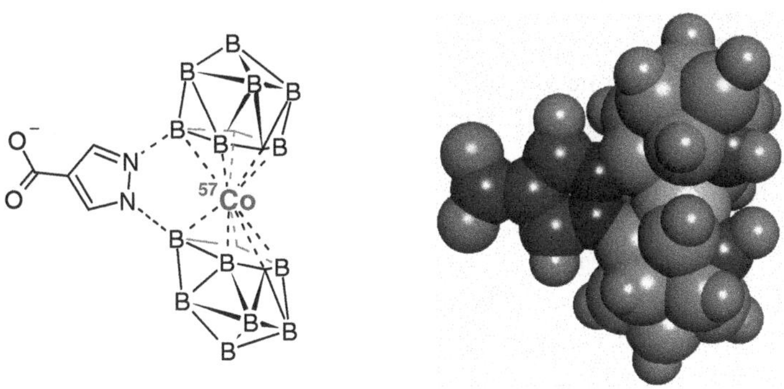

Figure 5.111 The radiolabeled Venus flytrap cluster (VFC, left, hydrogen atoms were removed for clarity) and its X-ray structure (right).

were achieved with a 1:4 template:monomer ratio utilizing methacrylic acid as a functional comonomer [244].

Venus flytrap clusters (VFCs) were proposed for the preparation of radiolabeled antibodies as delivery systems for various diagnostic and therapeutic applications. The cluster contains two carborane moieties attached to the pyrazole spacer. The radioactive atom ^{57}Co is located between the faces of the two carborane pincers (Figure 5.111). The pyrazole spacer unit contains a carboxylic group, which serves as an antibody linker. These radiolabeled antibodies have demonstrated clinical utility in radioimmunodetection and radioimmunotherapy [245, 246].

Examples of molecular tweezer applications presented here can only serve as an overview of this quickly growing field of applied sciences. With dozens of patent applications and journal articles published each year, it is safe to conclude that the molecular tweezer concept is one of the most advanced design ideas among other technomimetic molecules. It can also be envisioned that commercially viable tweezer molecules will appear on the market in the current decade.

References

1 Klärner, F.-G. and Schrader, T. (2013) Aromatic interactions by molecular tweezers and clips in chemical and biological systems. *Acc. Chem. Res.*, **46**, 967–978.
2 Harmata, M. (2004) Chiral molecular tweezers. *Acc. Chem. Res.*, **37**, 862–873.
3 Klärner, F.-G. and Kahlert, B. (2003) Molecular tweezers and clips as synthetic receptors. Molecular recognition and dynamics in receptor–substrate complexes. *Acc. Chem. Res*, **36**, 919–932.
4 Hardouin-Lerouge, M., Hudhomme, P., and Sallé, M. (2010) Molecular clips and tweezers hosting neutral guests. *Chem. Soc. Rev.*, **40**, 30–43.
5 Leblond, J. and Petitjean, A. (2011) Molecular tweezers: concepts and applications. *ChemPhysChem*, **12**, 1043–1051.
6 Warrener, R.N., Margetic, D., Amarasekara, A.S. *et al.* (1999) Building BLOCK strategies for the synthesis of molecular clefts with inside functionality. *Org. Lett.*, **1**, 199–202.
7 Shimizu, K.D., Dewey, T.M., and Rebek, J.J. (1994) Convergent functional groups. 15. Synthetic and structural studies of large and rigid molecular clefts. *J. Am. Chem. Soc.*, **116**, 5145–5149.
8 Banwell, M., Gable, R., Greenwood, R. *et al.* (1997) Ring-fused *gem*-dibromocyclopropanes as precursors to "tunable" molecular clefts possessing convergent functional groups. *Synlett*, **1997**, 953–955.

9 Reek, J.N.H., Elemans, J.A.A.W., de Gelder, R. *et al.* (2003) Self-association and self-assembly of molecular clips in solution and in the solid state. *Tetrahedron*, **59**, 175–185.
10 Wu, A., Chakraborty, A., Fettinger, J.C. *et al.* (2002) Molecular clips that undergo heterochiral aggregation and self-sorting. *Angew. Chem. Int. Ed.*, **41**, 4028–4031.
11 Sun, D., Tham, F.S., Reed, C.A. *et al.* (2002) Supramolecular fullerene–porphyrin chemistry. Fullerene complexation by metalated "jaws porphyrin" hosts. *J. Am. Chem. Soc.*, **124**, 6604–6612.
12 Habata, Y., Taniguchi, A., Ikeda, M. *et al.* (2013) Argentivorous molecules bearing two aromatic side-arms: Ag^+–π and CH–π interactions in the solid state and in solution. *Inorg. Chem.*, **52**, 2542–2549.
13 Chen, C.W. and Whitlock, H.W. (1978) Molecular tweezers: a simple model of bifunctional intercalation. *J. Am. Chem. Soc.*, **100**, 4921–4922.
14 Sygula, A. and Collier, W.E. (2011) Molecular clips and tweezers with corannulene pincers, in *Fragments of Fullerenes and Carbon Nanotubes: Designed Synthesis, Unusual Reactions, and Coordination Chemistry* (eds M.A. Petrukhina and L.T. Scott), John Wiley & Sons, pp. 1–40.
15 Hunter, C.A., Lawson, K.R., Perkins, J., and Urch, C.J. (2001) Aromatic interactions. *J. Chem. Soc. Perkin Trans.*, **2**, 651–669.
16 Zimmerman, S.C. (1991) Molecular tweezers: synthetic receptors for π-sandwich complexation of aromatic substrates, in *Bioorganic Chemistry Frontiers* (ed. P.H. Dugas), Springer, Berlin/Heidelberg, pp. 33–71.
17 Zimmerman, S.C. (1993) Rigid molecular tweezers as hosts for the complexation of neutral guests, in *Supramolecular Chemistry I—Directed Synthesis and Molecular Recognition*, Topics in Current Chemistry, vol. 165, Springer, Berlin/Heidelberg, pp. 71–102.
18 Biedermann, F., Nau, W.M., and Schneider, H.-J. (2014) The hydrophobic effect revisited—studies with supramolecular complexes imply high-energy water as a noncovalent driving force. *Angew. Chem. Int. Ed.*, **53**, 11158–11171.
19 Blake, J.F. and Jorgensen, W.L. (1990) Origin of the strong binding of adenine to a molecular tweezer. *J. Am. Chem. Soc.*, **112**, 7269–7278.
20 Fery-Forgues, S. and Al-Ali, F. (2004) Bis(azacrown ether) and bis(benzocrown ether) dyes: butterflies, tweezers and rods in cation binding. *J. Photochem. Photobiol. C Photochem. Rev.*, **5**, 139–153.
21 Marchi, E., Baroncini, M., Bergamini, G. *et al.* (2012) Photoswitchable metal coordinating tweezers operated by light-harvesting dendrimers. *J. Am. Chem. Soc.*, **134**, 15277–15280.
22 Takeshita, M. and Irie, M. (1998) Photoresponsive cesium ion tweezers with a photochromic dithienylethene. *Tetrahedron Lett.*, **39**, 613–616.

23 Irie, M. and Kato, M. (1985) Photoresponsive molecular tweezers. Photoregulated ion capture and release using thioindigo derivatives having ethylenedioxy side groups. *J. Am. Chem. Soc.*, **107**, 1024–1028.

24 Takeshita, M., Uchida, K., and Irie, M. (1996) Novel saccharide tweezers with a diarylethene photoswitch. *Chem. Commun.*, 1807–1808.

25 Iordache, A., Retegan, M., Thomas, F. *et al.* (2012) Redox-responsive porphyrin-based molecular tweezers. *Chem. Eur. J.*, **18**, 7648–7653.

26 Pochorovski, I. and Diederich, F. (2014) Development of redox-switchable resorcin[4]arene cavitands. *Acc. Chem. Res.*, **47**, 2096–2105.

27 Gao, C., Silvi, S., Ma, X. *et al.* (2012) Chiral supramolecular switches based on (*R*)-binaphthalene–bipyridinium guests and cucurbituril hosts. *Chem. Eur. J.*, **18**, 16911–16921.

28 Polavarapu, P., Melander, H., Langer, V. *et al.* (2008) Modulation and binding properties of extended glycoluril molecular clips. *New J. Chem.*, **32**, 643–651.

29 Petitjean, A., Khoury, R.G., Kyritsakas, N., and Lehn, J.-M. (2004) Dynamic devices. Shape switching and substrate binding in ion-controlled nanomechanical molecular tweezers. *J. Am. Chem. Soc.*, **126**, 6637–6647.

30 Parac, M., Etinski, M., Peric, M., and Grimme, S. (2005) A theoretical investigation of the geometries and binding energies of molecular tweezer and clip host–guest systems. *J. Chem. Theory Comput.*, **1**, 1110–1118.

31 Hermida-Ramón, J.M. and Estévez, C.M. (2007) Towards the design of neutral molecular tweezers for anion recognition. *Chem. Eur. J.*, **13**, 4743–4749.

32 Huang, H. and Drueckhammer, D.G. (2006) A modular molecular tweezer designed using CAVEAT. *Chem. Commun.*, 2995–2997.

33 Lee, S.K., Shin, S.Y., Lee, S. *et al.* (2001) Conformation and binding properties of polymethylene-linked bisviologens–2-naphthol complexes. *J. Chem. Soc. Perkin Trans.*, **2**, 1983–1988.

34 Phillips, M.D., Fyles, T.M., Barwell, N.P., and James, T.D. (2009) Carbohydrate sensing using a fluorescent molecular tweezer. *Chem. Commun.*, 6557–6559.

35 Užarević, K., Halasz, I., Đilović, I. *et al.* (2013) Dynamic molecular recognition in solid state for separating mixtures of isomeric dicarboxylic acids. *Angew. Chem. Int. Ed.*, **52**, 5504–5508.

36 Wallentin, C.-J., Wixe, T., Wendt, O.F. *et al.* (2010) Synthesis and self-aggregation of enantiopure and racemic molecular tweezers based on the bicyclo[3.3.1]nonane framework. *Chem. Eur. J.*, **16**, 3994–4002.

37 Lohr, A., Grüne, M., and Würthner, F. (2009) Self-assembly of bis(merocyanine) tweezers into discrete bimolecular π-stacks. *Chem. Eur. J.*, **15**, 3691–3705.

38 Hisamatsu, Y. and Aihara, H. (2010) Diverse dimerization of molecular tweezers with a 2,4,6-triphenyl-1,3,5-triazine spacer in the solid state. *Chem. Commun.*, **46**, 4902–4904.

39 Haino, T., Fujii, T., and Fukazawa, Y. (2006) Guest binding and new self-assembly of bisporphyrins. *J. Org. Chem.*, **71**, 2572–2580.

40 Peng, X.-X., Lu, H.-Y., Han, T., and Chen, C.-F. (2007) Synthesis of a novel triptycene-based molecular tweezer and its complexation with paraquat derivatives. *Org. Lett.*, **9**, 895–898.

41 Colquhoun, H.M., Zhu, Z., and Williams, D.J. (2003) Extreme complementarity in a macrocycle–tweezer complex. *Org. Lett.*, **5**, 4353–4356.

42 Kim, S.K., Lim, J.M., Pradhan, T. *et al.* (2014) Self-association and nitroaromatic-induced deaggregation of pyrene substituted pyridine amides. *J. Am. Chem. Soc.*, **136**, 495–505.

43 Krebs, F.C. and Jørgensen, M. (2001) A new versatile receptor platform. *J. Org. Chem.*, **66**, 6169–6173.

44 Lämsä, M., Raitamaa, K., and Pursiainen, J. (1999) Synthesis of podands bearing aromatic end groups and complex formation with tropylium tetrafluoroborate in 1,2-dichloroethane. *J. Phys. Org. Chem.*, **12**, 557–563.

45 Lämsä, M., Kiviniemi, S., Kettukangas, E.-R. *et al.* (2001) Synthesis and characterization of new aromatic tweezers and complex formation with tropylium ion in 1,2-dichloroethane. *J. Phys. Org. Chem.*, **14**, 551–558.

46 Legouin, B., Uriac, P., Tomasi, S. *et al.* (2009) Novel chiral molecular tweezer from (+)-usnic acid. *Org. Lett.*, **11**, 745–748.

47 Legouin, B., Gayral, M., Uriac, P. *et al.* (2010) Molecular tweezers: synthesis and formation of host–guest complexes. *Eur. J. Org. Chem.*, **2010**, 5503–5508.

48 Jaunet-Lahary, T., Jacquemin, D., Legouin, B. *et al.* (2015) Dissymmetric molecular tweezers in host–guest complexes: internal or external complexation? *J. Phys. Chem. C*, **119**, 3771–3779.

49 Otsuki, J., Oya, T., Lee, S.-H., and Araki, K. (1995) Design of organic superstructure based on a preorganised donor–acceptor unit. *J. Chem. Soc. Chem. Commun.*, 2193–2194.

50 Lee, S.-H., Imamura, K., Otsuki, J. *et al.* (1996) Characterisation of "sandwich" charge-transfer complexes composed of bis-anthracene donors and aromatic acceptors. *J. Chem. Soc. Perkin Trans.*, **2**, 847–852.

51 Maitra, U. and D'Souza, L.J. (1994) Bile acid based semi-rigid molecular tweezers. *J. Chem. Soc. Chem. Commun.*, 2793–2795.

52 D'Souza, L.J. and Maitra, U. (1996) Design, synthesis, and evaluation of bile acid-based molecular tweezers. *J. Org. Chem.*, **61**, 9494–9502.

53 Potluri, V.K. and Maitra, U. (2000) Bile acid-derived molecular tweezers: study of solvent effects in binding, and determination of thermodynamic parameters by an extraction-based protocol. *J. Org. Chem.*, **65**, 7764–7769.

54 Lima, C.F., Fernandes, A.M., Melo, A. *et al.* (2015) Diarylferrocene tweezers for cation binding. *Phys. Chem. Chem. Phys.*, **17**, 23917–23923.

55 Grimm, B., Santos, J., Illescas, B.M. *et al.* (2010) A new exTTF-crown ether platform to associate fullerenes: cooperative n–π and π–π effects. *J. Am. Chem. Soc.*, **132**, 17387–17389.

56 Biet, T. and Avarvari, N. (2014) Electroactive tetrathiafulvalene based pyridine-mono and -bis(1,2,3-triazoles) click ligands: synthesis, crystal structures and coordination chemistry. *CrstEngComm*, **16**, 6612–6620.

57 Mulla, K., Shaik, H., Thompson, D.W., and Zhao, Y. (2013) TTFV-based molecular tweezers and macrocycles as receptors for fullerenes. *Org. Lett.*, **15**, 4532–4535.

58 Ulrich, S. and Lehn, J.-M. (2009) Adaptation and optical signal generation in a constitutional dynamic network. *Chem. Eur. J.*, **15**, 5640–5645.

59 Ulrich, S., Petitjean, A., and Lehn, J.-M. (2010) Metallo-controlled dynamic molecular tweezers: design, synthesis, and self-assembly by metal-ion coordination. *Eur. J. Inorg. Chem.*, **2010**, 1913–1928.

60 Plante, J.P. and Glass, T.E. (2006) Shape-selective fluorescent sensing ensemble using a tweezer-type metalloreceptor. *Org. Lett.*, **8**, 2163–2166.

61 Doistau, B., Tron, A., Denisov, S.A. *et al.* (2014) Terpy(Pt–salphen)$_2$ switchable luminescent molecular tweezers. *Chem. Eur. J.*, **20**, 15799–15807.

62 Doistau, B., Rossi-Gendron, C., Tron, A. *et al.* (2015) Switchable platinum-based tweezers with Pt–Pt bonding and selective luminescence quenching. *Dalton Trans.*, **44**, 8543–8551.

63 Doistau, B., Cantin, J.-L., Chamoreau, L.-M. *et al.* (2015) Mechanical switching of magnetic interaction by tweezers-type complex. *Chem. Commun.*, **51**, 12916–12919.

64 Tanaka, Y., Wong, K.M.-C., and Yam, V.W.-W. (2012) Phosphorescent molecular tweezers based on alkynylplatinum(II) terpyridine system: turning on of NIR emission *via* heterologous Pt···M interactions (M = Pt^{II}, Pd^{II}, Au^{III} and Au^{I}). *Chem. Sci.*, **3**, 1185–1191.

65 Tanaka, Y., Wong, K.M.-C., and Yam, V.W.-W. (2013) Host–guest interactions of phosphorescent molecular tweezers based on an alkynylplatinum(II) terpyridine system with polyaromatic hydrocarbons. *Chem. Eur. J.*, **19**, 390–399.

66 Tian, Y.-K., Shi, Y.-G., Yang, Z.-S., and Wang, F. (2014) Responsive supramolecular polymers based on the bis[alkynylplatinum(II)] terpyridine molecular tweezer/arene recognition motif. *Angew. Chem. Int. Ed.*, **53**, 6090–6094.

67 Tanaka, Y., Wong, K.M.-C., and Yam, V.W.-W. (2013) Platinum-based phosphorescent double-decker tweezers: a strategy for extended heterologous metal–metal interactions. *Angew. Chem.*, **125**, 14367–14370.

68 Alvarez, C.M., García-Escudero, L.A., Garcia-Rodriguez, R. *et al.* (2014) Enhanced association for C_{70} over C_{60} with a metal complex with corannulene derivate ligands. *Dalton Trans.*, **43**, 15693–15696.

69 Valderrey, V., Aragay, G., and Ballester, P. (2014) Porphyrin tweezer receptors: binding studies, conformational properties and applications. *Coord. Chem. Rev.*, **258–259**, 137–156.

70 Borovkov, V.V., Lintuluoto, J.M., Sugiura, M. *et al.* (2002) Remarkable stability and enhanced optical activity of a chiral supramolecular bis-porphyrin tweezer in both solution and solid state. *J. Am. Chem. Soc.*, **124**, 11282–11283.

71 Li, X., Burrell, C.E., Staples, R.J., and Borhan, B. (2012) Absolute configuration for 1,*n*-glycols: a nonempirical approach to long-range stereochemical determination. *J. Am. Chem. Soc.*, **134**, 9026–9029.

72 Borovkov, V.V., Fujii, I., Muranaka, A. *et al.* (2004) Rationalization of supramolecular chirality in a bisporphyrin system. *Angew. Chem. Int. Ed.*, **43**, 5481–5485.

73 Brahma, S., Asif Ikbal, S., and Rath, S.P. (2011) *Syn–anti* conformational switching: synthesis and X-ray structures of tweezer and *anti* form in a zinc porphyrin dimer induced by axial ligands. *Inorg. Chim. Acta*, **372**, 62–70.

74 Chaudhary, A. and Rath, S.P. (2011) Efficient complexation of pyrrole-bridged dizinc(II) bisporphyrin with fluorescent probe pyrene: synthesis, structure, and photoinduced singlet–singlet energy transfer. *Chem. Eur. J.*, **17**, 11478–11487.

75 Murphy, R.B., Pham, D.-T., Lincoln, S.F., and Johnston, M.R. (2013) Molecular tweezers with freely rotating linker and porphyrin moieties. *Eur. J. Org. Chem.*, **2013**, 2985–2993.

76 Mukherjee, S., Banerjee, S., Bauri, A.K., and Bhattacharya, S. (2011) Synthesis, photophysical investigations and molecular structure of the supramolecular complexes of a newly designed diporphyrin receptor with fullerenes C_{60} and C_{70} in solution. *J. Mol. Struct.*, **1004**, 13–25.

77 Dudič, M., Lhoták, P., Stibor, I. *et al.* (2004) (Thia)calix[4]arene–porphyrin conjugates: novel receptors for fullerene complexation with C_{70} over C_{60} selectivity. *New J. Chem.*, **28**, 85–90.

78 Lee, C.-H., Yoon, H., and Jang, W.-D. (2009) Biindole-bridged porphyrin dimer as allosteric molecular tweezers. *Chem. Eur. J.*, **15**, 9972–9976.

79 Gomila, R.M., Quiñonero, D., Rotger, C. *et al.* (2002) Predicting experimental complexation-induced changes in ^{1}H NMR chemical shift for complexes between zinc-porphyrins and amines using the ab initio/GIAO-HF methodology. *Org. Lett.*, **4**, 399–401.

80 Liu, H., Wu, J., Xu, Y.-X. *et al.* (2007) Complexation of hydrogen bonding-driven preorganized di- and hexacationic bisporphyrin receptors for C_{60} in aqueous and DMSO media. *Tetrahedron Lett.*, **48**, 7327–7331.

81 Wu, Z.-Q., Shao, X.-B., Li, C. *et al.* (2005) Hydrogen-bonding-driven preorganized zinc porphyrin receptors for efficient complexation of C_{60}, C_{70}, and C_{60} derivatives. *J. Am. Chem. Soc.*, **127**, 17460–17468.
82 Li, C., Zhao, X., Gao, X. *et al.* (2013) Foldamer-derived preorganized bi- and tri-zinc porphyrin tweezers for a pentafluorobenzene-bearing pyridine guest: the binding pattern study. *Chin. J. Chem.*, **31**, 582–588.
83 Suzaki, Y., Tsuchido, Y., and Osakada, K. (2011) Tweezers-like aromatic molecules and their luminescent properties depending on the structures. *Tetrahedron Lett.*, **52**, 3883–3885.
84 Tsuchido, Y., Suzaki, Y., Ide, T., and Osakada, K. (2014) Dynamic properties of molecular tweezers with a bis(2-hydroxyphenyl)pyrimidine backbone. *Chem. Eur. J.*, **20**, 4762–4771.
85 Zimmerman, S.C. and VanZyl, C.M. (1987) Rigid molecular tweezers: synthesis, characterization, and complexation chemistry of a diacridine. *J. Am. Chem. Soc.*, **109**, 7894–7896.
86 Zimmerman, S.C., Mrksich, M., and Baloga, M. (1989) Highly efficient complexation of a π-acceptor by a molecular tweezer containing two π-donors: the role of preorganization. *J. Am. Chem. Soc.*, **111**, 8528–8530.
87 Zimmerman, S.C., VanZyl, C.M., and Hamilton, G.S. (1989) Rigid molecular tweezers: preorganized hosts for electron donor–acceptor complexation in organic solvents. *J. Am. Chem. Soc.*, **111**, 1373–1381.
88 Meyer, E.A., Castellano, R.K., and Diederich, F. (2003) Interactions with aromatic rings in chemical and biological recognition. *Angew. Chem. Int. Ed.*, **42**, 1210–1250.
89 Pace, C.J. and Gao, J. (2013) Exploring and exploiting polar–π interactions with fluorinated aromatic amino acids. *Acc. Chem. Res.*, **46**, 907–915.
90 Korenaga, T., Kawauchi, Y., Kosaki, T. *et al.* (2005) Synthesis of a molecular tweezer containing pentafluorophenyl groups and investigation of the π–π stacking interaction for a pentafluorophenyl group in a polar organic solvent. *Bull. Chem. Soc. Jpn.*, **78**, 2175–2179.
91 Korenaga, T., Kosaki, T., Kawauchi, Y. *et al.* (2006) Synthesis of molecular tweezers bearing pentafluorophenyl and several pendant aryl groups: effective application of the modified Suzuki–Miyaura coupling reaction. *J. Fluorine Chem.*, **127**, 604–609.
92 Etzkorn, M., Timmerman, J.C., Brooker, M.D. *et al.* (2010) Preparation, structures and preliminary host–guest studies of fluorinated *syn*-bis-quinoxaline molecular tweezers. *Beilstein J. Org. Chem.*, **6**.
93 Sommer, R.D., Rheingold, A.L., Goshe, A.J., and Bosnich, B. (2001) Supramolecular chemistry: molecular recognition and self-assembly using rigid spacer-chelators bearing cofacial terpyridyl palladium(II) complexes separated by 7 Å. *J. Am. Chem. Soc.*, **123**, 3940–3952.
94 Goshe, A.J., Steele, I.M., and Bosnich, B. (2003) Supramolecular recognition. Terpyridyl palladium and platinum molecular clefts and their

association with planar platinum complexes. *J. Am. Chem. Soc.*, **125**, 444–451.

95 Crowley, J.D., Goshe, A.J., and Bosnich, B. (2003) Molecular recognition. Electrostatic effects in supramolecular self-assembly. *Chem. Commun.*, 392–393.

96 Zimmerman, S.C. and Wu, W. (1989) A rigid molecular tweezers with an active site carboxylic acid: exceptionally efficient receptor for adenine in an organic solvent. *J. Am. Chem. Soc.*, **111**, 8054–8055.

97 Zimmerman, S.C. and Zeng, Z. (1990) Improved binding of adenine by a synthetic receptor. *J. Org. Chem.*, **55**, 4789–4791.

98 Zimmerman, S.C., Wu, W., and Zeng, Z. (1991) Complexation of nucleotide bases by molecular tweezers with active site carboxylic acids: effects of microenvironment. *J. Am. Chem. Soc.*, **113**, 196–201.

99 Kim, D., Lee, S., Gao, G. *et al.* (2010) A molecular-clip-based approach to cofacial zinc–porphyrin complexes. *J. Organomet. Chem.*, **695**, 111–119.

100 Nemoto, H., Kawano, T., Ueji, N. *et al.* (2000) A conformationally reliable spacer for molecular tweezers. *Org. Lett.*, **2**, 1015–1017.

101 Kurebayashi, H., Sakaguchi, M., Okajima, T. *et al.* (1999) Molecular tweezers based on dioxa[2.2]orthocyclophane skeleton. *Tetrahedron Lett.*, **40**, 5545–5548.

102 Kurebayashi, H., Haino, T., and Fukazawa, Y. (2000) Layered donor–acceptor–acceptor arrangement in complex of molecular tweezer based on dioxa[2.2]orthocyclophane. *Tetrahedron Lett.*, **41**, 477–480.

103 Kurebayashi, H., Haino, T., Usui, S., and Fukazawa, Y. (2001) Structure of supramolecular complex of flexible molecular tweezers and planar guest in solution. *Tetrahedron*, **57**, 8667–8674.

104 Nishiuchi, T., Kuwatani, Y., Nishinaga, T., and Iyoda, M. (2009) Dynamic molecular tweezers composed of dibenzocyclooctatetraene units: synthesis, properties, and thermochromism in host–guest complexes. *Chem. Eur. J.*, **15**, 6838–6847.

105 Sygula, A., Fronczek, F.R., Sygula, R. *et al.* (2007) A double concave hydrocarbon buckycatcher. *J. Am. Chem. Soc.*, **129**, 3842–3843.

106 Le, V.H., Yanney, M., McGuire, M. *et al.* (2014) Thermodynamics of host–guest interactions between fullerenes and a buckycatcher. *J. Phys. Chem. B*, **118**, 11956–11964.

107 Sygula, A., Yanney, M., Henry, W.P. *et al.* (2014) Inclusion complexes and solvates of buckycatcher, a versatile molecular host with two corannulene pincers. *Cryst. Growth Des.*, **14**, 2633–2639.

108 Katz, J.L., Geller, B.J., and Foster, P.D. (2007) Oxacalixarenes and oxacyclophanes containing 1,8-naphthyridines: a new class of molecular tweezers with concave-surface functionality. *Chem. Commun.*, 1026–1028.

109 Kleyn, A., Jacobs, T., and Barbour, L.J. (2011) Solid-state structural studies of oxacalix[2]arene[2]naphthalene as a molecular tweezer. *CrstEngComm*, **13**, 3175–3180.

110 Rowan, A.E., Elemans, J.A.A.W., and Nolte, R.J.M. (1999) Molecular and supramolecular objects from glycoluril. *Acc. Chem. Res.*, **32**, 995–1006.

111 Isaacs, L. (2014) Stimuli responsive systems constructed using cucurbit[*n*]uril-type molecular containers. *Acc. Chem. Res.*, **47**, 2052–2062.

112 Smeets, J.W.H., Sijbesma, R.P., Niele, F.G.M. *et al.* (1987) Novel concave building block for the synthesis of organic hosts. *J. Am. Chem. Soc.*, **109**, 928–929.

113 Sijbesma, R.P., Kentgens, A.P.M., and Nolte, R.J.M. (1991) Binding of dihydroxybenzenes in a synthetic molecular clip. Effect of hydrogen bonding and π-stacking. *J. Org. Chem.*, **56**, 3199–3201.

114 Sijbesma, R.P., Kentgens, A.P.M., Lutz, E.T.G. *et al.* (1993) Binding features of molecular clips derived from diphenylglycoluril. *J. Am. Chem. Soc.*, **115**, 8999–9005.

115 Wang, Z.-G., Wang, Z.-H., and Teng, L. (2008) A facile one-pot synthesis and the crystal structure of a molecular clip derived from diphenylglycoluril. *Monatsh. Chem. Chem. Mon.*, **139**, 949–952.

116 Murray, B.A. and Whelan, G.S. (1996) Molecular recognition of dihydroxyaromatics with bis-*o*-xylyleneglycoluril hosts. *Pure Appl. Chem.*, **68**, 1561–1567.

117 Reek, J.N.H., Priem, A.H., Engelkamp, H. *et al.* (1997) Binding features of molecular clips. Separation of the effects of hydrogen bonding and π–π interactions. *J. Am. Chem. Soc.*, **119**, 9956–9964.

118 Sijbesma, R.P., Bosnian, W.P., and Nolte, R.J.M. (1991) Structurally related hosts with remarkably different binding features. *J. Chem. Soc. Chem. Commun.*, 885–886.

119 Sijbesma, R.P., Wijmenga, S.S., and Nolte, R.J.M. (1992) A molecular clip that binds aromatic guests by an induced-fit mechanism. *J. Am. Chem. Soc.*, **114**, 9807–9813.

120 Reek, J.N.H., Elemans, J.A.A.W., and Nolte, R.J.M. (1997) Synthesis, conformational analysis, and binding properties of molecular clips with two different side walls. *J. Org. Chem.*, **62**, 2234–2243.

121 Elemans, J.A., Rowan, A.E., Nolte, R.J., and de Gelder, R. (1998) Bipyridine functionalized molecular clips. Self-assembly of their ruthenium complexes in water. *Chem. Commun.*, 1553–1554.

122 Cheng, P.-N., Chiang, P.-T., and Chiu, S.-H. (2005) A switchable macrocycle–clip complex that functions as a NOR logic gate. *Chem. Commun.*, 1285–1287.

123 Chiang, P.-T., Cheng, P.-N., Lin, C.-F. *et al.* (2006) A macrocycle/molecular-clip complex that functions as a quadruply controllable molecular switch. *Chem. Eur. J.*, **12**, 865–876.

124 Cotelle, Y., Allain, M., Legoupy, S., and Hudhomme, P. (2014) Fused glycoluril-tetrathiafulvalene molecular clips as receptors for neutral electron acceptor guests. *Org. Lett.*, **16**, 2590–2593.

125 Hardouin-Lerouge, M., Cotelle, Y., Legoupy, S., and Hudhomme, P. (2014) Synthesis of glycoluril-tetrathiafulvalene molecular clips for electron-deficient neutral guests through a straightforward Diels–Alder strategy. *New J. Chem.*, **38**, 5341–5348.

126 Cotelle, Y., Hardouin-Lerouge, M., Legoupy, S. *et al.* (2015) Glycoluril–tetrathiafulvalene molecular clips: on the influence of electronic and spatial properties for binding neutral accepting guests. *Beilstein J. Org. Chem.*, **11**, 1023–1036.

127 Düker, M.H., Gómez, R., Vande Velde, C.M.L., and Azov, V.A. (2011) Upper rim tetrathiafulvalene-bridged calix[4]arenes. *Tetrahedron Lett.*, **52**, 2881–2884.

128 Düker, M.H., Schäfer, H., Zeller, M., and Azov, V.A. (2013) Rationally designed calix[4]arene–pyrrolotetrathiafulvalene receptors for electron-deficient neutral guests. *J. Org. Chem.*, **78**, 4905–4912.

129 Korsching, K.R., Schäfer, H., Schönborn, J. *et al.* (2015) Enhancement of electron-donating character in alkylated monopyrrolo-tetrathiafulvalene derivatives. *RSC Adv.*, **5**, 82633–82637.

130 Norrehed, S., Polavarapu, P., Yang, W. *et al.* (2013) Conformational restriction of flexible molecules in solution by a semirigid bis-porphyrin molecular tweezer. *Tetrahedron*, **69**, 7131–7138.

131 Norrehed, S., Johansson, H., Grennberg, H., and Gogoll, A. (2013) Improved stereochemical analysis of conformationally flexible diamines by binding to a bisporphyrin molecular clip. *Chem. Eur. J.*, **19**, 14631–14638.

132 Reek, J.N.H., Rowan, A.E., Crossley, M.J., and Nolte, R.J.M. (1999) Synthesis and photophysical properties of porphyrin-functionalized molecular clips. *J. Org. Chem.*, **64**, 6653–6663.

133 Wang, Z.-G., Zhou, B.-H., Chen, Y.-F. *et al.* (2006) Substituent effects control the self-association of molecular clips in the crystalline state. *J. Org. Chem.*, **71**, 4502–4508.

134 Liu, Q., Wang, J., and Zhao, G. (2012) A new glycoluril clip and its self-association in solid state. *J. Chem. Crystallogr.*, **42**, 727–732.

135 Yin, G., Wang, Z., Chen, Y. *et al.* (2006) Synthesis of novel molecular clips based on diethoxycarbonyl glycoluril. *Synlett*, **1**, 0049–0052.

136 Li, Y., Yin, G., Guo, H. *et al.* (2006) One-pot synthesis of novel molecular clips and macrocyclic polyamines derived from bis(ethoxycarbonyl)glycoluril. *Synthesis*, **2006**, 2897–2902.

137 Jansen, R.J., de Gelder, R., Rowan, A.E. *et al.* (2001) Molecular clips based on propanediurea. Exceptionally high binding affinities for resorcinol guests. *J. Org. Chem.*, **66**, 2643–2653.

138 Hu, S.-L., She, N.-F., Yin, G.-D. *et al.* (2007) Synthesis, structural characterization, and fluorescent chemosensory properties of novel molecular clips based on diethoxycarbonyl glycoluril. *Tetrahedron Lett.*, **48**, 1591–1594.

139 Wu, A., Chakraborty, A., Witt, D. *et al.* (2002) Methylene-bridged glycoluril dimers: synthetic methods. *J. Org. Chem.*, **67**, 5817–5830.

140 Zhou, B.H., Cao, L.P., Yin, G.D. *et al.* (2007) X-ray structures and binding properties of molecular clips based on diethoxycarbonyl glycoluril. *Can. J. Chem.*, **85**, 586–591.

141 Zhou, B., Xu, B., Ma, L. *et al.* (2014) Synthesis, crystal structure of the receptor derived from diethoxycarbonyl glycoluril for aromatic amide guest. *Chin. J. Chem.*, **32**, 1111–1115.

142 Stancl, M., Hodan, M., and Sindelar, V. (2009) Glycoluril trimers: selective synthesis and supramolecular properties. *Org. Lett.*, **11**, 4184–4187.

143 Ma, D., Zavalij, P.Y., and Isaacs, L. (2010) Acyclic cucurbit[*n*]uril congeners are high affinity hosts. *J. Org. Chem.*, **75**, 4786–4795.

144 Klärner, F.-G. and Kuchenbrandt, M.C. (2008) Synthesis of molecular tweezers and clips by the use of a molecular lego set and their supramolecular functions, in *Strategies and Tactics in Organic Synthesis*, vol. 7 (ed. M. Harmata), Academic Press, pp. 99–153.

145 Klärner, F.-G., Panitzky, J., Bläser, D., and Boese, R. (2001) Synthesis and supramolecular structures of molecular clips. *Tetrahedron*, **57**, 3673–3687.

146 Klärner, F.-G., Kahlert, B., Boese, R. *et al.* (2005) Synthesis and supramolecular properties of molecular clips with anthracene sidewalls. *Chem. Eur. J.*, **11**, 3363–3374.

147 Marchioni, F., Juris, A., Lobert, M. *et al.* (2005) Luminescent host–guest complexes involving molecular clips and tweezers and tetracyanobenzene. *New J. Chem.*, **29**, 780–784.

148 Branchi, B., Balzani, V., Ceroni, P. *et al.* (2008) Molecular clips with extended aromatic sidewalls as receptors for electron-acceptor molecules. Synthesis and NMR, photophysical, and electrochemical properties. *J. Org. Chem.*, **73**, 5839–5851.

149 Klärner, F.-G., Lobert, M., Naatz, U. *et al.* (2003) Synthesis and supramolecular properties of trimethylene-bridged clips. *Chem. Eur. J.*, **9**, 5036–5047.

150 Klärner, F.-G., Burkert, U., Kamieth, M. *et al.* (1999) Molecular tweezers as synthetic receptors: molecular recognition of electron-deficient aromatic and aliphatic substrates. *Chem. Eur. J.*, **5**, 1700–1707.

151 Lobert, M., Bandmann, H., Burkert, U. *et al.* (2006) Dynamics in host–guest complexes of molecular tweezers and clips. *Chem. Eur. J.*, **12**, 1629–1641.

152 Dutt, S., Wilch, C., Gersthagen, T. *et al.* (2013) Molecular tweezers with varying anions: a comparative study. *J. Org. Chem.*, **78**, 6721–6734.

153 Jasper, C., Schrader, T., Panitzky, J., and Klärner, F.-G. (2002) Selective complexation of *N*-alkylpyridinium salts: recognition of NAD^+ in water. *Angew. Chem. Int. Ed.*, **41**, 1355–1358.
154 Fokkens, M., Jasper, C., Schrader, T. *et al.* (2005) Selective complexation of *N*-alkylpyridinium salts: binding of NAD^+ in water. *Chem. Eur. J.*, **11**, 477–494.
155 Fokkens, M., Klärner, F.-G., and Schrader, T. (2006) Nucleoside sensing. *Sensors*, **6**, 860–873.
156 Klärner, F.-G., Kahlert, B., Nellesen, A. *et al.* (2006) Molecular tweezer and clip in aqueous solution: unexpected self-assembly, powerful host–guest complex formation, quantum chemical ^1H NMR shift calculation. *J. Am. Chem. Soc.*, **128**, 4831–4841.
157 Fokkens, M., Schrader, T., and Klärner, F.-G. (2005) A molecular tweezer for lysine and arginine. *J. Am. Chem. Soc.*, **127**, 14415–14421.
158 Chou, T.-C., Lin, K.-C., and Wu, C.-A. (2009) Quinoxaline-based U-shaped septuple-bridged [7,7]orthocyclophanes: synthesis, solid-state structure, and self-assembly. *Tetrahedron*, **65**, 10243–10257.
159 Yanney, M., Fronczek, F.R., and Sygula, A. (2015) A 2:1 receptor/C_{60} complex as a nanosized universal joint. *Angew. Chem. Int. Ed.*, **54**, 11153–11156.
160 Abeyratne Kuragama, P.L., Fronczek, F.R., and Sygula, A. (2015) Bis-corannulene receptors for fullerenes based on Klärner's tethers: reaching the affinity limits. *Org. Lett.*, **17**, 5292–5295.
161 Harmata, M. and Barnes, C.L. (1990) Molecular clefts 2. An analogue of Kagan's ether as a molecular cleft: synthesis and clathrate formation with ethyl acetate. *Tetrahedron Lett*, **31**, 1825–1828.
162 Harmata, M. and Barnes, C.L. (1990) Molecular clefts. 3. The crystal structure of a chiral molecular tweezer and its guest. *J. Am. Chem. Soc.*, **112**, 5655–5657.
163 Harmata, M., Kahraman, M., Tyagarajan, S. *et al.* (1998) Molecular clefts derived from Kagan's ether. Synthesis and solid state inclusion complexes of a chiral molecular tweezer, in *Molecular Recognition and Inclusion* (ed. A.W. Coleman), Springer, The Netherlands, pp. 109–116.
164 Dolenský, B., Havlík, M., and Král, V. (2012) Oligo Tröger's bases—new molecular scaffolds. *Chem. Soc. Rev.*, **41**, 3839–3858.
165 Mas, T., Pardo, C., Salort, F. *et al.* (2004) A new entry to bis-Tröger's bases. *Eur. J. Org. Chem.*, **2004**, 1097–1104.
166 Masa, T., Pardo, C., and Elguero, J. (2004) A shorter synthesis of symmetrical 2,11-dimethyl-bis-Tröger's bases. A new molecular tweezer. *Arkivoc*, **4**, 86–93.
167 Havlík, M., Král, V., Kaplánek, R., and Dolenský, B. (2008) One-pot reaction as an efficient method for rigid molecular tweezers. *Org. Lett.*, **10**, 4767–4769.

168 Parchaňský, V., Matějka, P., Dolenský, B. *et al.* (2009) Modeling of a Tröger's tweezer and its complexation properties. *J. Mol. Struct.*, **934**, 117–122.

169 Pardo, C., Sesmilo, E., Gutiérrez-Puebla, E. *et al.* (2001) New chiral molecular tweezers with a bis-Tröger's base skeleton. *J. Org. Chem.*, **66**, 1607–1611.

170 Mas, T., Pardo, C., and Elguero, J. (2004) Synthesis of new chiral molecular tweezers with a tris-Tröger's base skeleton. *Mendeleev Commun.*, **14**, 235–237.

171 Dolenský, B., Valík, M., Sýkora, D., and Král, V. (2005) Synthetic routes to linear oligo-Tröger's bases. *Org. Lett.*, **7**, 67–70.

172 Dolenský, B., Kessler, J., Jakubek, M. *et al.* (2013) Synthesis and characterisation of a new naphthalene tris-Tröger's base derivative—a chiral molecular clip. *Tetrahedron Lett.*, **54**, 308–311.

173 Wilcox, C.S., Greer, L.M., and Lynch, V. (1987) Synthesis of chiral molecular clefts. New armatures for biomimetic systems. *J. Am. Chem. Soc.*, **109**, 1865–1867.

174 Li, P.-F., Han, Y., and Chen, C.-F. (2015) Synthesis and structures of triptycene-derived Tröger's base molecular clips. *Chin. Chem. Lett.*, **26**, 839–842.

175 Wixe, T., Wallentin, C.-J., Johnson, M.T. *et al.* (2013) Synthesis of an orthogonal topological analogue of helicene. *Chem. Eur. J.*, **19**, 14963–14969.

176 Fuchter, M.J., Schaefer, J., Judge, D.K. *et al.* (2012) [7]-Helicene: a chiral molecular tweezer for silver(I) salts. *Dalton Trans*, **41**, 8238–8241.

177 DiPalma, J.R., McMichael, R., and DiPalma, M. (1966) Touch receptor of Venus flytrap, *Dionaea muscipula*. *Science*, **152**, 539–540.

178 Bessis, A.-S., Rondard, P., Gaven, F. *et al.* (2002) Closure of the Venus flytrap module of mGlu8 receptor and the activation process: insights from mutations converting antagonists into agonists. *Proc. Natl. Acad. Sci. U.S.A.*, **99**, 11097–11102.

179 Mun, H.-C., Franks, A.H., Culverston, E.L. *et al.* (2004) The Venus fly trap domain of the extracellular Ca^{2+}-sensing receptor is required for L-amino acid sensing. *J. Biol. Chem.*, **279**, 51739–51744.

180 Mano, I., Lamed, Y., and Teichberg, V.I. (1996) A Venus flytrap mechanism for activation and desensitization of α-amino-3-hydroxy-5-methyl-4-isoxazole propionic acid receptors. *J. Biol. Chem.*, **271**, 15299–15302.

181 Hennrich, G. and Anslyn, E.V. (2002) 1,3,5-2,4,6-Functionalized, facially segregated benzenes—exploitation of sterically predisposed systems in supramolecular chemistry. *Chem. Eur. J.*, **8**, 2218–2224.

182 Schmuck, C. and Schwegmann, M. (2005) A molecular flytrap for the selective binding of citrate and other tricarboxylates in water. *J. Am. Chem. Soc.*, **127**, 3373–3379.

183 Schmuck, C. and Schwegmann, M. (2006) A naked-eye sensing ensemble for the selective detection of citrate—but not tartrate or malate—in water based on a tris-cationic receptor. *Org. Biomol. Chem.*, **4**, 836–838.

184 Ghosh, K., Sarkar, A.R., Sarkar, T. *et al.* (2014) Progress of 3-aminopyridinium-based synthetic receptors in anion recognition. *RSC Adv.*, **4**, 20114–20130.

185 Turner, D.R., Paterson, M.J., and Steed, J.W. (2006) A conformationally flexible, urea-based tripodal anion receptor: solid-state, solution, and theoretical studies. *J. Org. Chem.*, **71**, 1598–1608.

186 Abouderbala, L.O., Belcher, W.J., Boutelle, M.G. *et al.* (2002) Cooperative anion binding and electrochemical sensing by modular podands. *Proc. Natl. Acad. Sci. U.S.A.*, **99**, 5001–5006.

187 Watat, M.-L.L., Dülcks, T., Kemken, D., and Azov, V.A. (2014) Tripodal pyrrolotetrathiafulvalene receptors for recognition of electron-deficient molecular guests. *Tetrahedron Lett.*, **55**, 741–744.

188 Ni, X.-L., Jin, C.-C., Jiang, X.-K. *et al.* (2013) Tri-substituted hexahomotrioxacalix[3]arene derivatives bearing imidazole units: synthesis and extraction properties for cations and chromate anions. *Org. Biomol. Chem.*, **11**, 5435–5442.

189 Yanney, M. and Sygula, A. (2013) Tridental molecular clip with corannulene pincers: is three better than two? *Tetrahedron Lett.*, **54**, 2604–2607.

190 Habata, Y., Ikeda, M., Yamada, S. *et al.* (2012) Argentivorous molecules: structural evidence for Ag^+–π interactions in solution. *Org. Lett.*, **14**, 4576–4579.

191 Pochorovski, I., Ebert, M.-O., Gisselbrecht, J.-P. *et al.* (2012) Redox-switchable resorcin[4]arene cavitands: molecular grippers. *J. Am. Chem. Soc.*, **134**, 14702–14705.

192 Park, J.S., Bejger, C., Larsen, K.R. *et al.* (2012) Synthesis and recognition properties of higher order tetrathiafulvalene (TTF) calix[*n*]pyrroles ($n = 4–6$). *Chem. Sci.*, **3**, 2685–2689.

193 Davis, C.M., Lim, J.M., Larsen, K.R. *et al.* (2014) Ion-regulated allosteric binding of fullerenes (C_{60} and C_{70}) by tetrathiafulvalene-calix[4]pyrroles. *J. Am. Chem. Soc.*, **136**, 10410–10417.

194 Fukuzumi, S., Ohkubo, K., Kawashima, Y. *et al.* (2011) Ion-controlled on–off switch of electron transfer from tetrathiafulvalene calix[4]pyrroles to $Li^+@C_{60}$. *J. Am. Chem. Soc.*, **133**, 15938–15941.

195 Ruiz-Botella, S., Vidossich, P., Ujaque, G. *et al.* (2015) A tetraferrocenyl-resorcinarene cavitand as a redox-switchable host of ammonium salts. *Chem. Eur. J.*, **21**, 10558–10565.

196 Zimmerman, S.C., Saionz, K.W., and Zeng, Z. (1993) Chemically bonded stationary phases that use synthetic hosts containing aromatic binding clefts: HPLC analysis of nitro-substituted polycyclic aromatic hydrocarbons. *Proc. Natl. Acad. Sci. U.S.A.*, **90**, 1190–1193.

197 Elisa, L., Baldini, F., Giannetti, A. *et al.* (2010) Solid-supported Zn(II) porphyrin tweezers as optical sensors for diamines. *Chem. Commun.*, **46**, 3678–3680.

198 Kamieth, M., Burkert, U., Corbin, P.S. *et al.* (1999) Molecular tweezers as synthetic receptors: molecular recognition of electron-deficient aromatic substrates by chemically bonded stationary phases. *Eur. J. Org. Chem.*, **1999**, 2741–2749.

199 Shanmugaraju, S. and Mukherjee, P.S. (2015) π-Electron rich small molecule sensors for the recognition of nitroaromatics. *Chem. Commun*, **51**, 16014–16032.

200 Lee, Y.H., Liu, H., Lee, J.Y. *et al.* (2010) Dipyrenylcalix[4]arene—a fluorescence-based chemosensor for trinitroaromatic explosives. *Chem. Eur. J.*, **16**, 5895–5901.

201 Kim, S.-B., Lee, E.-B., Choi, J.-H., and Cho, D.-G. (2013) Simple fluorescent chemosensors for TNT: one-step synthesis. *Tetrahedron*, **69**, 4652–4656.

202 Zhang, F., Luo, L., Sun, Y. *et al.* (2013) Synthesis of a novel fluorescent anthryl calix[4]arene as picric acid sensor. *Tetrahedron*, **69**, 9886–9889.

203 Burattini, S., Colquhoun, H.M., Greenland, B.W. *et al.* (2009) Pyrene-functionalised, alternating copolyimide for sensing nitroaromatic compounds. *Macromol. Rapid Commun.*, **30**, 459–463.

204 Roy, B., Bar, A.K., Gole, B., and Mukherjee, P.S. (2013) Fluorescent tris-imidazolium sensors for picric acid explosive. *J. Org. Chem.*, **78**, 1306–1310.

205 Kumar, R., Sandhu, S., Singh, P. *et al.* (2014) Tripodal fluorescent sensor for encapsulation-based detection of picric acid in water. *Asian J. Org. Chem.*, **3**, 805–813.

206 Larsen, K.R., Nielsen, K.A., Sessler, J.L., and Jeppesen, J.O. (2013) Detection of nitroaromatic explosives using tetrathiafulvalene-calix[4]pyrroles, in *Organic Synthesis and Molecular Engineering* (ed. M.B. Nielsen), John Wiley & Sons, pp. 257–283.

207 Nielsen, K.A., Cho, W.-S., Jeppesen, J.O. *et al.* (2004) Tetra-TTF calix[4]pyrrole: a rationally designed receptor for electron-deficient neutral guests. *J. Am. Chem. Soc.*, **126**, 16296–16297.

208 Park, J.S., Le Derf, F., Bejger, C.M. *et al.* (2010) Positive homotropic allosteric receptors for neutral guests: annulated tetrathiafulvalene–calix[4]pyrroles as colorimetric chemosensors for nitroaromatic explosives. *Chem. Eur. J.*, **16**, 848–854.

209 Kim, D.-S., Lynch, V.M., Nielsen, K.A. *et al.* (2009) A chloride-anion insensitive colorimetric chemosensor for trinitrobenzene and picric acid. *Anal. Bioanal. Chem.*, **395**, 393–400.

210 Swinburne, A.N., Paterson, M.J., Beeby, A., and Steed, J.W. (2010) Fluorescent "twist-on" sensing by induced-fit anion stabilisation of a planar chromophore. *Chem. Eur. J.*, **16**, 2714–2718.

211 Lee, H.J., Yoon, I.J., Yoo, C.L. *et al.* (2000) Potentiometric evaluation of solvent polymeric carbonate-selective membranes based on molecular tweezer-type neutral carriers. *Anal. Chem.*, **72**, 4694–4699.

212 Choi, Y.S., Lvova, L., Shin, J.H. *et al.* (2002) Determination of oceanic carbon dioxide using a carbonate-selective electrode. *Anal. Chem.*, **74**, 2435–2440.

213 Zyryanov, G.V., Palacios, M.A., and Anzenbacher, P. (2007) Rational design of a fluorescence-turn-on sensor array for phosphates in blood serum. *Angew. Chem. Int. Ed.*, **46**, 7849–7852.

214 Guha, S. and Saha, S. (2010) Fluoride ion sensing by an anion–π interaction. *J. Am. Chem. Soc.*, **132**, 17674–17677.

215 Liu, X., Yang, X., Fu, Y. *et al.* (2011) Novel fluorescent sensor for Ag^+ and Hg^{2+} based on the BINOL-pyrene derivative via click reaction. *Tetrahedron*, **67**, 3181–3186.

216 Velmurugan, K., Raman, A., Easwaramoorthi, S., and Nandhakumar, R. (2014) Pyrene pyridine-conjugate as Ag selective fluorescent chemosensor. *RSC Adv.*, **4**, 35284–35289.

217 Nativi, C., Cacciarini, M., Francesconi, O. *et al.* (2007) Pyrrolic tripodal receptors effectively recognizing monosaccharides. Affinity assessment through a generalized binding descriptor. *J. Am. Chem. Soc.*, **129**, 4377–4385.

218 Yoon, H., Lee, C.-H., and Jang, W.-D. (2012) Absolute stereochemical determination of chiral carboxylates using an achiral molecular tweezer. *Chem. Eur. J.*, **18**, 12479–12486.

219 Brahma, S., Ikbal, S.A., and Rath, S.P. (2014) Synthesis, structure, and properties of a series of chiral tweezer–diamine complexes consisting of an achiral zinc(II) bisporphyrin host and chiral diamine guest: induction and rationalization of supramolecular chirality. *Inorg. Chem.*, **53**, 49–62.

220 Brahma, S., Ikbal, S.A., Dhamija, A., and Rath, S.P. (2014) Highly enhanced bisignate circular dichroism of ferrocene-bridged Zn(II) bisporphyrin tweezer with extended chiral substrates due to well-matched host–guest system. *Inorg. Chem.*, **53**, 2381–2395.

221 Palomino-Schätzlein, M., Virgili, A., Gil, S., and Jaime, C. (2006) Di-(*R*,*R*)-1-[10-(1-hydroxy-2,2,2-trifluoroethyl)-9-anthryl]-2,2,2-trifluoroethyl muconate: a highly chiral cavity for enantiodiscrimination by NMR. *J. Org. Chem.*, **71**, 8114–8120.

222 Gil, S., Palomino-Schätzlein, M., Burusco, K.K. *et al.* (2010) Molecular tweezers for enantiodiscrimination in NMR: di-(*R*,*R*)-1-[10-(1-hydroxy-2,2,2-trifluoroethyl)-9-anthryl]-2,2,2-trifluoroethyl benzenedicarboxylates. *Chirality*, **22**, 548–556.
223 Colquhoun, H.M., Zhu, Z., Cardin, C.J. *et al.* (2007) Sterically controlled recognition of macromolecular sequence information by molecular tweezers. *J. Am. Chem. Soc.*, **129**, 16163–16174.
224 Yang, F., Anilkumar, P., Anderson, A. *et al.* (2012) Facile and effective post-production separation of single-walled carbon nanotubes with paired aromatic molecules: a molecular tweezers approach. *J. Phys. Chem. C*, **116**, 6800–6804.
225 Peng, X., Komatsu, N., Kimura, T., and Osuka, A. (2007) Improved optical enrichment of SWNTs through extraction with chiral nanotweezers of 2,6-pyridylene-bridged diporphyrins. *J. Am. Chem. Soc.*, **129**, 15947–15953.
226 Shen, C., Ma, D., Meany, B. *et al.* (2012) Acyclic cucurbit[*n*]uril molecular containers selectively solubilize single-walled carbon nanotubes in water. *J. Am. Chem. Soc.*, **134**, 7254–7257.
227 Waring, M.J. (1993) Echinomycin and related quinoxaline antibiotics, in *Molecular Aspects of Anticancer Drug–DNA Interactions*, vol. **1** (eds S. Neidle and M.J. Waring), CRC Press, pp. 213–242.
228 Chen, H. and Patel, D.J. (1995) Solution structure of a quinomycin bisintercalator–DNA complex. *J. Mol. Biol.*, **246**, 164–179.
229 Gao, Q., Williams, L.D., Egli, M. *et al.* (1991) Drug-induced DNA repair: X-ray structure of a DNA–ditercalinium complex. *Proc. Natl. Acad. Sci. U.S.A.*, **88**, 2422–2426.
230 Van Vliet, L.D., Ellis, T., Foley, P.J. *et al.* (2007) Molecular recognition of DNA by rigid [*n*]-polynorbornane-derived bifunctional intercalators: synthesis and evaluation of their binding properties. *J. Med. Chem.*, **50**, 2326–2340.
231 Sinha, S., Lopes, D.H.J., Du, Z. *et al.* (2011) Lysine-specific molecular tweezers are broad-spectrum inhibitors of assembly and toxicity of amyloid proteins. *J. Am. Chem. Soc.*, **133**, 16958–16969.
232 Sinha, S., Du, Z., Maiti, P. *et al.* (2012) Comparison of three amyloid assembly inhibitors: the sugar scyllo-inositol, the polyphenol epigallocatechin gallate, and the molecular tweezer CLR01. *ACS Chem. Neurosci.*, **3**, 451–458.
233 Prabhudesai, S., Sinha, S., Attar, A. *et al.* (2012) A novel "molecular tweezer" inhibitor of α-synuclein neurotoxicity in vitro and in vivo. *Neurotherapeutics*, **9**, 464–476.
234 Bier, D., Rose, R., Bravo-Rodriguez, K. *et al.* (2013) Molecular tweezers modulate 14-3-3 protein–protein interactions. *Nat. Chem.*, **5**, 234–239.

235 Dutt, S., Wilch, C., Gersthagen, T. *et al.* (2013) Linker effects on amino acid and peptide recognition by molecular tweezers. *Eur. J. Org. Chem.*, **2013**, 7705–7714.

236 Zieringer, M., Casas Cartagena, M., Burakowska, E. *et al.* (2013) Polyglycerol-tagged molecular clips as receptors in protic solvents. *Eur. J. Org. Chem.*, **2013**, 362–367.

237 Ma, D., Hettiarachchi, G., Nguyen, D. *et al.* (2012) Acyclic cucurbit[*n*]uril molecular containers enhance the solubility and bioactivity of poorly soluble pharmaceuticals. *Nat. Chem.*, **4**, 503–510.

238 Hettiarachchi, G., Samanta, S.K., Falcinelli, S. *et al.* (2016) Acyclic cucurbit[*n*]uril-type molecular container enables systemic delivery of effective doses of albendazole for treatment of SK-OV-3 xenograft tumors. *Mol. Pharm.*, **13**, 809–818.

239 Ganapati, S., Zavalij, P.Y., Eikermann, M., and Isaacs, L. (2016) *In vitro* selectivity of an acyclic cucurbit[*n*]uril molecular container towards neuromuscular blocking agents relative to commonly used drugs. *Org. Biomol. Chem.*, **14**, 1277–1287.

240 Ma, D., Zhang, B., Hoffmann, U. *et al.* (2012) Acyclic cucurbit[*n*]uril-type molecular containers bind neuromuscular blocking agents in vitro and reverse neuromuscular block in vivo. *Angew. Chem. Int. Ed.*, **51**, 11358–11362.

241 Hoffmann, U., Grosse-Sundrup, M., Eikermann-Haerter, K. *et al.* (2013) Calabadion: a new agent to reverse the effects of benzylisoquinoline and steroidal neuromuscular-blocking agents. *Anesthesiology*, **119**, 317–325.

242 Haerter, F., Simons, J.C.P., Foerster, U. *et al.* (2015) Comparative effectiveness of calabadion and sugammadex to reverse non-depolarizing neuromuscular-blocking agents. *Anesthesiology*, **123**, 1337–1349.

243 Ma, D., Glassenberg, R., Ghosh, S. *et al.* (2012) Acyclic cucurbituril congener binds to local anaesthetics. *Supramol. Chem.*, **24**, 325–332.

244 Davidson, L., Blencowe, A., Drew, M.G.B. *et al.* (2003) Synthesis and evaluation of a solid supported molecular tweezer type receptor for cholesterol. *J. Mater. Chem.*, **13**, 758–766.

245 Paxton, R.J., Beatty, B.G., Hawthorne, M.F. *et al.* (1991) A transition metal complex (Venus flytrap cluster) for radioimmunodetection and radioimmunotherapy. *Proc. Natl. Acad. Sci. U.S.A.*, **88**, 3387–3391.

246 Hawthorne, M.F., Varadarajan, A., Knobler, C.B. *et al.* (1990) Radiometallacarboranes as tumor imaging reagents. *J. Am. Chem. Soc.*, **112**, 5365–5366.

6

Concluding Remarks

I felt obliged to write some concluding remarks, since the very topic of this book touches upon the subject of accelerating technological progress – which is, with the advent of technomimetics, now entering the molecular domain. This expansion might have far-reaching consequences and therefore should be considered in a broader context beyond the boundaries of traditional science and technology. As a natural conclusion of the book, but with a different view angle, these remarks are presented here as a postscript.

There are two fundamental approaches that dominate the literature regarding the impact and meaning of exponential technological growth. One asserts that the human race embracing accelerating technological progress as a primary survival tool is essentially a plague of the planet, if not the universe itself. The opposite one implies that our technological civilization is the crown jewel of almost 4 billion years of evolution of life on Earth and a critical, if not essential, part of the whole universe development. Obviously, both approaches are based on emotion and cannot substitute an objective scientific analysis.

A rational analysis of the expanding technological civilization is a complicated task, exacerbated by the fact that no comparative studies can be made due to the apparent solitude of our civilization in the observable part of the universe. Nonetheless, some general considerations are possible.

First and foremost, it is a mistake to consider our civilization a weird fluke waiting to be eradicated by the reversion to the mean. This is by no means a statistical aberration, but rather the normal behavior of certain thermodynamically open subsystems (including Earth itself) within the broader trend of overall development of the universe, characterized by the expansion and entropy increase.[1]

1 "As entropy increases, the universe, and all closed systems in the universe, tend naturally to deteriorate and lose their distinctiveness, to move from the least to the most probable state … But while the universe as a whole tends to run down, there are local enclaves of whose direction seems opposed to that of the universe at large and in which there is a limited and temporary tendency for organization to increase. Life finds its home in these enclaves" Ref. [1].

Molecular Devices: An Introduction to Technomimetics and Its Biological Applications,
First Edition. Andrei A. Gakh.

A convoluted play of these development forces provides a ground for the emergence of the self-sustainable information processes first manifesting themselves on Earth in their rudimental forms as biological life processes about 4 billion years ago. These information processes rely on the replication and transformation of information packages – discrete sets of functionally homomorphic elements in which specific ordering can be trans-coded from one physical carrier to another. These information processes are governed by their own dynamics and have little, if any, resemblance to the other "natural" processes. Because of this, it is reasonable to assume that a new reality – an information domain – was established on Earth about 4 billion years ago.

Currently, the information domain has three distinct subdomains: biological, cognitive, and computer-based. However, the rapid advent of molecular genetics reversing the original flow of information from genotype to phenotype and the progress in computer–human interfaces coupled with social network technologies make it increasingly clear that these three subdomains are closely interconnected and uniform in their general core.

It is customary to present the information reality as a secondary, subordinate part of the physical reality, but the actual interplay between the physical and information domains is far more complex. Although all the objects of the information world would have to be written (coded) on the physical carriers, and thus obey all constrains and limitations of the physical domain, the information processes themselves – replications and transformations of information packages – easily defy many of these physical limitations. More importantly, the information reality has its own set of laws, such as the laws of evolution, which do not exist in the physical reality. The two realities are also fundamentally different in other basic features. For example, the physical domain has only one trajectory and resolves any potential paradoxes to achieve a unique solution for any given initial setup. Furthermore, no process in the physical reality can be affected by future events. The physical domain also does not allow for any miracles violating basic conservation or limitation laws, such as the law of the conservation of energy or the speed of light limitation. On the contrary, the information domain allows for multiple trajectories, with many variants of the future coexisting and selectable at will. More importantly, the state of information objects can be affected by the perceived events of probable future scenarios. Paradoxes and apparent miracles are possible.

Although it is true that the preponderance of observable physical objects have their imprints (projections) in the information world, the information domain also contains objects that do not exist in the physical world. At some point in evolutionary development, the information reality witnessed the emergence of complex life forms capable of not only adapting to the physical reality but also invoking changes in the physical environment to satisfy their needs. Subsequent progress yielded self-aware information systems capable of creating sophisticated information models of the physical reality with predictive features specifically aimed at optimizing the required future actions in pursuit of the most desirable outcome. At this point, the exchange dynamics between the two worlds engrosses the massive intrusion of the information domain into the physical domain facilitated by sophisticated functional devices controlled by the information models with predictive capabilities. This can be described as an emergence of a new fundamental development factor, which, with the advent of modern science and technology, quickly gains prominence on Earth and near-Earth environments. Its actions are governed by collective intelligent choices – features that exist only in the information domain. More importantly, planning for the most desirable outcomes among all available choices becomes a hallmark of our civilization.

It is customary to assume that the effects of information reality are insignificant in comparison to the other fundamental forces, such as gravitation or electromagnetism, and can simply be ignored in grand cosmological models. While this is definitely correct at the current level of development, the ultimate forecast of an interdependent future of the information and physical domains is far from certain, given the exponential growth tendencies of the information realm.

The advent of the information reality irreversibly changed the trajectory of the physical reality, which now has to be considered as existing in a *mixed* state. Before the emergence of the information reality, the trajectory of the physical reality was always directed toward the most *probable* future. The information reality, to the contrary, is gravitating to the most *desirable* future. This is possible due to modeling, which entails creating artificial dynamic projections of the physical reality within the information domain, exploring various scenarios using these models until the best variant of the future is attained, and then acting accordingly. Both primitive and advanced biological life forms follow the same script, just using different approaches.

Primitive biological life operates primarily on the chemical level by creating coded heteropolymers with specific sets of instructions written and executed with the use of several types of monomers – nucleosides (in DNA and RNA) and amino acids (in peptides). Its progress, or evolution, is a relatively slow process employing random search algorithms to achieve the best desirable outcome: the survival and propagation of the biological species.

Contrary to primitive biological life forms, advanced life forms utilize specialized neuron-based networks executing sophisticated, nonrandom hierarchical algorithms within complex information models in search of the preferred outcome. These life forms are capable of constructing dynamic neuron-based models for real-time operations based on past experience and of developing empirical constructs reflecting important event interconnections of the physical reality in the form of gene-coded persistent correlations (instincts). Although evolutionarily effective, these advanced life forms have one major disadvantage: there is no easy, direct way to preserve the accumulated individual information at the genome level. As a result, maintaining the best behavioral models for future generations in the form of instincts is a slow and convoluted process.

Civilizations, which are networks (or societies) of advanced intelligent life forms, can overcome this limitation by creating technological means of storing and processing the information outside the biological subdomain. With the advent of ancient civilizations, information processes began to spread into the physical environment. However, only modern, advanced civilizations are capable of transforming the environment well beyond the boundaries of their natural habitat areas. There are no physical limitations to the outer space penetration of intelligent life forms by technological means, but before we dwell on such a technically achievable possibility, it would be prudent to ask a simple question – why should we do that? In other words, what do we (collectively, as a human civilization) actually want long term beyond simple survival and propagation? Even more importantly, should we even try to answer this question?

Perhaps the easiest way of long-term planning entails making no specific plans at all – and letting the natural forces take the lead in defining our future. Unfortunately, the obvious outcome of this "no-planning" approach is mean reversion and restoration of the natural, most *probable* trajectory of the physical reality, which is currently grossly distorted by the presence of the information domain. Due to the probabilistic character of the natural self-balancing forces, it is impossible to predict whether eradication of our civilization will occur due a catastrophic technological error, or a natural disaster, or social apathy and degeneration, or a combination of thereof. What is certain is the inevitability of this outcome as long as we follow this hand-off approach and stay on Earth for too long. In any case, this is definitely not the most *desirable* scenario.

It is fair to say that there is actually no shortage of long-term planning ideas in written history. The earliest global models were based on faith. Many of them are several millennia-old, but are still in demand even now. Due to their self-consistent nature, they are essentially infallible from the outside, but suffer one major drawback. All of them rely on some sort of supernatural intervention, which is beyond the reach of humanity. As a result, none of these plans can be fully executed in the physical reality by rational means.

This fundamental flaw of faith-based planning gave rise to various utopian ideas. However, so far, every attempt to execute purely utopian plans has led to ruinous results with no exception. Utopian and apocalyptic scenarios appear to be the same, outcome-wise. While the exact analysis of this convergence lies beyond the scope of this postscript, one obvious observation is that traditional utopian scenarios do not reflect the disruptive nature of exponential technological progress and simply assume that human capabilities will remain intact in years to come.

Recent futuristic models attempt to remove these limitations, with an acknowledgment that our human civilization is not the ultimate form of technological evolution. One of the most popular scenarios (which can be traced back to Ulam – von Neumann's pioneering discussions more than a half century ago)[2] is depicted in Figure 6.1. While the first four steps of the presented scenario are well within confirmed historical trends, the last two steps – the merging of biological-, cognitive-, and computer-based information processes followed by the spread of the information domain well beyond the current human habitat boundaries – require further elaboration.

The merger of computers, brains, and biology is essentially upon us. Technology enthusiasts place the timing of this event within the reach of the current generation, but more conservative estimates coalesce within a several hundred year time frame, which is still a very near future event in the historical context of human civilization. The exact timing of this event is critical, since the actual contributions of the three subdomains during the merger will be greatly affected by this time factor. The biology subdomain is

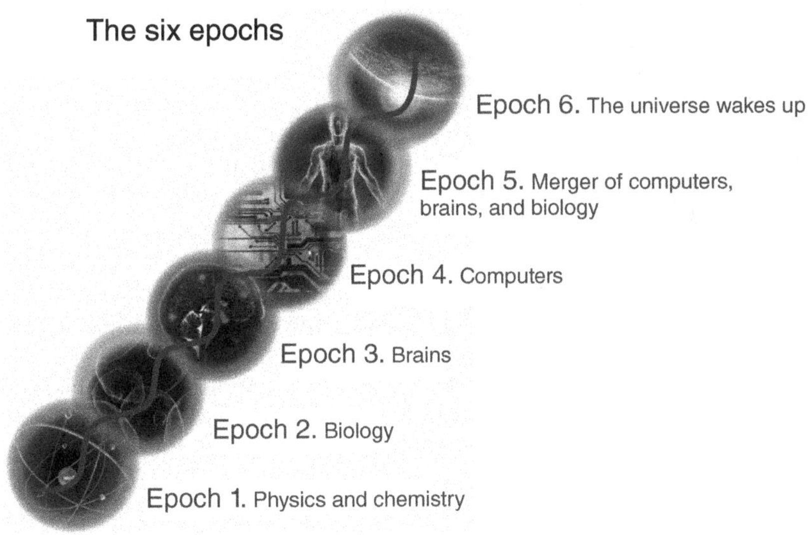

Figure 6.1 Evolution of information processes. *Credits*: Idea – R. Kurzweil; graphics – R. Lubin, R. Denning, N. Staubach.

2 "One conversation centered on the ever accelerating progress of technology and changes in the mode of human life, which gives the appearance of approaching some essential singularity in the history of the race beyond which human affairs, as we know them, could not continue" Ref. [2].

the oldest and the most developed one, but at the same time is the slowest evolution-wise, even in the age of the genetic revolution. In addition, it is tightly optimized for the Earth's physical conditions, and adaptation of the current biological ecosystem to extraterrestrial conditions is a formidable task. On the contrary, computer evolution is very fast. Driven by Moorse's law (which is actually closer to a self-fulfilling prophecy than an ordinary evolutionary law),[3] computer information processing capacity is capable of doubling every 18–24 months. In addition, current solid-state electronic devices are easily adaptable to a variety of environmental conditions. Last but not least, the clear separation of software and hardware makes information transcoding a relatively straightforward task compared to the brain or any other biological cell network systems.

Given the circumstances, it would be desirable to facilitate an early merger and put a particular emphasis on the development of human-centric molecular technologies, including bio- and technomimetics. In this case, the core essence of human experience will be adequately represented during the merger event. This desirable scenario is far from automatic and can be attained only with significant effort and willful redirection of available resources within the science and technology areas, since human-centric technologies lie outside the natural direction of current investment trends heavily favoring most profitable, quickly evolving computer technologies. These willful interventions would have to be made at the policy or ideology levels.

Most prominent ideologies offer some sort of perceived benefits beyond pure economic value. These perceived benefits could be social, as in utopian ideas, or psychological, such as peace-of-mind or sense of purpose. Another prized benefit is hope for a second life, which is designed to circumvent the fear of death deeply embedded in the set of basic instincts of all complex biological organisms. Technically, all these attractive features, including peace-of-mind and sense of purpose, can be easily incorporated in the framework of human-centric technologies and make them attractive enough for capital redirection. The biggest selling point, hope for a second life, is also attainable.

Similar to the reconstruction of DNA sequences in ancient organisms, the autonomic information processes of human consciousness can be captured and then transferred to a different physical carrier. This is a highly complicated, but technically achievable, procedure, provided, of course, that there is an adequately preserved source of an individual digital imprint. It remains to be seen how much of the original functionality can be retained after this

3 "...nowadays we speak of 'Moore's Law', as if it were a law of Nature. The validity of this law cannot be understood from the technical procedures by which the chips are made. The fact that the law holds so well is an effect of the way actors (in industry, in science and in government) judge their own and each others' accomplishments with respect to what Moore's Law predicts" [3].

transfer, but, with the proper technology, adequate functionality (with modest improvements) is not out of the question.

It is naïve to assume that a second life will be a given chance for everybody regardless of circumstances. The laws of the information reality are different from the laws of physical reality, but nevertheless have the same objective features. For any information object to survive the endless onslaught of entropy, the overall benefits of preservation must exceed the expenses. Therefore, only competitively functional information objects will eventually survive in this new environment.

It is difficult to speculate what will happen to the human experience after the completion of the merger and the progression toward the final stages of the information evolution (Figure 6.1). It is commonly assumed that the subsequent development of supercivilizations will stay on the traditional path, with a relatively slow formation of a few stationary centers followed by a targeted development of new centers at the periphery of the ever expanding information habitat area. This scenario favors the retention of human-centric values for many millennia. However, the final outcome of the last epoch, "Awakening of the Universe", is far from certain. We do not see any signs of such a development in our galaxy, even though the probability of it should be quite high due to the fact that our galaxy contains well over 100 billions of stars. One may argue that we simply cannot comprehend the powers of super-intelligent life forms and therefore are simply unable to detect any signs of their existence. What we do know, however, is that there is another variant of the final epoch that would not require the development of superintelligence.

The current exponential technological progress has all of the distinctive characteristics of a detonation process. In late development stages, it allows for the construction of technological tools capable of acting as "smart" energy catalysts by facilitating the release of the energy preserved behind the natural potential barriers, such as nuclear energy. Unlike common self-balanced habitats that attain essentially homeostatic conditions by receiving the constant flow of free energy from external sources, such as the Sun, and then releasing equal amounts of it in the form of thermal radiation, any large-scale use of the internal potential energy would inevitably result in an environment with quickly raising temperatures and radiation. Since no known information processes can operate effectively in such an environment, long-lasting information habitats can use the internal energy sources on a very limited scale. This, in turn, limits the overall expansion to a slow process of hopping from one homeostatic planetary system to another. An alternative scenario, with the heavy use of all available energy sources, entails the eventual formation of a highly dynamic information process constantly moving forward and leaving a very hot and uninhabitable world behind the front of this information shockwave. Only primitive forms of life are capable of such locust-type modes of action.

Any controls will be lost once an exponentially growing civilization reaches critical conditions with the formation of a self-propagating technological shockwave. Although it is impossible to know in advance when the process will cross a point of no return, this is quite likely in the next few centuries, if not sooner. Little comfort can be attained from the absence of any signs of these information-based shockwave bubbles in our observable space–time, since we might be alone, at least in our galaxy. Obviously, it would be best to avoid this detonation scenario, in which the bulk of human experience will inevitably be lost.

The question of what we really want from our future, and how to achieve it, remains a paramount consideration. This question cannot be answered by science alone without invoking the "active intelligent choice" concept. This concept entails the recognition of the necessity to choose, understanding all available options and their probabilities, and, finally, selecting and implementing the best route toward the most desirable outcome. Unfortunately, there is no way of knowing all the available choices and all intended and sometimes unintended consequences of our actions. However, even if the best route would not lead to the desired goal, acting is still better than doing nothing.

A popular assumption is that individuals have no particular roles in shaping our common future since it is a cumulative product, a resulting vector of many driving forces. This assumption, although generally correct, is not entirely accurate. Like other information processes, our civilization creates a so-called active environment. In this active environment, the correlation between individual events and their consequences is no longer linear, and even small local perturbations can grow into global phenomena. History provides ample examples of individuals making a difference and affecting the society in a very profound, sometimes revolutionary way. In addition, a peculiarity in intelligence development on Earth yielded a system of individual autonomic information processes instead of one unified global consciousness. There are many reasons why our collective intelligence operates on an individual level (lack of natural effective communication tools is probably the major one), but the direct consequence of this peculiarity is the existence of individual free will. With this freedom, even one individual can offer and pursue alternative venues of future development.

The search for hidden opportunities beyond simple survival and propagation remains one of the fundamental goals of science. Of course, there is a plethora of already developed scenarios. Some of them are technically achievable, others are not. However, the main effort should not be directed at exploring the existing ideas, but rather at creating new, better ones. We have plenty of time ahead of us to do that, but this time should not be wasted in vain. After all, the only reliable way to build a better future is to invent it, because the future is the only thing in this world that can be changed at will.

References

1 Wiener, N. (1950) *The Human Use of Human Beings: Cybernetics and Society*, Houghton Mifflin Company/The Riverside Press, Preface, p. 12 (in 1988 edition).
2 Ulam, S. (1958) Tribute to John von Neumann. *Bull. Amer. Math. Soc.*, **64** (3), 1–49.
3 van Lente, H. and Rip, A. (1998) Expectations in technological developments: an example of prospective structures to be filled in by agency. In *Getting New Technologies Together. Studies in Making Sociotechnical Order* (eds C. Disco and B. van der Meulen), Walter de Gruyter, Berlin, pp. 203–231.

Index

Molecular Devices: An Introduction to Technomimetics and Its Biological Applications,
First Edition. Andrei A. Gakh.